AF355964

Development and Application of Starter Cultures

Development and Application of Starter Cultures

Guest Editors

Roberta Comunian
Luigi Chessa

Basel • Beijing • Wuhan • Barcelona • Belgrade • Novi Sad • Cluj • Manchester

Guest Editors

Roberta Comunian
Agris Sardegna
Agricultural Research Agency
of Sardinia
Sassari
Italy

Luigi Chessa
Agris Sardegna
Agricultural Research Agency
of Sardinia
Sassari
Italy

Editorial Office
MDPI AG
Grosspeteranlage 5
4052 Basel, Switzerland

This is a reprint of the Special Issue, published open access by the journal *Fermentation* (ISSN 2311-5637), freely accessible at: www.mdpi.com/journal/fermentation/special_issues/Starter_Cultures.

For citation purposes, cite each article independently as indicated on the article page online and using the guide below:

Lastname, A.A.; Lastname, B.B. Article Title. *Journal Name* **Year**, *Volume Number*, Page Range.

ISBN 978-3-7258-2978-1 (Hbk)
ISBN 978-3-7258-2977-4 (PDF)
https://doi.org/10.3390/books978-3-7258-2977-4

Contents

About the Editors

Roberta Comunian

Dr. Roberta Comunian holds a master's degree in Biological Sciences and a Ph.D. in Production and Safety of Food of Animal Origin from the Department of Veterinary Medicine. With over 27 years of experience in food microbiology, she has been leading the Microbiology Unit at Agris Sardegna, the Agricultural Research Agency of Sardinia (Italy), since 2007. Her expertise includes phenotypic/genotypic identification and technological/safety characterization of microbial strains and communities of agri-food interest. She oversees the Agris MBDS-BNSS Culture Collection, focusing on the long-term preservation of microbial communities crucial to the quality and safety of traditional fermented food products. In her leadership role, she coordinates the Microbiology Unit's research projects, actively contributes to the scientific literature, and shares her findings through conferences and seminars. She collaborates with the School of Specialization in Inspection of Food of Animal Origin of the Veterinary Medicine Department (University of Sassari), teaching dairy microbiology. She is a member of the Scientific Technical Committee of the Italian Ministry of Agriculture, contributing to the National Plan on Biodiversity of Agricultural Relevance; is one of the Italian experts in the Standing Committee on Methods for Dairy Microorganisms of the International Dairy Federation (FIL-IDF); and serves on the Scientific Committee of MIRRI.it, the Italian Network of Microbial Culture Collections.

Luigi Chessa

Luigi Chessa (Ph.D.), a researcher at the Microbiology Laboratory of Agris Sardegna, the Agricultural Research Agency of Sardinia (Italy), holds a master's degree in Agricultural and Environmental Biotechnology and has a Ph.D. in Microbial Biotechnology, with additional certification of Doctor Europaeus. He was a visiting Ph.D. student at the Julius Kühn-Institut, Bundesforschungsinstitut für Kulturpflanzen—Institut für Epidemiologie und Pathogendiagnostik in Braunschweig (Germany), and a post-doctoral fellow at the University of Sassari (Italy), and is currently a member of the Italian Society for Agricultural Food and Environmental Microbiology (SIMTREA). As a passionate teacher of dairy microbiology at the School of Specialization in Inspection of Food of Animal Origin at the Department of Veterinary Medicine (University of Sassari), his expertise has earned him invitations to speak at numerous international and national conferences. His work has also been showcased through poster presentations on several prestigious platforms. His research journey began with a focus on veterinary antibiotics and the dynamic world of soil microbial communities. Currently, his cutting-edge research delves into dairy microbiology, exploring foods derived from various matrices like milk, meat, and vegetables. He is particularly committed to the characterization, preservation, and innovative use of bacteria and natural, undefined starter cultures that play a crucial role in the art and science of food production.

Preface

In recent decades, starter cultures have been developed to improve raw material processing and produce a variety of fermented foods. Well-designed selected starters offer a convenient solution for easy and safe fermentation, especially when the natural microbiota colonizing the production environment and the raw material is insufficient or when natural starter cultures are difficult to obtain and manage.

There are two main approaches for developing a starter culture:

a) Selecting a limited and well-defined number of species or strains based on their proven aptitude to perform the biochemical processes required for each production technology, as well as their ease of cultivation under laboratory conditions.

b) Replicating autochthonous biodiverse natural cultures, where a wide variety of species and strains, both starter and non-starter, coexist in equilibrium and contribute equally to the fermentation and maturation processes.

Both approaches have their advantages and disadvantages. Selected starters are easy to reproduce at the laboratory scale, in high concentration, making them technologically effective and ensuring consistent quality standardization. As a result, they are widely applied in industrial-scale production. In contrast, natural starters are complex and biodiverse microbial communities with a unique composition of strains that cannot be replicated outside their place of origin. These cultures can enhance foods with distinctive sensory features tied to their territory of production and help to preserve local microbial diversity. Autochthonous natural starter cultures are typically found in the most traditional and high-quality agri-food products. However, their technological performance is not standardized, and their use is not without risks. Along with useful autochthonous microorganisms, pathogen or spoilage organisms can be inadvertently introduced into the product. The goal of this Special Issue is to present innovative research or review articles addressing the challenge of developing starter cultures that can be applied at artisanal, pilot, or industrial scales, ensuring safety, consistency in quality, and technological performance while also preserving biodiversity and the unique sensory characteristics associated with traditional products.

We thank all the authors of the 10 papers contributing to making this Special Issue successful.

Roberta Comunian and Luigi Chessa
Guest Editors

fermentation

Editorial

Development and Application of Starter Cultures

Roberta Comunian * and Luigi Chessa *

Agris Sardegna, Agricultural Research Agency of Sardinia, Associated Member of the JRUMIRRI-IT, Loc. Bonassai SS 291 km 18.600, 07100 Sassari, Italy
* Correspondence: rcomunian@agrisricerca.it (R.C.); lchessa@agrisricerca.it (L.C.)

Citation: Comunian, R.; Chessa, L. Development and Application of Starter Cultures. *Fermentation* **2024**, *10*, 512. https://doi.org/10.3390/fermentation10100512

Received: 3 October 2024
Accepted: 4 October 2024
Published: 8 October 2024

Driven by the imperative of reconciling food safety with the preservation of traditional sensory profiles, the landscape of starter culture research is evolving. This Special Issue on "Development and Application of Starter Cultures" has brought to light cutting-edge research that, across various fermented food and beverage applications, not only advances our understanding of microbial fermentation but also challenges long-held paradigms in the field.

The work of Chessa et al. [1] on undefined starter cultures for traditional dairy products, while emphasizing the importance of biodiversity in natural starters, serves as a clarion call for a re-evaluation of our approach to microbial safety. Their findings raise crucial questions about the coexistence of QPS (Qualified Presumption of Safety) and non-QPS microorganisms in traditional artisanal production, pushing us to ask whether current regulatory frameworks, safety protocols, and scientific risk assessments sufficiently account for the complexity of natural microbial communities.

In their meticulous screening of acetic acid bacteria for Kombucha production, Lee et al. [2] showcase the untapped potential residing in diverse microbial ecosystems. Their research, alongside a comprehensive review by de Oliveira Hosken et al. [3] on lactic acid bacteria (LAB) from Brazilian artisanal cheeses, opens a new frontier in starter culture development. This study expands our understanding of the potential applications of acetic acid bacteria beyond traditional vinegar production, challenging us to look beyond conventional sources, and suggesting that revolutionary strains could be found in previously unexplored niches.

Further advancing the field, Česnik et al. [4] provide a sophisticated analysis of the use of *Saccharomyces* yeasts in cider production, underscoring the significance of strain-level characterization and marking a significant leap towards what we might term "precision fermentation". Their work on aroma compound production, linked to amino acid metabolism, illustrates the complex interplay between starter culture metabolism and final product quality, paving the way for tailor-made starter cultures designed to achieve precise sensory outcomes. The implications of this approach extend far beyond cider, potentially revolutionizing how we craft fermented foods across the board.

Building on ancient fermentation practices, the research of Huang et al. [5] on optimizing *Monascus purpureus* fermentation bridges ancient fermentation practices and modern biotechnology. The success of optimizing both pigment production and saccharification through the precise manipulation of growth conditions demonstrates the potential for technological advancements and nutritional enhancements in traditional fermented foods, all without sacrificing authenticity. Other studies, including those by Cecchi et al. [6] on Taggiasca olives, Tolu et al. [7] on sourdough, and a review by Rădoi-Encea et al. [8] on Romanian wine yeasts, underscore the crucial role of indigenous microbiota in traditional fermented products. These works emphasize the potential of harnessing local microbial biodiversity for product differentiation and quality improvement. There is immense value in preserving and leveraging local microbial biodiversity, highlighting the vast and largely untapped microbial resources associated with traditional fermented products around the world. The challenge lies in integrating these traditional resources with cutting-edge biotechnology.

Iosca et al. [9], in their development of bioprotective cultures based on lactic acid bacteria to combat bread rope spoilage, exemplify the trend towards "clean label" biopreservatives. This research demonstrates the dual functionality of starter cultures as both fermentation agents and natural preservatives, addressing consumer demand for fewer additives in food products.

Lastly, Neviani et al. [10] provide a critical review of natural whey starters, advocating for a paradigm shift in how we view starter cultures: not as isolated strains, but as complex, interacting microbial ecosystems. This "microbiome approach" to starter culture development could lead to more robust, adaptable fermentation systems capable of navigating the complexities of microbial interactions for improved food production.

Together, these studies point to an exciting future for starter culture research, one that embraces both tradition and innovation to push the boundaries of fermentation science.

Charting the course forward, the contributions to this Special Issue represent not just incremental progress, but a fundamental shift in how we approach starter culture development and application. We are standing at the threshold of a new era in fermentation technology, one that promises unprecedented control over fermentation outcomes while honoring the rich microbial heritage of traditional foods. The future of starter culture research hinges on our ability to balance standardization with biodiversity, safety with complexity, and efficiency with authenticity.

Future research in food fermentation should focus on the following areas:

1. Metagenomics and strain-level characterization: leveraging advanced genomic and metagenomic tools to characterize natural starter cultures at the strain level, allowing for more precise applications of microorganisms.
2. Microbial interactions: more in-depth studies on the interactions between microorganisms in mixed cultures are needed to develop more effective and stable multi-strain starter cultures.
3. Functional metabolites: continuing to explore bioactive compounds and functional metabolites produced by starter cultures could enhance the nutritional and health-promoting properties of fermented foods.
4. Biopreservation: developing starter cultures with bioprotective properties could provide more natural methods of food preservation.
5. Adapting to climate change: research on the resilience and adaptability of starter cultures to changing environmental conditions will be crucial in the context of climate change.
6. Biotechnological applications: exploring the potential of starter microorganisms beyond food fermentation, such as in bioremediation or the production of high-value compounds.
7. Regulatory frameworks: updating regulatory approaches to ensure food safety by considering both individual microbial species and strains and the natural microbial communities used in traditional fermented products.

This Special Issue showcases the rich diversity and potential of starter cultures across various fermented food applications. As we move forward, starter culture development and applications offer exciting opportunities for innovation in food technology, preservation, and the creation of products with enhanced quality, safety, and functionality. By embracing interdisciplinary approaches and fostering collaboration among microbiologists, food technologists, and sensory scientists, the full potential of microbial fermentation can be unlocked.

As we close this Special Issue, it is evident that the field of starter culture research is not merely evolving but undergoing a revolution. The challenges ahead are significant, but so are the opportunities. By addressing these challenges and pursuing the research directions outlined here, we can ensure that starter cultures remain pivotal in shaping the future of food production, meeting the evolving needs of both industry and consumers.

Conflicts of Interest: The authors declare no conflict of interest.

References

1. Chessa, L.; Daga, E.; Dupré, I.; Paba, A.; Fozzi, M.C.; Dedola, D.G.; Comunian, R. Biodiversity and Safety: Cohabitation Experimentation in Undefined Starter Cultures for Traditional Dairy Products. *Fermentation* **2024**, *10*, 29. [CrossRef]
2. Lee, D.-H.; Kim, S.-H.; Lee, C.-Y.; Jo, H.-W.; Lee, W.-H.; Kim, E.-H.; Choi, B.-K.; Huh, C.-K. Screening of Acetic Acid Bacteria Isolated from Various Sources for Use in Kombucha Production. *Fermentation* **2024**, *10*, 18. [CrossRef]
3. Hosken, B.d.O.; Melo Pereira, G.V.; Lima, T.T.M.; Ribeiro, J.B.; Magalhães Júnior, W.C.P.d.; Martin, J.G.P. Underexplored Potential of Lactic Acid Bacteria Associated with Artisanal Cheese Making in Brazil: Challenges and Opportunities. *Fermentation* **2023**, *9*, 409. [CrossRef]
4. Česnik, U.; Martelanc, M.; Øvsthus, I.; Radovanović Vukajlović, T.; Hosseini, A.; Mozetič Vodopivec, B.; Butinar, L. Functional Characterization of Saccharomyces Yeasts from Cider Produced in Hardanger. *Fermentation* **2023**, *9*, 824. [CrossRef]
5. Huang, Y.; Chen, J.; Chen, Q.; Yang, C. Effects of Main Nutrient Sources on Improving Monascus Pigments and Saccharifying Power of Monascus purpureus in Submerged Fermentation. *Fermentation* **2023**, *9*, 696. [CrossRef]
6. Cecchi, G.; Di Piazza, S.; Rosa, E.; De Vecchis, F.; Silvagno, M.S.; Rombi, J.V.; Tiso, M.; Zotti, M. Autochthonous Microbes to Produce Ligurian Taggiasca Olives (Imperia, Liguria, NW Italy) in Brine. *Fermentation* **2023**, *9*, 680. [CrossRef]
7. Tolu, V.; Fraumene, C.; Carboni, A.; Loddo, A.; Sanna, M.; Fois, S.; Roggio, T.; Catzeddu, P. Dynamics of Microbiota in Three Backslopped Liquid Sourdoughs That Were Triggered with the Same Starter Strains. *Fermentation* **2022**, *8*, 571. [CrossRef]
8. Rădoi-Encea, R.-Ș.; Pădureanu, V.; Diguță, C.F.; Ion, M.; Brîndușe, E.; Matei, F. Achievements of Autochthonous Wine Yeast Isolation and Selection in Romania—A Review. *Fermentation* **2023**, *9*, 407. [CrossRef]
9. Iosca, G.; Fugaban, J.I.I.; Özmerih, S.; Wätjen, A.P.; Kaas, R.S.; Hà, Q.; Shetty, R.; Pulvirenti, A.; De Vero, L.; Bang-Berthelsen, C.H. Exploring the Inhibitory Activity of Selected Lactic Acid Bacteria against Bread Rope Spoilage Agents. *Fermentation* **2023**, *9*, 290. [CrossRef]
10. Neviani, E.; Levante, A.; Gatti, M. The Microbial Community of Natural Whey Starter: Why Is It a Driver for the Production of the Most Famous Italian Long-Ripened Cheeses? *Fermentation* **2024**, *10*, 186. [CrossRef]

Article

Biodiversity and Safety: Cohabitation Experimentation in Undefined Starter Cultures for Traditional Dairy Products

Luigi Chessa *, Elisabetta Daga, Ilaria Dupré, Antonio Paba, Maria C. Fozzi, Davide G. Dedola and Roberta Comunian

Agris Sardegna, Servizio Ricerca Prodotti di Origine Animale, Associated Member of the JRUMIRRI-IT, Loc. Bonassai SS 291 km 18.600, 07100 Sassari, Italy; edaga@agrisricerca.it (E.D.); idupre@agrisricerca.it (I.D.); apaba@agrisricerca.it (A.P.); mfozzi@agrisricerca.it (M.C.F.); dgdedola@agrisricerca.it (D.G.D.); rcomunian@agrisricerca.it (R.C.)
* Correspondence: lchessa@agrisricerca.it

Abstract: Natural starter cultures, characterised by undefined microbiota, can contribute to the technological process, giving peculiar characteristics to artisanal fermented foods. Several species have a long history of safe use and have obtained Qualified Presumption of Safety (QPS) status from the European Food Safety Authority (EFSA), whereas others (non-QPS) could represent a potential risk for consumers' health and must undergo a safety assessment. In this work, the biodiversity, at species and strain level, by pulsed-field gel electrophoresis (PFGE) and (GTG)$_5$ rep-PCR, of an undefined natural starter culture, in frozen and lyophilized form, obtained from ewe's raw milk avoiding thermal treatment or microbial selection, was investigated. The culture was constituted by different biotypes of *Enterococcus durans*, *Enterococcus faecium*, *Enterococcus faecalis*, and *Lacticaseibacillus paracasei*. *Streptococcus oralis* and *Streptococcus salivarius* were also found, over species belonging to the *Streptococcus bovis*–*Streptococcus equinus* complex (SBSEC), like *Streptococcus gallolyticus* subsp. *macedonicus*, *Streptococcus lutetiensis*, and *Streptococcus equinus*. Molecular investigation on virulence and antibiotic resistance genes, as well as minimum inhibitory concentration (MIC) determination, revealed that all the non-QPS strains can be considered safe in the perspective of using this culture for cheesemaking. The obtainment of a natural culture directly from ewe's raw milk bypassing thermal treatment and selection of pro-technological bacteria can be advantageous in terms of biodiversity preservation, but non-QPS microorganisms can be included in the natural starter and also in cheeses, especially in traditional ones obtained from fermenting raw milk. Following EFSA guidelines, artisanal factories should not be allowed to produce starter cultures by themselves from raw milk, running the risk of including some non-QPS species in their culture, and only selected starters could be used for cheesemaking. A revision of the criteria of QPS guidelines should be necessary.

Keywords: natural starters; cheesemaking; food safety; biodiversity; microbial fingerprint

1. Introduction

Starter cultures are used to aid raw material processing with the aim of easily and safely carrying out fermentation and obtaining different types of fermented foods. In cheesemaking, commercial starters consisting of a few (one to three) selected species/strains, or natural starter cultures, can be used [1,2]. Selected starters represent a suitable solution to perform fermentation if the microbiota of raw material and production environment is inadequate, or natural starter cultures are difficult to obtain and manage [3]. They are widely added at high concentrations in industrial production processes, becoming dominant in the food microbiota, causing a dramatic decrease in microbial diversity and a loss of peculiar sensory characteristics of foods of particular geographic niches [4]. On the contrary, natural cultures are complex microbial communities that, having a strain composition mostly undefined [5], are not reproducible in any place other than their origin.

Their use in food production contributes to preserving microbial biodiversity, enriching artisanal products with peculiar sensory features that link them to the territory of production [6]. Indeed, autochthonous natural starter cultures usually characterise the most typical and high-quality agri-food products. Natural cultures are usually reproduced daily by cheesemakers, but after repetitive passages of reproduction, they are not able to properly accomplish their technological role (i.e., acidification ability) any longer.

The use of these natural cultures could not be risk free since, together with useful autochthonous microorganisms, even pathogenic or spoilage ones could be potentially inoculated and can contaminate the product [2]. Several microorganisms can be used for food production, such as lactic acid bacteria, whereas others could represent a potential risk for the consumers' health, e.g., enterococci, some streptococci, and staphylococci. In recent decades, the European Food Safety Authority (EFSA) introduced the definition of Qualified Presumption of Safety (QPS) for long-safe-history microorganisms that can be used for food and feed production without prior safety assessments [7]. The QPS list is published every three years, and the updates are based on a two-year assessment carried out by the Biological Hazards (BIOHAZ) Panel [8]. The QPS status is the result of a pre-assessment that covers safety concerns for humans, animals, and the environment. During this process, experts assess the taxonomic identity of the microorganism and the related body of knowledge. Microorganisms that are not well defined, and for which it is not possible to conclude whether they pose a safety concern, are not considered suitable for QPS status and must undergo a full safety assessment. However, also some non-QPS microorganisms can be used after the ascertainment of safety, e.g., *Enterococcus faecium* [9,10]. A possible solution to overcome these problems (obtainment, management, safety, and technological ability loss) could be the use of natural cultures in lyophilized form [2].

The aim of this work was to evaluate the biodiversity, at species and strain level, of an undefined natural starter culture, obtained from raw ewe's milk, avoiding any thermal treatment or microbial selection, as well as to assess the safety of non-QPS strains in the perspective of use of this culture, in frozen or lyophilised form, for cheesemaking.

2. Materials and Methods

2.1. Experimental Plan

A natural starter culture for cheesemaking, in frozen (NSC) and lyophilized (LNSC) form, was characterised for its microbial diversity, at the strain level, and safety. NSC was obtained directly from raw ewe milk in a study described by Chessa et al. [11], where the composition in microbial groups (i.e., mesophilic cocci and bacilli, thermophilic cocci and bacilli, and enterococci) and the technological performances in acidification were assessed. Moreover, NSC was lyophilized by Veneto Agricoltura (Thiene, Italy) as external service. In this study, both the frozen (NSC) and lyophilized (LNSC) forms of the natural starter culture were characterised for their microbial diversity. Moreover, the evaluation of safety for non-QPS bacteria isolated from NSC and LNSC was also performed. Operatively, microbial counts in NSC, after thawing, and LNSC, after lyophilisation, were performed, and a total of 157 microbial colonies were picked up from the different elective media used for the enumeration of microbial groups. The microbial isolates were identified, first, by species-specific PCR and then molecularly typed by $(GTG)_5$ rep-PCR or pulsed-field gel electrophoresis (PFGE). Molecular and phenotypic tests for the detection of antibiotic resistance and virulence genes, and for the determination of minimum inhibitory concentration (MIC), were also performed.

2.2. Microbial Counts and Isolation

Microbial counts on both NSC and LNSC for the enumeration of thermophilic cocci and lactobacilli, mesophilic cocci and lactobacilli, enterococci, citrate-fermenting bacteria, staphylococci, and coliforms were performed by spreading 0.1 mL of each serial 10-fold dilution on agar plates, as described by Chessa et al. [11]. From each media used for microbial counts, 10 colonies were picked up from the lowest countable dilution (Table S1), and each colony was purified by 3 repetitive passages on the agar media of origin. Morphology of isolates was also checked using Axio-Phot optic microscope (Carl Zeiss, Oberkochen, Germany) equipped with Objective EC Plan-Neofluar 1009/1.30 OilPol M27.

2.3. Microbial Identification and Biodiversity Evaluation

2.3.1. Species Identification

The identification of 157 bacterial isolates was performed by genus/species-specific PCR and DNA sequencing analysis. Genomic DNA was extracted from purified cocci and bacilli, isolated from NSC and LNSC, using PrepMan Ultra Sample Preparation reagent (Applied Biosystems—Thermo Fisher Scientific, Rodano, Italy), according to the manufacturer's instructions. The extracted DNA was diluted 1:50 in sterile Milli-Q water and used as template for genus/species determination with the primers listed in Table 1. PCR products were separated on agarose gel, supplemented with $1\times$ SYBR SAFE (Invitrogen— Thermo Fisher Scientific) in Tris-acetate buffer, and gel images were acquired by the UV transilluminator FireReader V4 (UVITec, Warwickshire, UK). The microbial isolates for which it was not possible to identify by genus/species-specific PCR were investigated via DNA sequencing analysis targeting 16S rRNA and *sod*A genes using the universal primers *p27f* and *p765r* [12], and *d1* and *d2* [13] targeting *16S rRNA* (about 750 bp) and *sodA*, respectively (Table 1). Sequences were edited and aligned with BioEdit (v. 7.2) using the ClustalW algorithm. Consensus sequences were built for each sample and, for species determination, were compared to those deposited in the nucleotide database of the National Center for Biotechnology Information by means of NCBI BLAST. Identification was deemed reliable if values for sequence similarities were >99%.

Table 1. Molecular primers used for microbial identification.

Genus/Species	Primer	Gene Target	Primer Sequence (5′–3′)	Annealing (°C)	Size (bp)	Reference
Lactococcus lactis	LcLspp-F	*16S rRNA*	GTTGTATTAGCTAGTTGGTGAGGTAAA	55	387	[14]
	Lc-R		GTTGAGCCACTGCCTTTTAC			
Lactobacillus delbrueckii lactis	Lac-LACTIS-F733	*dppE*	TGCCAAGCTCTACTCCGTTT	58	217	[15]
	Lac-LACTIS-R949		GTCAAGCGGCATAGTGTCAA			
Lactobacillus delbrueckii bulgaricus	Lac-BULG-F391	*lacZ*	GGAAGACTCCGTTTTGGTCA	58	395	[15]
	Lac-BULG-R785		AGTTCAAGTCTGCCCCATTG			
Lactobacillus helveticus	Lac-HELV-F73	*prt*H	GGCGGGGAAAGAGGTAACTA	58	509	[15]
	Lac-HELV-R581		TGACGCAAACTTAATGAACCA			
Limosilactisbacillus fermentum	Lac-FER-F753	*Arc*D	CCAGATCAGCCAACTTCACA	58	310	[15]
	Lac-FER-R1062		GGCAAACTTCAAGAGGACCA			
Limosilactobacillus reuteri	REUT1	*16S rRNA*	TGAATTGACGATGGATCACCAGTG	65	1000	[16]
	LOWLAC		CGACGACCATGAACCACCTGT			
Lacticaseibacillus paracasei	Y2	*16S rRNA*	CCCACTGCTGCCTCCCGTAGGAGT	55	290	[17]
	PARA		CACCGAGATTCAACATGG			
Lactiplantibacillus plantarum	planF	*recA*	CCGTTTATGCGGAACACCTA	56	318	[18]
	pREV		TCGGGATTACCAAACATCAC			
Streptococcus thermophilus	Str-THER-F2116	*lacZ*	GCTTGTGTTCTGAGGGAAGC	58	577	[15]
	Str-THER-R2693		CTTTCTTCTGCACCGTATCCA			
Enterococcus	ENT1	*Tuf*	TACTGACAAACCATTCATGATG	59	112	[19]
	ENT2		AACTTCGTCACCAACGCGAAC			
Enterococcus faecium	FM1	*sod*A	GAAAAAACAATAGAAGAATTAT	55	215	[20]
	FM2		TGCTTTTTTGAATTCTTCTTTA			

Table 1. *Cont.*

Genus/Species	Primer	Gene Target	Primer Sequence (5′–3′)	Annealing (°C)	Size (bp)	Reference
Enterococcus faecalis	FL1	*sod*A	ACTTATGTGACTAACTTAACC	55	360	[20]
	FL2		TAATGGTGAATCTTGGTTTGG			
Universal	p27f	*16S rRNA*	GAGAGTTTGATCCTGGCTCAG	58	≈750	[12]
	p765r		CTGTTTGCTCCCCACGCTTTC			
Degenerate	d1	*sod*A	CCITAYICITAYGAYGCIYTIGARCC	37	≈480	[13]
	d2		ARRTARTAIGCARRTARTAIGCRTGYTCCCAIACRTC			

N = A, C, G, and T; R = A and G; W = A and T; Y = C and T.

2.3.2. Molecular Biotyping

The intraspecific identification of all the bacteria isolated from NSC and LNSC (*Enterococcus durans, Enterococcus faecium, Enterococcus faecalis, Streptococcus gallolyticus* subsp. *macedonicus, Streptococcus equinus, Streptococcus lutetiensis, Streptococcus oralis, Streptococcus salivarius*, and *Lacticaseibacillus paracasei*) was performed by (GTG)$_5$ rep-PCR microbial fingerprint, as described by Chessa et al. [21], using an FTA® Disc for DNA analysis (GE Healthcare, Chicago, IL, USA) as template. PCR products were separated on agarose gel (1.8% *w/v*) with 1× SYBR Safe (Invitrogen—Thermo Fisher Scientific), at 100 V (222 V/h) in Tris-acetate buffer.

Only the isolates identified as *S. equinus* and *S. lutetiensis* were genotyped by pulsed-field gel electrophoresis (PFGE). Genomic DNA extraction was prepared according to the protocol described by Graves and Swaminathan [22], then digested with 25 U of *Sma*I (BioLabs, Heidelberg, Germany) for 4 h at 25 °C. Electrophoresis was carried out in a contour-clamped homogeneous electric field (CHEF)-Mapper apparatus (Bio-Rad Laboratories, Hercules, CA, USA) at 14 °C in 0.5× TBE buffer. DNA fragments were separated at 6 V/cm gradient with 120° angle for 16 h. The running time was divided into 3 blocks: block 1 of 5 h, initial switch time 1 s final switch time 20 s; block 2 of 5 h, initial switch time 1 s final switch time 5 s; block 3 for 6 h, initial switch time 10 s final switch time 40 s. Gels were stained with 1× SYBR Safe.

Gel images of both (GTG)$_5$ rep-PCR and PFGE gels were acquired with the UV transilluminator FireReader V4 (UVITec) and elaborated by BioNumerics (v. 6.6.11; Applied Maths, Sint-Martens-Latem, Belgium). Cluster analysis was performed by unweighted pair group method with arithmetic averages (UPMGA); then, Pearson and Dice similarity correlation indexes for (GTG)$_5$ rep-PCR and PFGE profiles, respectively, were used. The isolates sharing ≥93% similarity among the (GTG)$_5$ rep-PCR profiles and 100% similarity among the PFGE profiles analysed according to Tenover et al. [23] were considered the same biotype.

2.4. Safety Assessment

2.4.1. Antibiotic Resistance and Virulence Gene Detection

For safety assessment, PCR reactions were carried out using primers and annealing temperatures listed in Table S2. Total-community DNA from NSC and LNSC, extracted following the protocol described by Paba et al. [24], and DNA from isolates, previously extracted for microbial identification, were used as template. PCR products were separated on 1.5% (*w/v*) agarose gel with 1× SYBR Safe in Tris-acetate buffer.

For the assessment of safety of *E. faecium* isolated from the cultures, the detection of the pathogenic-related genes hyl_{Efm}, *esp*, and IS*16* was carried out, according to the European Food and Safety Authority (EFSA) guidelines [10]. Same protocol was applied for the *E. durans* and *E. faecalis* isolates. *E. faecalis* was also tested for the presence of tetracycline resistance genes *tet*M, *tet*K, *tet*L, *tet*S.

The *Streptococcus* isolates were checked for antibiotic resistance genes for macrolides (*ermA, ermB, ermC*, and *mefA*), lincosamides (*lnuC*), tetracyclines (*tet*M, *tet*K, *tet*L, *tet*S, and the putative conjugative Tn*916*-like transposon), and for the presence of potential virulence genes: *scpB, hylB, bca, bac, emm, smeZ, speA, speG*, and *ssa*.

The natural starter cultures NSC and LNSC *in toto* were tested for all the genes listed above.

2.4.2. Antibiotic Susceptibility

The isolates belonging to *Enterococcus* and *Streptococcus* genera, not currently included in the QPS list of EFSA [9], were phenotypically tested, by broth micro-dilution method, for their susceptibility to several antibiotics. Ampicillin (Sigma Aldrich, Milan, Italy) minimum inhibitory concentration (MIC) determination for *E. faecium* was performed using homemade trays prepared according to the ISO 20776-1:2019 [25], and the MIC breakpoint ≤ 2 mg/L indicated by EFSA for safety evaluation was applied [10]. For each tray, 7 *E. faecium* strains were tested. For each strain, positive (strain DSM 2570, equivalent to ATCC 29212) and negative (not inoculated) control wells were also included. Plates were inoculated with 100 µL of Mueller-Hinton Broth (Thermo Fisher Scientific, Rodano, Italy) at 1×10^5 CFU/mL final concentration, then incubated at 37 °C overnight before visual examination of microbial growth.

The *E. durans* (from LNSC) and *E. faecalis* (1 from NSC and 2 from LNSC) isolates were tested for antimicrobial susceptibility by the Sensititre™ EU Surveillance Enterococcus EUVENC AST Plate (Thermo Fisher Scientific).

For *Streptococcus* isolates, Sensititre™ STP6F (Thermo Fisher Scientific) plates for MIC determination were used. Operatively, 100 µL of reconstitution volume in Cation-Adjusted Mueller-Hinton Broth with Lysed Horse Blood (Thermo Fisher Scientific) at 1×10^5 CFU/mL final concentration of the well was used, and although the plate was intended for 50 µL, the resulting dilutions were twice the lower dilution. The antibiotics used in this study for MIC determination are listed in Table 3.

The interpretation of MIC breakpoints was based, for *Enterococcus*, on the Clinical and Laboratory Standards Institute (CLSI) [26], and for *Streptococcus*, on The European Committee on Antimicrobial Susceptibility Testing [27] (EUCAST), applied for most of the antibiotics tested, except for tetracycline, erythromycin, and chloramphenicol, for which the breakpoints indicated in Clinical and Laboratory Standards Institute (CLSI) [26] were considered (Table S3).

2.5. Statistical Analysis

Differences in microbial counts about the concentrations of the microbial groups investigated between NSC and LNSC were compared using the Student *t* test using the software SPSS Statistics (v. 21.0; IBM Corp., Armonk, NY, USA). Molecular fingerprints performed by (GTG)$_5$ rep-PCR and PFGE were elaborated by BioNumerics (v. 5.0; Applied Maths, Sint-Martens-Latem, Belgium). Cluster analysis for PFGE and (GTG)$_5$-rep fingerprints was performed using deep significance clustering (DICE) and Pearson's correlation index through the unweighted pair group method using arithmetic averages (UPGMA), respectively.

The Simpson's diversity index (DI) was calculated for each microbial species found in NSC and LNSC, where the number of biotypes was ≥ 2, according to Hunter and Gaston [28], using Equation (1), where N is the total number of isolates, S the total number of biotypes identified, and nj the number of isolates belonging to each biotype.

$$DI = 1 - \frac{1}{N(N-1)} \sum_{j=1}^{S} nj(nj - 1) \tag{1}$$

3. Results

3.1. Microbial Counts of the Natural Starter Cultures

Microbial counts performed on the starter culture object of this study (NSC) and its lyophilized form (LNSC) revealed that the concentration of viable cells in LNSC was always

1.2–2.2 Log CFU/g higher than in NSC (Figure 1). Indeed, mesophilic and thermophilic cocci, mesophilic and thermophilic bacilli, as well as enterococci, were always significantly ($p < 0.05$) higher in LNSC. Staphylococci and coliforms were not detected in NSC nor in LNSC.

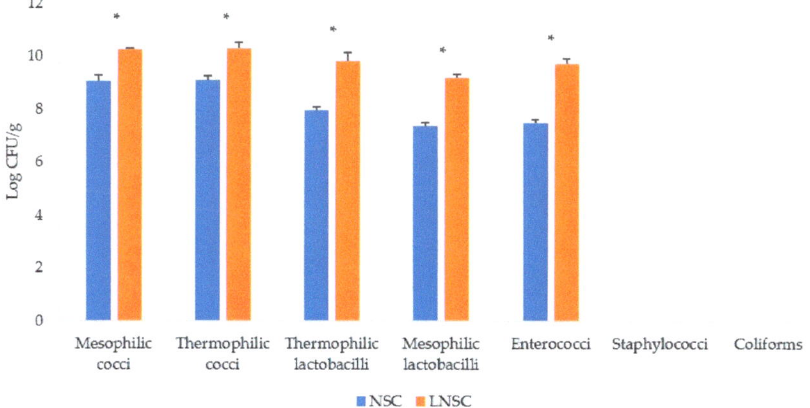

Figure 1. Microbial counts of presumptive thermophilic cocci and lactobacilli, mesophilic cocci and lactobacilli, enterococci, staphylococci, and coliforms in the frozen natural starter culture (NSC) and the lyophilized NSC (LNSC). Microbial counts were expressed as Log CFU/g $\pm$ standard deviation. *, significant difference ($p < 0.05$) in the Log CFU/g between NSC and LNSC, according to the Student *t* test.

3.2. Biodiversity Evaluation

3.2.1. Biodiversity at Species Level

A total of 157 isolates from the natural starter cultures NSC and LNSC (53 and 104, respectively) were molecularly identified at the species level. In particular, 135 isolates were identified by genus/species-specific PCR (36 from NSC and 99 from LNSC) and the remaining 22 isolates by 16S rRNA sequencing analysis (17 from NSC and 5 from LNSC).

The characterisation by genus/species-specific PCR revealed the presence of different microbial species in the two starter cultures (Figure 2). In particular, 25 *E. durans* isolates, only in the lyophilized culture (LNSC) in M17 (incubated at 22 and 45 °C) and MRS media, at dilution -6 were found. *E. faecium* was found in both cultures. In NSC, 10 isolates, all coming from the KAA medium at dilution -4, were found, whereas in LNSC, the *E. faecium* isolates characterised were 45: 23 from KAA at dilution -5, 17 from MRS at dilution -6, and 5 from M17 incubated at 22 and 45 °C, at dilution -6. Three *E. faecalis* were isolated, one from NSC (in KAA medium at dilution -4) and two from LNSC (in M17 incubated at 30 °C, at dilution -6). Furthermore, 35 colonies of *L. paracasei* were isolated from the FH medium, 12 from NSC, at dilution -3, and 23 from LNSC, at dilution -5. *S. gallolyticus* subsp. *macedonicus* was isolated more frequently from NSC than LNSC (13 versus 4 isolates, respectively), from the M17 medium at dilutions -5 and -6. Among streptococci, another four species were isolated and identified: 3 *S. oralis* (1 from NSC and 2 from LNSC) isolated from M17 at 45 °C, 1 *S. salivarius* isolated only from NSC in KAA at dilution -4, and 18 isolates belonging to the *S. bovis/equinus* complex (SBSEC), 15 from NSC and 3 from LNSC. For this latter microbial group, the *sod*A gene, encoding for the manganese-dependent superoxide dismutase, one of the most reliable biomarkers for SBSEC [29,30], was used for the identification at the species level. All 18 isolates were confirmed to belong to SBSEC; 13 of them were *S. equinus* (isolated from NSC in M17 30 °C and 45 °C, and MRS) and 5 were *S. lutetiensis* (2 from NSC and 3 from LNSC) (Figure 2).

(a)

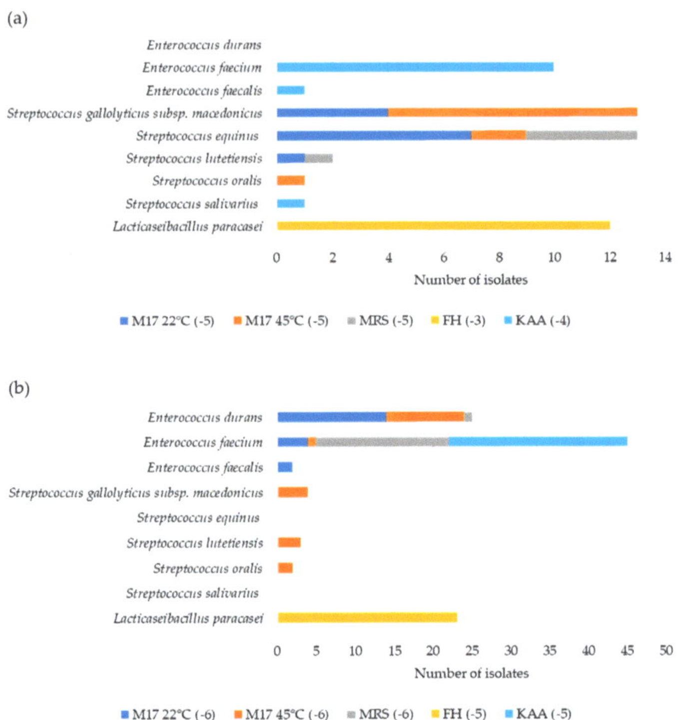

(b)

Figure 2. Bacterial species isolated from the natural cultures NSC (**a**) and LNSC (**b**). From each medium (M17 incubated at 22 °C, M17 incubated at 45 °C, MRS, FH, and KAA) and serial 10-fold dilution (in brackets), the numbers of isolates are indicated.

3.2.2. Biodiversity at Strain Level

The microbial isolates picked up from NSC and LNSC were also characterised at the strain level by $(GTG)_5$-rep and PFGE fingerprinting to calculate the number of microbial biotypes in the two forms of the natural starter culture investigated and to assess the biodiversity level. The calculation of the Simpson's diversity index (DI) revealed that *E. durans* (25 isolates), found only in LNSC (Table 2), was represented by eight rep-PCR profiles, with a DI of 0.85. The other *Enterococcus* species, *faecium* and *faecalis*, were instead found in both NSC and LNSC. In particular, six *E. faecium* biotypes were found in NSC (from 10 total isolates) and nine in LNSC (from 45 isolates), with one biotype in common between the two forms of the natural culture. Moreover, a DI decrease after the freeze-drying process, from 0.87 in NSC to 0.62 in LNSC, was observed. Two *E. faecalis* biotypes were calculated using the Pearson's correlation of $(GTG)_5$-rep fingerprints, one in NSC and one in LNSC, and the biotype found was in common between the cultures. However, DI calculation for *E. faecalis* was not possible since a single biotype was found both in NSC and LNSC. In addition, *L. paracasei* was also found both in NSC (12 isolates) and LNSC (23 isolates), with one and five biotypes detected, respectively. The DI for *L. paracasei* in LNSC was 0.78, whereas the diversity calculation was not possible in NSC. *S. gallolyticus* subsp. *macedonicus*, found both in NSC and LNSC, was represented by one and two biotypes, respectively, and the DI accountable, only for LNSC, was 0.50. Two *S. oralis* biotypes were found, one in NSC and one in LNSC, and none were in common, whereas only one *S. salivarius* from NSC was found. For the species belonging to the SBSEC complex, one molecular biotype of *L. lutetiensis* from NSC was found, and the same biotype was found in LNSC, whereas *S. equinus* was isolated only from NSC, and only one biotype (from 13 isolates) was found.

Table 2. Microbial biotypes for each species in the natural starter culture (NSC) and in the lyophilized NSC (LNSC).

Species	NSC			LNSC			NSC + LNSC			
	Isolates No.	rep-PCR Profiles No.	PFGE Profiles No.	Isolates No.	rep-PCR Profiles No.	PFGE Profiles No.	Isolates No.	rep-PCR Profiles No.	rep-PCR Profiles in Common	PFGE Profiles in Common
Enterococcus durans	0	n.d.	n.d.	25	8	n.d.	25	8	n.d.	n.d.
Enterococcus faecium	10	6	n.d.	45	9	n.d.	55	14	1	n.d.
Enterococcus faecalis	1	1	n.d.	2	1	n.d.	3	2	1	n.d.
Lacticaseibacillus paracasei	12	1	n.d.	23	5	n.d.	35	6	n.d.	n.d.
Streptococcus gallolyticus subsp. *macedonicus*	13	1	n.d.	4	2	n.d.	17	2	1	n.d.
Streptococcus oralis	1	1	n.d.	2	1	n.d.	3	2	0	n.d.
Streptococcus salivarius	1	1	n.d.	0	n.d.	n.d.	1	1	0	n.d.
Streptococcus lutetiensis	2	1	1	3	1	2	5	1	1	1
Streptococcus equinus	13	1	3	0	n.d.	n.d.	13	1	0	0
Total isolates	53			104			157			

Rep-PCR, (GTG)5 rep-PCR; PFGE, pulse field gel electrophoresis; n.d., not determined.

3.3. Safety Assessment

All of the 55 *E. faecium* isolates, 10 from NSC and 45 from LNSC, showing ampicillin MIC ≤ 2 mg/L (Table 2) and absence of the virulence-associated genes hyl_{Efm}, esp, and $IS16$, were considered safe to be used as food additives since they met the standards of compliance with the safety requirements indicated by the EFSA [31].

Results about the safety of the other microbial species isolated are reported below.

3.3.1. Antibiotic Resistance and Virulence Genes Investigation

The cultures NSC and LNSC, *in toto*, and the microbial isolates belonging to the genus *Streptococcus*, were analysed for the detection of resistance genes for macrolides (*erm*A, *erm*B, *erm*C, and *mef*A), lincosamides (*lnu*C), tetracyclines (the putative conjugative Tn916-like transposon), and for the presence of potential virulence genes. *mef*A, encoding for an efflux pump, the activity of which, driven by proton motive force, involves only macrolides that determine the M phenotype [32] was detected in *S. oralis* and *S. salivarius*, both phenotypically resistant to azithromycin and erythromycin. The cultures NSC and LNSC, *in toto*, were positive for the Tn916-like transposon, whereas they, and the isolates, were negative for the antibiotic resistance genes listed above. LNSC cultures were negative for pathogenic-related genes hyl_{Efm}, esp, and $IS16$, while NSC was positive for esp due to the presence of *E. faecalis*.

3.3.2. Minimum Inhibitory Concentration Determination

The MIC of the antibiotics included in Table 3 was determined for all the isolates except those belonging to *E. faecium*. All of the 25 *E. durans* isolates (from LNSC) were susceptible to chloramphenicol, tetracycline, erythromycin, ampicillin, linezolid, tigecycline, vancomycin, teicoplanin, and ciprofloxacin (Table 3), using the cut off established by EUCAST or CLSI, as indicated in Table 3.

Table 3. Antibiotic susceptibility of *Enterococcus* isolates from the natural cultures.

Bacteria Tested	Culture	Antibiotics Tested								
		Penicillins AMP	Macrolides ERY	Amphenicols CHL	Oxazolidinones LZD	Tetracyclines TET	Glycycyclines TGC	Glycopeptides VAN	TEI	Fluoroquinolones CIP
E. faecium	NSC	S [1]	n.t.	n.t.	n.t.	n.t.	n.t.	n.t.	n.t.	n.t.
	LNSC	S [1]	n.t.	n.t.	n.t.	n.t.	n.t.	n.t.	n.t.	n.t.
E. faecalis	NSC	S [1]	I [2]	S [2]	S [1]	R [2]	S [1]	S [1]	S [1]	S [1]
	LNSC	S [1]	I [2]	S [2]	S [1]	R [2]	S [1]	S [1]	S [1]	S [1]
E. durans	NSC	abs.	abs.	abs.	abs.	abs.	abs.	abs.	abs.	abs.
	LNSC	S [1]	S [2]	S [2]	S [1]	S [2]	S [1]	S [1]	S [1]	S [1]

NSC, natural starter culture; LNSC, Lyophilized natural starter culture (LNSC). AMP, Ampicillin; PEN, Penicillin; AZI, Azithromycin; ERY, Erythromycin; FEP, Cefepime; FOT, Cefotaxime; AXO, Ceftriaxone; CHL, Chloramphenicol; CLI, Clindamycin; ETP, Ertapenem; MERO, Meropenem; LEVO, Levofloxacin; LZD, Linezolid; TET, Tetracycline; TGC, Tigecycline; VAN, Vancomycin; TEI, Teicoplanin; SYN, Quinupristin/Dalfopristin; CIP, Ciprofloxacin; S, sensitive; R, resistant; I, intermediate; n.t., not tested; abs., the species was absent in the culture. [1] Breakpoint by EUCAST 2023. [2] Breakpoint by CLSI 2020.

Among the enterococci investigated, the *E. faecalis* isolates, one from NSC and two from LNSC, were resistant to tetracycline and intermediate to erythromycin (CLSI cut off) [26] but susceptible to ampicillin, chloramphenicol, linezolid, tigecycline, vancomycin, teicoplanin, and ciprofloxacin (Table 3).

The 13 *S. gallolyticus* subsp. *macedonicus* isolated from NSC, and the 4 isolates from LNSC, were susceptible to penicillin, azithromycin, erythromycin, cefepime, cefotaxime, ceftriaxone, chloramphenicol, clindamycin, ertapenem, meropenem, levofloxacin, linezolid, tetracycline, and vancomycin. The same results were found for all the *S. lutetiensis* isolates (2 from NSC and 3 from LNSC) and the 13 *S. equinus* isolated only from NSC. *S. oralis* (one isolate from NSC and two from LNSC) was resistant to azithromycin and erythromycin. The only *S. salivarius* isolated from NSC was resistant to azithromycin and erythromycin, and it showed an intermediate profile for penicillin susceptibility (Table 4).

Table 4. Antibiotic susceptibility of *Streptococcus* isolates from the natural cultures.

Bacteria Tested	Culture	Antibiotics Tested															
		Penicillins		Macrolides		Cephalosporins			Amphenicols	Lincosamides	Carbapenems		Fluoroquinolones	Oxazolidinones	Tetracyclines	Glycopeptides	Streptogramins
		AMP	PEN	AZI	ERY	FEP	FOT	AXO	CHL	CLI	EPT	MERO	LEVO	LZD	TET	VAN	SYN
S. gallolyticus macedonicus	NSC	n.t.	S^{2}*	S^{2}	S^{2}	S^{2}	S^{2}	S^{2}	S^{2}	S^{2}	S^{2}	S^{2}	S^{2}	S^{2}	S^{2}	S^{2}	n.t.
	LNSC	n.t.	S^{2}	S^{2}	S^{2}	S^{2}	S^{2}	S^{2}	S^{2}	S^{2}	S^{2}	S^{2}	S^{2}	S^{2}	S^{2}	S^{2}	n.t.
S. equinus	NSC	n.t.	S^{2}	S^{2}	S^{2}	S^{2}	S^{2}	S^{2}	S^{2}	S^{2}	S^{2}	S^{2}	S^{2}	S^{2}	S^{2}	S^{2}	n.t.
	LNSC	abs.	abs.	abs.	abs.	abs.	abs.	abs.	abs.	abs.	abs.	abs.	abs.	abs.	abs.	abs.	abs.
S. lutetiensis	NSC	n.t.	S^{2}	S^{2}	S^{2}	S^{2}	S^{2}	S^{2}	S^{2}	S^{2}	S^{2}	S^{2}	S^{2}	S^{2}	S^{2}	S^{2}	n.t.
	LNSC	n.t.	S^{2}	S^{2}	S^{2}	S^{2}	S^{2}	S^{2}	S^{2}	S^{2}	S^{2}	S^{2}	S^{2}	S^{2}	S^{2}	S^{2}	n.t.
S. oralis	NSC	S^{1}	S^{2}	R^{2}	R^{2}	S^{2}	S^{2}	S^{2}	S^{2}	S^{2}	S^{2}	S^{2}	S^{2}	S^{2}	S^{2}	S^{2}	I^{2}
	LNSC	n.t.	S^{2}	R^{2}	R^{2}	S^{2}	S^{2}	S^{2}	S^{2}	S^{2}	S^{2}	S^{2}	S^{2}	S^{2}	S^{2}	S^{2}	n.t.
S. salivarius	NSC	S^{1}	I^{2}	R^{2}	R^{2}	S^{2}	S^{2}	S^{2}	S^{2}	S^{2}	S^{2}	S^{2}	S^{2}	S^{2}	S^{2}	S^{2}	S^{2}
	LNSC	abs.	abs.	abs.	abs.	abs.	abs.	abs.	abs.	abs.	abs.	abs.	abs.	abs.	abs.	abs.	abs.

NSC, natural starter culture; LNSC, Lyophilized natural starter culture (LNSC). AMP, Ampicillin; PEN, Penicillin; AZI, Azithromycin; ERY, Erythromycin; FEP, Cefepime; FOT, Cefotaxime; AXO, Ceftriaxone; CHL, Chloramphenicol; CLI, Clindamycin; ETP, Ertapenem; MERO, Meropenem; LEVO, Levofloxacin; LZD, Linezolid; TET, Tetracycline; TGC, Tigecycline; VAN, Vancomycin; TEI, Teicoplanin; SYN, Quinupristin/Dalfopristin; CIP, Ciprofloxacin; S, sensitive; R, resistant; I, intermediate; n.t., not tested; abs., the species was absent in the culture. * one isolate was intermediate. [1] Breakpoint by EUCAST 2023. [2] Breakpoint by CLSI 2020.

4. Discussion

In this study, a natural starter culture, in frozen and lyophilized form, obtained from raw ewe's milk was investigated for its biodiversity and safety, at the strain level, with the perspective of use as a food additive in cheesemaking. The novelty of this work follows the new approach of Chessa et al. [11], where a new method for the obtainment of a natural starter culture directly from raw ewe milk without applying any thermal treatment or selection of pro-technological starter strains, to recover as much of the microbial raw milk biodiversity, was described. The evaluation of the microbial biodiversity was performed both on the frozen natural starter culture (NSC) and on its lyophilized form (LNSC). The latter is easier to handle in cheesemaking at both artisanal- and industrial-scale dairy plants. The bacterial isolates were picked up from the Petri dishes used for plate counts performed in the previous work and revealed the presence of eight species belonging to three genera, with different biotypes for each species. *L. paracasei* was the sole representative of the *Lacticaseibacillus* genus, with one biotype found in NSC and five in LNSC. The genus *Enterococcus* was represented by three species: *E. durans*, with eight strains found only in LNSC; *E. faecalis*, with one biotype detected both in the frozen NSC and in the lyophilized LNSC; and *E. faecium*, of which several biotypes were found (six in NSC and nine in LSNSC), with one of them shared between the cultures' forms. Enterococci are not currently included in the QPS list provided by EFSA every three years and updated every six months, and, therefore, their use as a food additive is not yet recommended [9] since they can be involved in nosocomial infections and antibiotic resistance spread [33]. That being said, their presence in cheese is desirable, due to their well-known contribution to aroma development [34]. Although infections caused by enterococci in humans, outside the nosocomial environment, are not common, the frequency of *E. faecium* isolated from hospitalised patients (i.e., belonging to the clade A) and responsible for infections has increased in the last decades [31]. To exclude the origin of the *E. faecium* strains from clade A, and, therefore, the possibility that they might have an advantage in the gastrointestinal tract given by ampicillin, amoxicillin, or vancomycin resistance, the evaluation of ampicillin susceptibility must be carried out. The EFSA suggested that *E. faecium* isolates having ampicillin MIC ≤ 2 mg/L and negative for the presence of the genes IS*16* (a marker associated with nosocomial strains), *esp* (pathogenicity marker), and hyl_{Efm} (able to facilitate intestinal colonisation) can be considered safe and used as feed additives, assuming the microbial isolate tested is of environmental origin [31]. All the *E. faecium* biotypes investigated in this study complied with the requirements set by EFSA and can be safely used for cheesemaking. Although the assessment of safety for *E. faecium* is well described by the EFSA, this evaluation for other species is still not clear. Recently, the EFSA received notifications of re-assessment for the inclusion of several microorganisms in the QPS list. Among these, nine notifications for *E. faecium* were not evaluated, and this species is still outside the QPS list [8]. Moreover, other taxonomic units found in the starter culture investigated in this study have been evaluated for possible QPS status by the EFSA, concluding that *S. salivarius*, one isolate found only in NSC, is not recommended for the QPS list due to its ability to cause bacteraemia and systemic infection that results in a variety of morbidities [35], even though it is a commensal bacterium in the oral cavity and seems to contribute to human health, preventing biofilm formation [36]. Similarly, *S. oralis*, found both in NSC and LNSC, is a commensal species and opportunistic pathogen, showing low pathogenicity and virulence in immunocompromised patients [37]. Although its virulence mechanism is unclear, it is still not recommended for QPS status due to safety concerns [8]. Both *S. salivarius* and *S. oralis* were found in the starter cultures at dilution -4 for *S. salivarius* and at -5 and -6 for *S. oralis*; therefore, their presence in inoculated milk for cheesemaking is not that likely. Indeed, if the starter culture (10^{10} Log CFU/mL) is inoculated at a final concentration of 10^5 Log CFU/mL, up to 1 to 10 CFU/mL might be present in the inoculated milk. Presumably, their presence in ripened cheese and survival in the gastrointestinal tract after ingestion should not be a cause of concern. To gain a more detailed picture of these strains, for which the safety ascription is still of debate, an antibiotic resistance evaluation was performed in this study. *S. salivarius* and *S. oralis* were sensitive to all the antibiotics tested except for macrolides, revealing resistance at low concentrations (2–4 mg/L) of erythromycin and azithromycin (M-phenotype, 14–15 membered ring macrolides) [32]. At the molecular level, they were positive to the *mef*A gene, coding for an efflux pump, that is correlated to the M-phenotype. High rates (76%) of erythromycin resistance were found in commensal isolates belonging to different sequence types by multilocus sequence typing [36]. Macrolide resistance in the *viridans* group of streptococci from healthy people's oropharynx is reported in the literature [38]. Moreover, the other genes tested for macrolide (*erm*B) and lincosamide (*lnu*C) resistance were absent.

The other *Streptococcus* found in the culture were *S. gallolyticus* subsp. *macedonicus*, *S. equinus*, and *S. lutetiensis*, which are included in the *S. bovis-S. equinux* complex (SBSEC), the non-enterococcal group D streptococci, and, from a safety point of view, could deserve attention [39]. *S. gallolyticus* subsp. *macedonicus*, commonly found in several European cheeses, was isolated for the first time from naturally fermented Greek Kasseri cheese, and it can also be found in Italian cheeses [40]. It is moderately acidifying and proteolytic, potentially contributing to cheese ripening, and can be considered a multifunctional candidate as a nonstarter lactic acid bacterium and adjunct culture for dairy manufacturing since it is non-pathogenic [41]. *S. equinus* and *S. lutetiensis*, inhabitants of the rumen and gastrointestinal tracts of animals and humans, are associated with bovine mastitis [42]. Although considered potential human pathogens [43], they have been isolated from camel, buffalo, and bovine milk, and also from traditional fermented milks from Gambia and Ethiopia, obtained after natural fermentation of raw cow and camel milks [44,45], and from the traditional cheeses Darfiyeh, a Lebanese product from raw goat milk [46], and Mozzarella di Bufala Campana, an Italian protected designation-of-origin (PDO) cheese [47]. Similar to the *Enterococcus* strains isolated from the natural culture, all the *S. gallolyticus* subsp. *macedonicus* and *S. lutetiensis* were negative for antibiotic resistance at the molecular and phenotypic level and, since these species are commonly found in raw milks [33,48], it is believed that they can be used for cheesemaking, even if not included in the QPS list.

Potential pathogens can be introduced by food consumption, and their survival in the human and animal GIT is affected by diet. Therefore, oral, pyogenic, and other streptococci, like those causing mastitis, can be considered food-related microorganisms. The presence of streptococci such as *S. equinus* are considered to be indicators of faecal pollution of food because they have an advantage over coliforms as they are more resistant to most environmental stresses [41]. Therefore, the results of this study show that some *Streptococcus* and *Enterococcus* species were found in the natural starter culture investigated. Indeed, as described by Chessa et al. [11], the culture was obtained from raw milk without applying any thermal treatment but only a slightly acidic condition to remove potential pathogens, and spoilage and hygiene indicators, such as coliforms and staphylococci. Indeed, resilient microorganisms like streptococci or acidic-tolerant like enterococci, already present in the raw milk, survived and were detected in both the natural starter culture NSC and also, after lyophilisation, in LNSC. The potential pathogens found, *E. faecium* and the *Streptococcus* species, were present in the raw ewe milk used in this study and consequently found in the natural starter culture. Furthermore, some SBSEC members such as *S. gallolyticus* subsp. *macedonicus* and *S. lutetiensis* are part of the daily diet, also producing bacteriocins useful for food preservation [49], and can be considered safe, although some strains may be potentially pathogenic [39].

The natural culture investigated, both in frozen and lyophilized form, revealed good biodiversity, both at the species and strain level. Unlike commercial starters, built by few selected species/strains, usually one to three, this was obtained directly from raw ewe milk without any thermal treatment nor isolation to select pro-technological bacteria. Nevertheless, some species found were potentially pathogens and marked for attention by the EFSA as they are not yet included in the QPS list, thus requiring attention before being used as food additives. Nonetheless, all the tests, both at molecular and phenotypic levels, gave reassurances regarding their safety and suggest their suitability for cheesemaking.

5. Conclusions

The obtainment of a natural culture directly from raw ewe milk bypassing the thermal treatment and the selection of pro-technological bacteria may be advantageous in terms of microbial diversity. Conversely to the commercial starters composed of a few species and/or strains cultured separately then artificially mixed together, the natural culture is characterised by a variety of species and strains in a delicate equilibrium that can contribute to the uniqueness and typicity of artisanal products. The assessment of taxonomic identity of microorganisms was the first step in the evaluation of the microbial composition, at the strain level, but this work also aimed to focus attention in terms of food safety since non-QPS microorganisms can be included in the natural starter and can be found in cheeses, especially in traditional ones obtained from raw milk, where thermal treatment is not commonly applied or even not allowed by strict production regulations, like those for the PDOs. The current QPS list does not include some microbial species commonly present in raw milk, inevitably found in traditional products and even in the natural starter culture investigated. Nonetheless, it contained the most important characteristics in addition to technological abilities, i.e., biodiversity and safety. Following the actual EFSA guidelines, artisanal factories may not be able to provide the production of starter cultures from raw milk by themselves, running the risk of including some non-QPS species in their culture, and only commercial selected starters would be allowed for cheese production. Considering the findings of this study, a revision of the criteria for accession to the QPS list should be performed.

Supplementary Materials: The following supporting information can be downloaded at: https://www.mdpi.com/article/10.3390/fermentation10010029/s1, Table S1: Origin of the NSC and LNSC bacterial isolates picked out from the 10-fold dilution of the respective media; Table S2: Molecular primers used for antibiotic resistance genes and virulence genes detection; Table S3: Breakpoints used for the safety evaluation of non-QPS microorganisms.

Author Contributions: Conceptualization, L.C. and R.C.; methodology, L.C., A.P. and R.C.; formal analysis, L.C. and I.D.; investigation, E.D., I.D., M.C.F. and D.G.D.; data curation, L.C., E.D. and I.D.; writing—original draft preparation, L.C.; writing—review and editing, L.C. and R.C.; project administration, R.C.; funding acquisition, R.C. All authors have read and agreed to the published version of the manuscript.

Funding: This research was funded by the Regione Autonoma della Sardegna—Progetto ValIdeS—Valorizzazione e tutela dei sistemi di produzione agroalimentare Identitari del centro Sardegna—L.R.7/2007, "Promozione della ricerca scientifica e dell'innovazione tecnologica in Sardegna" annualità 2020.

Institutional Review Board Statement: Not applicable.

Informed Consent Statement: Not applicable.

Data Availability Statement: Data are contained within the article and Supplementary Materials.

Conflicts of Interest: The authors declare no conflicts of interest.

References

1. Bassi, D.; Puglisi, E.; Cocconcelli, P.S. Comparing natural and selected starter cultures in meat and cheese fermentations. *Curr. Opin. Food Sci.* **2015**, *2*, 118–122. [CrossRef]
2. Powell, I.B.; Broome, M.C.; Limsowtin, G.K.Y. Cheese | Starter Cultures: General Aspects. In *Encyclopedia of Dairy Sciences*, 2nd ed.; Fuquay, J.W., Ed.; Academic Press: San Diego, CA, USA, 2011; pp. 552–558. [CrossRef]
3. Durso, L.; Hutkins, R. Starter cultures. In *Encyclopedia of Food Sciences and Nutrition*, 2nd ed.; Caballero, B., Ed.; Academic Press: Oxford, UK, 2003; pp. 5583–5593. [CrossRef]
4. Chessa, L.; Paba, A.; Daga, E.; Dupré, I.; Piga, C.; Di Salvo, R.; Mura, M.; Addis, M.; Comunian, R. Autochthonous Natural Starter Cultures: A Chance to Preserve Biodiversity and Quality of Pecorino Romano PDO Cheese. *Sustainability* **2021**, *13*, 8214. [CrossRef]
5. Parente, E.; Cogan, T.M.; Powell, I. *Starter Cultures: General Aspects*; Academic Press: Cambridge, MA, USA, 2017; pp. 201–226. [CrossRef]

6. Abarquero, D.; Bodelón, R.; Manso, C.; Rivero, P.; Fresno, J.M.; Tornadijo, M.E. Effect of autochthonous starter and non-starter cultures on the physicochemical, microbiological and sensorial characteristics of Castellano cheese. *Int. J. Dairy Technol.* **2023**. [CrossRef]

7. European Food Safety Authority. The EFSA's 2nd Scientific Colloquium Report—Qps. *EFSA Support. Publ.* **2005**, *2*, 109E. [CrossRef]

8. EFSA Panel on Additives Products or Substances used in Animal Feed. Update of the list of qualified presumption of safety (QPS) recommended microorganisms intentionally added to food or feed as notified to EFSA. *EFSA J.* **2023**, *21*, e07747. [CrossRef]

9. EFSA BIOHAZ Panel (EFSA Panel on Biological Hazards); Koutsoumanis, K.; Allende, A.; Alvarez-Ordonez, A.; Bolton, D.; Bover-Cid, S.; Chemaly, M.; Davies, R.; De Cesare, A.; Hilbert, F.; et al. Statement on the update of the list of QPS-recommended biological agents intentionally added to food or feed as notified to EFSA 15: Suitability of taxonomic units notified to EFSA until September 2021. *EFSA J.* **2022**, *20*, 7045. [CrossRef]

10. EFSA Panel on Additives Products or Substances used in Animal Feed. Guidance on the characterisation of microorganisms used as feed additives or as production organisms. *EFSA J.* **2018**, *16*, e05206. [CrossRef]

11. Chessa, L.; Paba, A.; Dupré, I.; Daga, E.; Fozzi, M.C.; Comunian, R. A Strategy for the Recovery of Raw Ewe's Milk Microbiodiversity to Develop Natural Starter Cultures for Traditional Foods. *Microorganisms* **2023**, *11*, 823. [CrossRef]

12. Di Cello, F.; Pepi, M.; Baldi, F.; Fani, R. Molecular characterization of an n-alkane-degrading bacterial community and identification of a new species, Acinetobacter venetianus. *Res. Microbiol.* **1997**, *148*, 237–249. [CrossRef]

13. Poyart, C. Taxonomic dissection of the *Streptococcus bovis* group by analysis of manganese-dependent superoxide dismutase gene (*sod*A) sequences: Reclassification of 'Streptococcus infantarius subsp. coli' as Streptococcus lutetiensis sp. nov. and of Streptococcus bovis biotype II.2 as Streptococcus pasteurianus sp. nov. *Int. J. Syst. Evol. Microbiol.* **2002**, *52*, 1247–1255. [CrossRef]

14. Odamaki, T.; Yonezawa, S.; Kitahara, M.; Sugahara, Y.; Xiao, J.-Z.; Yaeshima, T.; Iwatsuki, K.; Ohkuma, M. Novel multiplex polymerase chain reaction primer set for identification of *Lactococcus* species. *Lett. Appl. Microbiol.* **2011**, *52*, 491–496. [CrossRef] [PubMed]

15. Cremonesi, P.; Vanoni, L.; Morandi, S.; Silvetti, T.; Castiglioni, B.; Brasca, M. Development of a pentaplex PCR assay for the simultaneous detection of *Streptococcus thermophilus, Lactobacillus delbrueckii* subsp. bulgaricus, L. delbrueckii subsp. lactis, L. helveticus, L. fermentum in whey starter for Grana Padano cheese. *Int. J. Food Microbiol.* **2011**, *146*, 207–211. [CrossRef] [PubMed]

16. Chagnaud, P.; Machinis, K.; Coutte, L.c.A.; Marecat, A.; Mercenier, A. Rapid PCR-based procedure to identify lactic acid bacteria: Application to six common *Lactobacillus* species. *J. Microbiol. Methods* **2001**, *44*, 139–148. [CrossRef] [PubMed]

17. Ward, L.J.; Timmins, M.J. Differentiation of *Lactobacillus casei, Lactobacillus paracasei* and *Lactobacillus rhamnosus* by polymerase chain reaction. *Lett. Appl. Microbiol.* **1999**, *29*, 90–92. [CrossRef] [PubMed]

18. Torriani, S.; Felis, G.E.; Dellaglio, F. Differentiation of *Lactobacillus plantarum, L. pentosus*, and *L. paraplantarum* by *rec*A gene sequence analysis and multiplex PCR assay with *rec*A gene-derived primers. *Appl. Environ. Microbiol.* **2001**, *67*, 3450–3454. [CrossRef] [PubMed]

19. Ke, D.; Picard, F.J.; Martineau, F.; Menard, C.; Roy, P.H.; Ouellette, M.; Bergeron, M.G. Development of a PCR assay for rapid detection of enterococci. *J. Clin. Microbiol.* **1999**, *37*, 3497–3503. [CrossRef] [PubMed]

20. Jackson, C.R.; Fedorka-Cray, P.J.; Barrett, J.B. Use of a Genus- and Species-Specific Multiplex PCR for Identification of Enterococci. *J. Clin. Microbiol.* **2004**, *42*, 3558–3565. [CrossRef]

21. Chessa, L.; Paba, A.; Daga, E.; Comunian, R. Effect of growth media on natural starter culture composition and performance evaluated with a polyphasic approach. *Int. J. Dairy Technol.* **2019**, *72*, 152–158. [CrossRef]

22. Graves, L.M.; Swaminathan, B. PulseNet standardized protocol for subtyping *Listeria monocytogenes* by macrorestriction and pulsed-field gel electrophoresis. *Int. J. Food Microbiol.* **2001**, *65*, 55–62. [CrossRef]

23. Tenover, F.C.; Arbeit, R.D.; Goering, R.V.; Mickelsen, P.A.; Murray, B.E.; Persing, D.H.; Swaminathan, B. Interpreting chromosomal DNA restriction patterns produced by pulsed-field gel electrophoresis: Criteria for bacterial strain typing. *J. Clin. Microbiol.* **1995**, *33*, 2233–2239. [CrossRef]

24. Paba, A.; Chessa, L.; Cabizza, R.; Daga, E.; Urgeghe, P.P.; Testa, M.C.; Comunian, R. Zoom on starter lactic acid bacteria development into oxytetracycline spiked ovine milk during the early acidification phase. *Int. Dairy J.* **2019**, *96*, 15–20. [CrossRef]

25. 20776-1:2019; Susceptibility Testing of Infectious Agents and Evaluation of Performance of Antimicrobial Susceptibility Test Devices. In Part 1: Broth Micro-Dilution Reference Method for Testing the In Vitro Activity of Antimicrobial Agents against Rapidly Growing Aerobic Bacteria Involved in Infectious Diseases. International Organization for Standardization (ISO): Geneva, Switzerland, 2019.

26. Clinical and Laboratory Standards Institute (CLSI). *M100—Performance Standards for Antimicrobial Susceptibility Testing, Clinical and Laboratory Standards Institute*, 30th ed.; Clinical and Laboratory Standards Institute (CLSI): Wayne, PA, USA, 2020.

27. The European Committee on Antimicrobial Susceptibility Testing. Breakpoint Tables for Interpretation of MICs and Zone Diameters. *Version 13.0.* 2023. Available online: http://www.eucast.org (accessed on 15 October 2023).

28. Hunter, P.R.; Gaston, M.A. Numerical index of the discriminatory ability of typing systems: An application of Simpson's index of diversity. *J. Clin. Microbiol.* **1988**, *26*, 2465–2466. [CrossRef] [PubMed]

29. Dekker, J.P.; Lau, A.F. An Update on the Streptococcus bovis Group: Classification, Identification, and Disease Associations. *J. Clin. Microbiol.* **2016**, *54*, 1694–1699. [CrossRef] [PubMed]

30. Hinse, D.; Vollmer, T.; Erhard, M.; Welker, M.; Moore, E.R.; Kleesiek, K.; Dreier, J. Differentiation of species of the *Streptococcus bovis/equinus*-complex by MALDI-TOF Mass Spectrometry in comparison to *sod*A sequence analyses. *Syst. Appl. Microbiol.* **2011**, *34*, 52–57. [CrossRef] [PubMed]

31. EFSA Panel on Additives Products or Substances used in Animal Feed. Guidance on the safety assessment of *Enterococcus faecium* in animal nutrition. *EFSA J.* **2012**, *10*, 2682. [CrossRef]

32. Leclercq, R. Mechanisms of resistance to macrolides and lincosamides: Nature of the resistance elements and their clinical implications. *Clin. Infect. Dis.* **2002**, *34*, 482–492. [CrossRef] [PubMed]

33. Jans, C.; Meile, L.; Kaindi, D.W.M.; Kogi-Makau, W.; Lamuka, P.; Renault, P.; Kreikemeyer, B.; Lacroix, C.; Hattendorf, J.; Zinsstag, J.; et al. African fermented dairy products—Overview of predominant technologically important microorganisms focusing on African *Streptococcus infantarius* variants and potential future applications for enhanced food safety and security. *Int. J. Food Microbiol.* **2017**, *250*, 27–36. [CrossRef]

34. Ogier, J.-C.; Serror, P. Safety assessment of dairy microorganisms: The *Enterococcus* genus. *Int. J. Food Microbiol.* **2008**, *126*, 291–301. [CrossRef]

35. Hazards, E. Panel o.B.; Koutsoumanis, K.; Allende, A.; Alvarez-Ordóñez, A.; Bolton, D.; Bover-Cid, S.; Chemaly, M.; Davies, R.; De Cesare, A.; Hilbert, F.; et al. Update of the list of QPS-recommended microbiological agents intentionally added to food or feed as notified to EFSA 16: Suitability of taxonomic units notified to EFSA until March 2022. *EFSA J.* **2022**, *20*, e07408. [CrossRef]

36. Chaffanel, F.; Charron-Bourgoin, F.; Libante, V.; Leblond-Bourget, N.; Payot, S. Resistance Genes and Genetic Elements Associated with Antibiotic Resistance in Clinical and Commensal Isolates of *Streptococcus salivarius*. *Appl. Environ. Microbiol.* **2015**, *81*, 4155–4163. [CrossRef]

37. Joyce, L.R.; Youngblom, M.A.; Cormaty, H.; Gartstein, E.; Barber, K.E.; Akins, R.L.; Pepperell, C.S.; Palmer, K.L. Comparative Genomics of *Streptococcus oralis* Identifies Large Scale Homologous Recombination and a Genetic Variant Associated with Infection. *mSphere* **2022**, *7*, e00509-22. [CrossRef] [PubMed]

38. Malhotra-Kumar, S.; Lammens, C.; Martel, A.; Mallentjer, C.; Chapelle, S.; Verhoeven, J.; Wijdooghe, M.; Haesebrouck, F.; Goossens, H. Oropharyngeal carriage of macrolide-resistant viridans group streptococci: A prevalence study among healthy adults in Belgium. *J. Antimicrob. Chemother.* **2004**, *53*, 271–276. [CrossRef] [PubMed]

39. Pompilio, A.; Di Bonaventura, G.; Gherardi, G. An Overview on *Streptococcus bovis/Streptococcus equinus* Complex Isolates: Identification to the Species/Subspecies Level and Antibiotic Resistance. *Int. J. Mol. Sci.* **2019**, *20*, 480. [CrossRef] [PubMed]

40. Blaiotta, G.; Sorrentino, A.; Ottombrino, A.; Aponte, M. Short communication: Technological and genotypic comparison between *Streptococcus macedonicus* and *Streptococcus thermophilus* strains coming from the same dairy environment. *J. Dairy Sci.* **2011**, *94*, 5871–5877. [CrossRef] [PubMed]

41. Gobbetti, M.; Calasso, M. STREPTOCOCCUS | Introduction. In *Encyclopedia of Food Microbiology*, 2nd ed.; Batt, C.A., Tortorello, M.L., Eds.; Academic Press: Oxford, UK, 2014; pp. 535–553. [CrossRef]

42. Chen, P.; Qiu, Y.; Liu, G.; Li, X.; Cheng, J.; Liu, K.; Qu, W.; Zhu, C.; Kastelic, J.P.; Han, B.; et al. Characterization of *Streptococcus lutetiensis* isolated from clinical mastitis of dairy cows. *J. Dairy Sci.* **2021**, *104*, 702–714. [CrossRef] [PubMed]

43. Yu, A.T.; Shapiro, K.; Beneri, C.A.; Wilks-Gallo, L.S. *Streptococcus lutetiensis* neonatal meningitis with empyema. *Access Microbiol.* **2021**, *3*, 000264. [CrossRef]

44. Fugl, A.; Berhe, T.; Kiran, A.; Hussain, S.; Laursen, M.F.; Bahl, M.I.; Hailu, Y.; Sørensen, K.I.; Guya, M.E.; Ipsen, R.; et al. Characterisation of lactic acid bacteria in spontaneously fermented camel milk and selection of strains for fermentation of camel milk. *Int. Dairy J.* **2017**, *73*, 19–24. [CrossRef]

45. Baldeh, E.; Bah, S.A.; Camara, S.; Fye, B.L.; Nakamura, T. Bacterial diversity of Gambian traditional fermented milk, "Kosam". *Anim. Sci. J.* **2022**, *93*, e13699. [CrossRef]

46. Alexandraki, V.; Kazou, M.; Angelopoulou, A.; Arena, M.P.; Capozzi, V.; Russo, P.; Fiocco, D.; Spano, G.; Papadimitriou, K.; Tsakalidou, E. Chapter 6—The Microbiota of Non-cow Milk and Products. In *Non-Bovine Milk and Milk Products*; Tsakalidou, E., Papadimitriou, K., Eds.; Academic Press: San Diego, CA, USA, 2016; pp. 117–159. [CrossRef]

47. Devirgiliis, C.; Barile, S.; Caravelli, A.; Coppola, D.; Perozzi, G. Identification of tetracycline- and erythromycin-resistant Gram-positive cocci within the fermenting microflora of an Italian dairy food product. *J. Appl. Microbiol.* **2010**, *109*, 313–323. [CrossRef]

48. Quigley, L.; O'Sullivan, O.; Stanton, C.; Beresford, T.P.; Ross, R.P.; Fitzgerald, G.F.; Cotter, P.D. The complex microbiota of raw milk. *FEMS Microbiol. Rev.* **2013**, *37*, 664–698. [CrossRef]

49. Tarrah, A.; Callegaro, S.; Pakroo, S.; Finocchiaro, R.; Giacomini, A.; Corich, V.; Cassandro, M. New insights into the raw milk microbiota diversity from animals with a different genetic predisposition for feed efficiency and resilience to mastitis. *Sci. Rep.* **2022**, *12*, 13498. [CrossRef] [PubMed]

fermentation

MDPI

Article

Screening of Acetic Acid Bacteria Isolated from Various Sources for Use in Kombucha Production

Dong-Hun Lee [1,†] , Su-Hwan Kim [2,†], Chae-Yun Lee [1], Hyeong-Woo Jo [1], Won-Hee Lee [1], Eun-Hye Kim [1], Byung-Kuk Choi [3] and Chang-Ki Huh [1,2,*]

[1] Department of Food Science and Technology, Sunchon National University, Suncheon 57922, Republic of Korea; glehdgns@naver.com (D.-H.L.); imgoy1536@naver.com (C.-Y.L.); cloud4502@naver.com (H.-W.J.); patawon759@naver.com (W.-H.L.); kimeunhye515@naver.com (E.-H.K.)

[2] Research Institute of Food Industry, Sunchon National University, Suncheon 57922, Republic of Korea; suhwan010@scnu.ac.kr

[3] Fermented Food Industry Support Center, Suncheon 57908, Republic of Korea; suncheon82@ffic.kr

* Correspondence: hck1008@sunchon.ac.kr; Tel.: +82-61-750-3251

† These authors contributed equally to this work.

Abstract: The objective of this study was to isolate and identify strains of *Acetobacter* suitable for use in the development of a complex microbial culture for producing Kombucha and to examine the fermentation characteristics for selection of suitable strains. A medium supplemented with calcium carbonate was used for isolation of acetic acid bacteria from 22 various sources. Colonies observed in the clear zone resulting from decomposition of calcium carbonate by acid produced by microorganisms were collected. Identification of the collected strains was based on biological and morphological characteristics, and the results of base sequence analysis. A total of 37 strains were identified, including six species in the *Acetobacter* genus: *Acetobacter pasteurianus, Acetobacter orientalis, Acetobacter cibinongensis, Acetobacter pomorum, Acetobacter ascendens*, and *Acetobacter malorum*, as well as one species in the *Gluconobacter* genus, *Gluconobacter oxydans*. Among thirty-seven strains, seven strains of acetic acid bacteria with exceptional acid and alcohol tolerance were selected, and an evaluation of their fermentation characteristics according to fermentation temperature and period was performed. The results showed a titratable acidity of 1.68% for the *Acetobacter pasteurianus* SFT-18 strain, and an acetic acid bacteria count of 9.52 log CFU/mL at a fermentation temperature of 35 °C. The glucuronic acid and gluconate contents for the *Gluconobacter oxydans* SFT-27 strain were 10.32 mg/mL and 25.49 mg/mL, respectively.

Keywords: Kombucha; *Acetobacter*; *Gluconobacter*; glucuronic acid; fermentation characteristics

check for updates

Citation: Lee, D.-H.; Kim, S.-H.; Lee, C.-Y.; Jo, H.-W.; Lee, W.-H.; Kim, E.-H.; Choi, B.-K.; Huh, C.-K. Screening of Acetic Acid Bacteria Isolated from Various Sources for Use in Kombucha Production. *Fermentation* **2024**, *10*, 18. https://doi.org/10.3390/fermentation10010018

Academic Editors: Luca Settanni, Roberta Comunian and Luigi Chessa

Received: 23 November 2023
Revised: 11 December 2023
Accepted: 22 December 2023
Published: 26 December 2023

1. Introduction

The oxidative fermentation capacity of acetic acid bacteria (AAB) is known to involve an incomplete oxidation process where the substrate is oxidized by dehydrogenase, leading to release of the resulting oxidized product [1]. Nineteen genera of AAB, including *Acetobacter, Gluconobacter, Gluconacetobacter*, and others, have been recognized based on the results of genetic analysis and their respective characteristics [2]. The presence of AAB has been detected in a variety of foods; *Acetobacter aceti* (*A. aceti*), *Acetobacter pasteurianus* (*A. pasteurianus*), *Acetobacter malorum* (*A. malorum*), and *Acetobacter pomorum* (*A. pomorum*) are the most frequently isolated species in the process of vinegar fermentation [3,4].

Growth of *Gluconobacter* (*Glu.*), a Gram-negative, rod-shaped acetic acid bacterium, can cause incomplete oxidization of a wide range of carbohydrates and alcohols, which can occur in highly concentrated sugar solutions and at low pH. *Gluconobacter* is used extensively in industrial processes in the production of gluconic acid from glucose and sorbose from L-sorbitol [5]. *A. aceti, A. pasteurianus, Glu. europaeus, Glu. hanseni*, and *Glu. oxydans* species have received approval from the Korean Ministry of Food and Drug

Safety as generally recognized as safe (GRAS) food materials for use in the production of vinegar [6,7].

Acetic acid bacteria and gluconic acid-producing bacteria, mainly *Komagataeibacter xylinus* (*K. xylinus*), *Bacterium. gluconicum*, *A. aceti*, *A. pasteurianus*, *A. musti*, *Glucobacter oxygendans* (*G. oxygendans*), and *Glu. potus*, are the dominant prokaryotes found in Kombucha cultures [8]. Among them, an association of *K. xylinus* with the production of cellulose biofilms floating on the surface of tea broth in Kombucha has been reported [9].

Kombucha, a fermented beverage, is produced by introducing a symbiotic culture of bacteria and yeast (SCOBY) into a mixture created by combining sugar with water brewed from green or black tea. This beverage was reportedly administered in Ancient China as a remedy for various infirmities during the period of Emperor Qin Shi Huangdi, and it is believed that it was first distributed from Russia to Eastern Europe, traveling by trade routes, and gained popularity in Germany during the 19th century and then expanded to European countries [10,11].

Metabolism of microorganisms by a SCOBY, a cellulose biofilm formed during fermentation of Kombucha, occurs in the production of a variety of functional substances during the process of Kombucha fermentation. In addition, use of SCOBY has been attempted in various fields of active research, not only for medical applications that better support high-water holding capacity and strength compared to the properties of plant cellulose, but also in a range of commercial applications through the synthesis of bioactive compounds containing bacterial cellulose with fine structures [12]. A SCOBY, composed of a mixture of bacteria and yeast used in the preparation of foods and beverages, contains particular genera of bacteria and yeasts, including *Gluconobacter*, *Acetobacter*, *Zygosaccharomyces*, *Saccharomyces* sp., and *Schizosaccharomyces* [13]. In the process of symbiotic fermentation, yeast is responsible for converting sugar into alcohol, while acetic acid bacteria utilize alcohol and sugar to produce acetic acid and gluconic acid [14,15].

The production of Kombucha involves fermentation through cooperation of specific bacteria and yeast, using a SCOBY composed of various species of bacteria and yeast [16]. The flavor profile of Kombucha is significantly influenced by the resulting microbial compositions and fermentation conditions. In addition, this process of fermentation can yield substances that include polyphenols, amino acids, organic acids (including acetic acid, gluconic acid, and glucuronic acid), minerals, vitamins, and D-saccharic acid 1,4-lactone (DSL), which contribute to its proven health benefits, including antioxidant effects, promotion of digestion, skin health, antimicrobial properties, and others [17,18].

Despite extensive research on the efficacy and marketability of Kombucha, focus on the development of standardized manufacturing methods has been limited. This includes use of fermentation techniques that can ensure consistent culture time, temperature, substrate, and additive parameters. In particular, the preparation of Kombucha is currently reliant on the use of imported Kombucha powder and SCOBY starter.

These are important considerations because various variables are dependent on microbial composition. Particularly during fermentation, the challenge of producing exceptional fermented products with consistent functionalities (including gluconic acid, glucuronic acid content, and antioxidant activity, etc.) is more complex. Therefore, the objective of this study was to identify isolated strains suitable for the composition of acetic acid bacteria among complex microbial cultures, which are major bacterial components.

2. Materials and Methods

2.1. Materials

Vinegar starter, plum extract, and wine (Suncheon, Republic of Korea) were supplied by the Food Fermentation Engineering Laboratory, Department of Food Engineering, Sunchon National University. Nine types of fruits (Suncheon, Republic of Korea) were obtained from Suncheon Agricultural Products Wholesale Market, and the collection of bacteria from the surface of fruit was performed using a 3M Pipette Swab Plus⁺ (3M Korea Ltd., Seoul, Republic of Korea). Eight types of commercial fruit vinegars (Jangseong and

Namwon, Republic of Korea) and commercially available Kombucha (Masontops, North York, ON, Canada) were purchased for use as samples in the isolation of acetic acid bacteria.

2.2. Reagents

Yeast extract (Life Technologies Co., Miami, FL, USA), D-(+)-Glucose (Sigma-Aldrich Co., Louis, MO, USA), $CaCO_3$ (Taekyung Bk Co., Seoul, Republic of Korea), mannitol (Junsei Chemical Co., Chuo-ku, Tokyo), peptone (Duksan Pure Chemical Co., Ansan-si, Republic of Korea), and ethyl alcohol anhydrous (Daejung, Siheung-si, Republic of Korea) were purchased for use in preparation of medium. The medium used for isolation and selection of acetic acid bacteria contained YGCE agar (1.0% Yeast extract, 5.0% Glucose, 2.5% $CaCO_3$, 4.0% Ethanol, 2.0% Agar,) and MA agar (0.5% Yeast extract, 2.5% Mannitol, 0.3% Peptone, 1.0% $CaCO_3$, 1.5% Agar). The medium used for screening the most suitable strains according to fermentation characteristics contained YGE broth (1% Yeast extract, 5% Glucose, 3% Ethanol).

2.3. Isolation, Screening, and Identification of the Most Suitable Strains of Acetic Acid Bacteria

2.3.1. Isolation of Acetic Acid Bacteria

The 22 collected samples diluted with 0.85% NaCl were spread on isolation plate medium (YGCE agar and MA agar), 200 µL each, followed by incubation at 30 °C for three days. Isolation of pure bacterial strains from subcultures was repeated three times. This process was based on formation of clear zones around the colony, which could be easily observed by the naked eye [19]. The isolated strains were transplanted onto slant agar medium (1.0% Yeast extract, 5.0% Glucose, 4.0% Ethanol, 2.0% Agar) and used in experiments for selection of the most suitable strain.

2.3.2. Screening and Identification of Acetic ACID Bacteria

Screening of acetic acid bacteria was based on morphological and biological characteristics. Gram staining and simple staining were performed for microscopic examination to determine the morphology [20]. For biological evaluation, one drop of $Fecl_3$ solution was added to 1 mL of strain culture solution for testing of gluconic acid based on change of color from yellow to dark brown (Figure 1). For the catalase test, bacterial isolates were obtained from the surface of a sterilized glass slide using a loop according to the Reiner [21] method, followed by addition of one drop of 3% hydrogen peroxide for detection of bubbles (O_2 + water = bubbles).

(A)   **(B)**

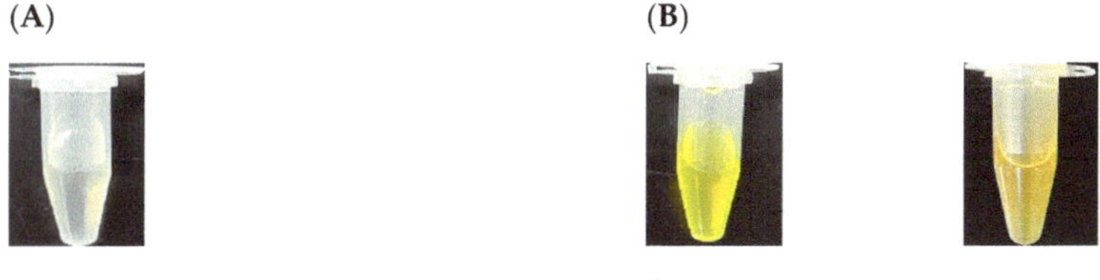

Figure 1. Changes in gluconic acid test of Gluconobacter strains. (**A**) Before reaction. (**B**) After reaction (+, positive; −, negative).

Colony PCR was performed according to the method reported by Wan et al. [22] using 785F (5′-GGA TTA GAT ACC CTG GTA-3′) and 907R (5′-CCG TCA ATT CMT TTR AGT TT-3′) primers, and sequencing of 16S ribosomal RNA in the screened bacterial colony was requested from Macrogen Inc. (Seoul, Republic of Korea) for confirmation of both forward (5′) and reverse (3′) directions. The analyzed DNA sequences were inserted into the BLAST (Basic Local Alignment Search Tool) program provided by the NCBI (National Center for Biotechnology Information (http://www.ncbi.nlm.nih.gov/)(accessed on 17 January 2023) for comparison with a search of the sequence database for identification of homologous sequences and to determine the systematical genetic relationship [23].

2.4. Selection of the Most Suitable Strains According to Fermentation Characteristics

2.4.1. Measurement of pH and Titratable Acidity (TTA)

The pH values for each strain were measured in 10 mL of culture media using a pH meter (HM-40X, Dkk-toa Co., Shinjuku-ku, Tokyo, Japan). The total amount of acid was shown as the percentage of acetic acid (%) after calculating the amount of solution used in neutralizing 2 mL of supernatant obtained from the centrifuged sample using 0.1 N NaOH until reaching a pH level of 8.3 after addition of 2–3 drops of 1% phenolphthalein [24].

2.4.2. Acid Resistance

YGE broth was used as the medium for determining the resistance level of the strains at various concentrations of acid. 1 N HCl and 1 N NaOH was added to the YGE broth for adjustment of the pH range to 4.0–8.0. pH-adjusted YGE broth was inoculated with 1% of each target strain, followed by culture at 30 °C. Absorbance was measured at 660 nm using a microplate reader (SPECTROstarNano, BMG Labtech, Ortenberg, Germany) for determination of growth rates according to incubation periods, which were presented as a percentage (%) compared to the controls.

2.4.3. Alcohol Tolerance

YG broth, consisting of 1% yeast extract and 5% glucose, was used as the medium for evaluating the tolerance level of the strains at various concentrations of alcohol. Ethyl alcohol anhydrous (Daejung, Siheung-si, Republic of Korea) was added for adjustment of the alcohol concentration of the YG broth to 2.0–10%. YG broth with the adjusted concentration of ethanol was inoculated with 1% of each target strain, followed by culture at 30 °C. Absorbance was measured at 660 nm using a microplate reader (SPECTROstarNano, BMG Labtech, Ortenberg, Germany) for determination of growth rates according to incubation periods, which were presented as a percentage (%) compared to the controls.

2.5. Viable Cell Count of Acetic Acid Bacteria

A standard plate count (SPC) was used for counting the number of viable cells in acetic acid bacteria according to fermentation temperature and period. Dilution of each sample with sterile diluent (0.85% NaCl) was performed in a step-by-step manner using the decimal dilution method, followed by plating of 1 mL of each diluted sample on YGE agar medium and incubation at 30 °C for three days. The average number of colonies was determined from the results of three independent experiments for calculation of the colony count, which was expressed as log CFU (colony forming units)/mL [25,26].

2.6. Content of Gluconate and Glucuronic Acid

Measurement of the gluconate and glucuronic acid content was performed using a modified version of the method reported by Ansari et al. [27]. The culture solution was centrifuged at 1000 rpm for 3 min (HA-1000-3, Hanil Science Industrial Co., Ltd., Incheon, Republic of Korea), followed by filtering of the supernatant through a 0.45 μm membrane filter (PVDF 25 mm, Chromdisc, Daegu, Republic of Korea), and analysis was then performed using HPLC (Waters 1525 and 717, Waters Co., Milford, MA, USA). A Supelcogel c-610h column (30 cm × 7.8 mm, Supelco, Bellefonte, PA, USA) was used with an oven temperature of 30 °C. The mobile phase was composed of 0.1% phosphoric acid with a flow rate of 0.5 mL/min. UV detection was measured at 210 nm using a Waters 996 detector (Waters Co., Milford, MA, USA). Sodium gluconate and D-glucuronic acid (Sigma-Aldrich Co., Louis, MO, USA) were used as standard reference materials (SRMs) and the content was presented using the external standard method.

2.7. Statistical Analysis

For statistical analyses, the experiments were repeated three times or more and analysis of the data was performed using IBM SPSS Statistics version 27 (IBM Corp., Armonk, NY,

USA). Calculations of Mean ± SD and testing for significant difference of mean values were performed using Duncan's multiple range test ($p < 0.05$).

3. Results and Discussion

3.1. Isolation, Screening, and Identification of the Most Suitable Strains for Production of Acetic Acid and Gluconic Acid Bacteria for Kombucha Fermentation

3.1.1. Isolation and Selection of Acetic Acid Bacteria

For isolation of acetic acid bacteria, the 22 collected samples were spread on YGCE and MA agar media and 42 pure strains were isolated according to the size of clear zones formed around the colony (Figure 2). The morphological and biological characteristics of the isolated pure strains are shown in Table 1. Morphologically, most of the isolated strains were Gram-negative bacilli. FPA-4, FPP-1, FPP-3, FPP-4, and FPS-3 were identified as Gram-positive streptococci. The color of colonies was brown in most strains, while a white color was observed in colonies produced by FPA-4, FPP-1, FPP-3, FPP-4, and FPS-3. Regarding biological characteristics, the negative decomposition ability of mannitol was observed in FPP-1, FPP-3, FPP-4, and FPS-3 and a positive result was obtained from the remaining strains. A negative result was obtained for biofilm formation and the catalase-test in FPA-4, FPP-1, FPP-3, FPP-4, and FPS-3, and a positive result was obtained from the remaining strains. Strains FPA-4, FPP-1, FPP-3, FPP-4, and FPS-3 were identified as lactic acid bacteria based on the white colony color and negative results on the Gram-positive and catalase test [28]. The remaining strains exhibited characteristics identical to those of *Acetobacter* sp. (Figure 3) including Gram-negative, bacillus, obligate aerobe, and biofilm formation [29].

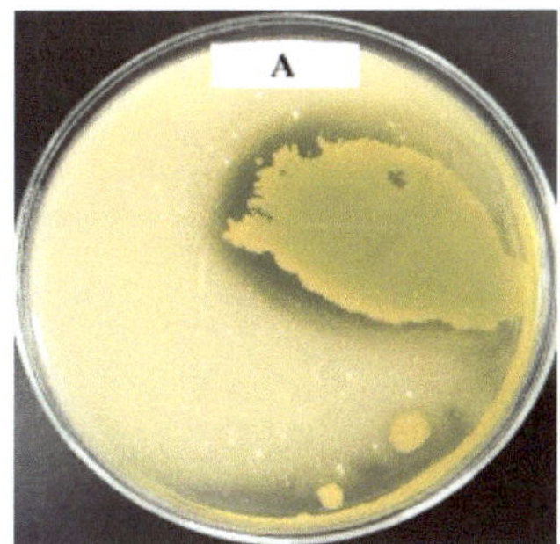
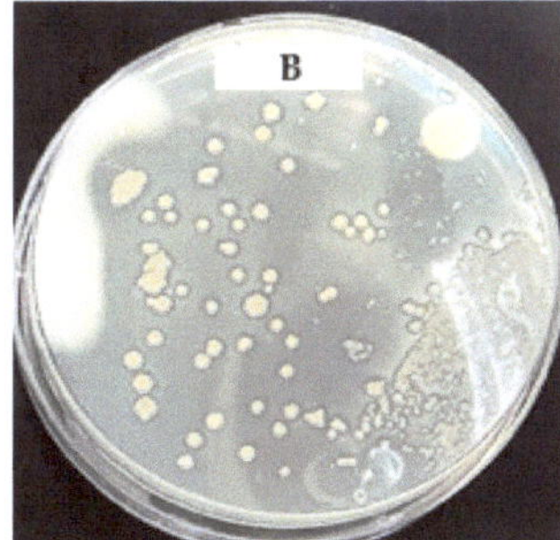

Figure 2. Isolated strains formed clear zones around the colony. (**A**) Clear zone of SMC-4 strain. (**B**) Clear zone of VVJ-2 strain.

The production of 5-keto- and 2-ketogluconic acids by strains of *Gluconobacter* is known to occur by partial oxidation of the carbon source (D-glucose) and alcohol. A dark yellow color was observed for gluconic acid, with Fe^{3+} oxidation-reduction in iron (II) ions of iron (III) chloride by the hydroxy group [30,31]. The result of the gluconic acid test was positive only for VVJ-1 and VVJ-2, which were identified as *Glucobobacter* sp. based on positive results on the catalase-test and the presence/absence of biofilm formation [32,33]. Following isolation and selection of acetic acid bacteria, 37 out of 42 species were confirmed as strains of acetic acid bacteria and sequencing of their 16S rDNA gene was performed.

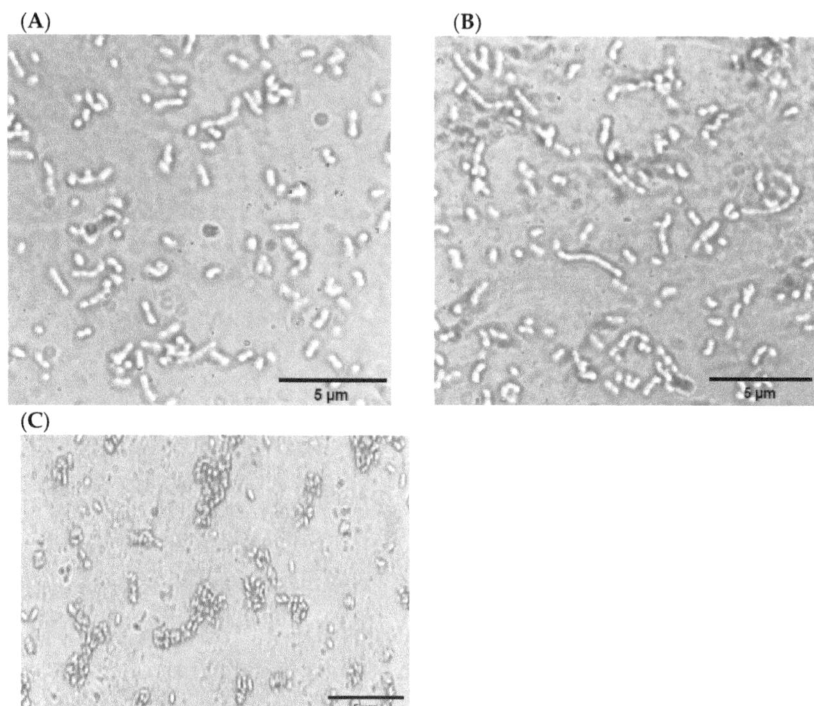

Figure 3. Cell characteristics of strains isolated from domestic fermented foods and produce. (**A**) Microscopic examination of SMC-4 strain, (**B**) microscopic examination of VVJ-2 strain, (**C**) Gram-negative reaction result of acetic acid bacteria.

Table 1. Morphological, biological, and fermentation characteristics of 42 strains isolated from collected samples.

Isolate	Morphological		Biological			D-Mannitol Assimilation
	Colony Morphology	Gram Staining	Gluconic Acid Test	Biofilm Formation	Catalase Test	
SVC-04	rod-shaped, light brown	−	−	+	+	+
SVC-12	rod-shaped, light brown	−	−	+	+	+
SVC-14	rod-shaped, light brown	−	−	+	+	+
SVC-22	rod-shaped, light brown	−	−	+	+	+
SVC-38	rod-shaped, light brown	−	−	+	+	+
SVC-49	rod-shaped, light brown	−	−	+	+	+
SVC-410	rod-shaped, light brown	−	−	+	+	+
SVC-54	rod-shaped, light brown	−	−	+	+	+
SMC-1	rod-shaped, light brown	−	−	+	+	+
SMC-2	rod-shaped, light brown	−	−	+	+	+
SMC-3	rod-shaped, light brown	−	−	+	+	+
SMC-4	rod-shaped, light brown	−	−	+	+	+
SMC-5	rod-shaped, light brown	−	−	+	+	+
FPA-1	rod-shaped, reddish brown	−	−	+	+	+
FPA-2	rod-shaped, reddish brown	−	−	+	+	+
FPA-3	rod-shaped, reddish brown	−	−	+	+	+
FPA-4	Coccus, white	+	−	−	−	+
JGV-1	rod-shaped, light brown	−	−	+	+	+
JGV-2	rod-shaped, light brown	−	−	+	+	+
FPP-1	Coccus, white	+	−	−	−	−
FPP-3	Coccus, white	+	−	−	−	−
FPP-4	Coccus, white	+	−	−	−	−
FPS-3	Coccus, white	+	−	−	−	−
FPS-4	rod-shaped, light brown	−	−	+	+	+

Table 1. *Cont.*

Isolate	Morphological			Biological			D-Mannitol Assimilation
	Colony Morphology	**Gram Staining**	**Gluconic Acid Test**	**Biofilm Formation**	**Catalase Test**		
ACJ-1	rod-shaped, light brown	−	−	+	+		+
ACJ-2	rod-shaped, light brown	−	−	+	+		+
MPV-1	rod-shaped, light brown	−	−	+	+		+
PVJ-1	rod-shaped, light brown	−	−	+	+		+
PVJ-4	rod-shaped, light brown	−	−	+	+		+
PVJ-5	rod-shaped, light brown	−	−	+	+		+
VVJ-1	rod-shaped, reddish brown	−	+	+	+		+
VVJ-2	rod-shaped, reddish brown	−	+	+	+		+
AVJ-3	rod-shaped, light brown	−	−	+	+		+
PEV-1	rod-shaped, light brown	−	−	+	+		+
PEV-4	rod-shaped, light brown	−	−	+	+		+
URV-1	rod-shaped, light brown	−	−	+	+		+
URV-2	rod-shaped, light brown	−	−	+	+		+
KS-1	rod-shaped, light brown	−	−	+	+		+
KS-2	rod-shaped, light brown	−	−	+	+		+
KS-3	rod-shaped, light brown	−	−	+	+		+
PS-1	rod-shaped, light brown	−	−	+	+		+
WS-1	rod-shaped, light brown	−	−	+	+		+

+: positive or activated, −: negative or inactive.

3.1.2. Identification of Isolated Strains

The results from the identification of 37 strains of acetic acid based on their 16S rDNA gene sequences are shown in Table 2. Sixteen strains in the genus *A. pasteurianus*, three strains in *A. orientalis*, one strain in *A. cibinongensis*, seven strains in *A. pomorum*, three strains in *A. ascendens*, and five strains in *A. malorum* were identified. Two strains of *Glu. oxydans* in the genus *Gluconobacter* were identified.

Table 2. Identification of acetic acid bacteria isolated from 22 samples.

Strains No.	Species	Identities (%)	Strain Distinction	Source
SVC-04		99.9	SFT-1	
SVC-12		99.7	SFT-2	
SVC-14		99.7	SFT-3	
SVC-22	*Acetobacter pasteurianus* [1]	99.9	SFT-4	Vinegar, Sunchon University b3 115, Republic of Korea
SVC-38		99.9	SFT-5	
SVC-49		99.9	SFT-6	
SVC-410		99.8	SFT-7	
SVC-54		99.5	SFT-8	
FPA-1		99.8	SFT-9	
FPA-2	*Acetobacter orientalis*	99.9	SFT-10	*Prunus armeniaca* (surface), Suncheon-si, Jeollanam-do, Republic of Korea
FPA-3		99.9	SFT-11	
FPS-4	*Acetobacter cibinongensis*	99.6	SFT-12	*Prunus salicina* (surface), Suncheon-si, Jeollanam-do, Republic of Korea
JGV-1	*Acetobacter pasteurianus* [1]	99.8	SFT-13	Vinegar (persimmon), Jangseong-gun, Jeollanam-do, Republic of Korea
JGV-2		99.6	SFT-14	
SMC-1		99.7	SFT-15	
SMC-2		99.5	SFT-16	
SMC-3	*Acetobacter pasteurianus* [1]	99.8	SFT-17	Maesil cheong, Sunchon University b3 115, Republic of Korea
SMC-4		99.6	SFT-18	
SMC-5		99.7	SFT-19	
ACJ-1	*Acetobacter pomorum*	99.4	SFT-20	Vinegar (*Ananas comosus*), Jangseong-gun, Jeollanam-do, Republic of Korea
ACJ-2		99.7	SFT-21	
MPV-1	*Acetobacter pomorum*	99.5	SFT-22	Vinegar (*Malus pumila*), Jangseong-gun, Jeollanam-do, Republic of Korea

Table 2. *Cont.*

Strains No.	Species	Identities (%)	Strain Distinction	Source
PVJ-1	*Acetobacter pomorum*	99.7	SFT-23	Vinegar (*persimmon*), Jangseong-gun, Jeollanam-do, Republic of Korea
PVJ-4		99.7	SFT-24	
PVJ-5	*Acetobacter pasteurianus* [1]	99.8	SFT-25	
VVJ-1	*Gluconobacter oxydans* [1]	99.7	SFT-26	Vinegar (*Vitis vinifera L.*), Jangseong-gun, Jeollanam-do, Republic of Korea
VVJ-2		99.9	SFT-27	
AVJ-3	*Acetobacter pomorum*	99.5	SFT-28	Vinegar (*Aronia melanocarpa*), Jangseong-gun, Jeollanam-do, Republic of Korea
PEV-1	*Acetobacter ascendens* [1]	99.9	SFT-29	Vinegar (*Passiflora edulis*), Jangseong-gun, Jeollanam-do, Republic of Korea
PEV-4		99.9	SFT-30	
URV-1	*Acetobacter ascendens* [1]	99.9	SFT-31	Vinegar (Brown rice), Namwon-si, Jeollabuk-do, Republic of Korea
URV-2	*Acetobacter pomorum*	99.7	SFT-32	
KS-1	*Acetobacter pomorum*	99.8	SFT-33	Kombucha (Masontops), North York, ON, Canada
KS-2		99.9	SFT-34	
KS-3		99.9	SFT-35	
PS-1	*Acetobacter pomorum*	99.8	SFT-36	Peach cheong, Sunchon University b3 115, Republic of Korea
WS-1	*Acetobacter pomorum*	99.9	SFT-37	Wine, Sunchon University b3 115, Republic of Korea

[1] List of 21 microorganisms approved by the Ministry of Food and Drug Safety for use as food ingredients.

In Korea, the use of acetic acid bacteria is mainly limited to acetic acid fermentation [6,7]. Of the 37 isolated strains, 21 strains are included on the list of relevant acetic acid bacteria. Therefore, the fermentation characteristics of 21 applicable strains and 10 strains with limited capacity for fermentation of acetic acid were compared for selection of the most suitable strains.

3.2. Fermentation Characteristics of Selected Strains

3.2.1. Titratable Acidity (TTA)

Changes in titratable acidity according to the incubation periods for acetic acid bacteria are shown in Table 3. A titratable acidity of 0.11% first showed an increasing trend ranging between 0.45 and 1.48%, with increasing incubation time on the third day. A titratable acidity of more than 1% was detected in six strains of *A. pasteurianus* (SFT-1, 6, 7, 16), *A. orientalis* SFT-10, and *A. ascendens* SFT-30).

The titratable acidity increased with increasing incubation time. However, variation in the range of increase was observed according to the strain. In agreement with the results of a previous study reported by Eom et al. [34], differences in changes in titratable acidity according to fermentation time were observed in four species of *A. pasteurianus*, even in the same species or genera of bacteria from the same source of isolation. The formation of organic acids, the metabolic products of acetic acid bacteria, can be assessed using the titratable acidity according to cultivation, which has been reported as an important indicator in the selection of acetic acid bacteria [35]. In addition, the findings of this study demonstrated the importance of acid resistance and ethanol tolerance as factors in the selection of acetic acid bacteria most suitable for use in symbiotic fermentation in the production of Kombucha.

Therefore, an experiment was conducted for the evaluation of the growth rate according to pH and alcohol concentration for a selection of exceptional strains.

Table 3. Comparative analysis of titratable acidity changes during the growth of select bacterial strains.

Sample	Titratable Acidity (%)			
	Fermentation Time (Days)			
	0	1	2	3
Acetobacter pasteurianus (SFT-1)		0.27 ± 0.00 [b]	0.64 ± 0.03 [d]	1.18 ± 0.00 [d]
Acetobacter pasteurianus (SFT-2)		0.22 ± 0.00 [ef]	0.42 ± 0.02 [hij]	0.75 ± 0.00 [hi]
Acetobacter pasteurianus (SFT-3)		0.24 ± 0.00 [c]	0.49 ± 0.00 [f]	0.87 ± 0.01 [f]
Acetobacter pasteurianus (SFT-4)		0.19 ± 0.01 [hi]	0.36 ± 0.00 [lmno]	0.55 ± 0.00 [no]
Acetobacter pasteurianus (SFT-5)		0.25 ± 0.01 [c]	0.45 ± 0.05 [gh]	0.85 ± 0.03 [f]
Acetobacter pasteurianus (SFT-6)		0.21 ± 0.01 [fg]	0.60 ± 0.01 [e]	1.07 ± 0.05 [e]
Acetobacter pasteurianus (SFT-7)		0.25 ± 0.00 [c]	0.73 ± 0.01 [b]	1.39 ± 0.01 [b]
Acetobacter pasteurianus (SFT-8)		0.23 ± 0.01 [de]	0.48 ± 0.02 [fg]	0.82 ± 0.03 [fg]
Acetobacter orientalis (SFT-10)		0.28 ± 0.02 [ab]	0.84 ± 0.02 [a]	1.48 ± 0.05 [a]
Acetobacter cibinongensis (SFT-12)		0.21 ± 0.00 [fg]	0.34 ± 0.01 [nop]	0.60 ± 0.02 [lmn]
Acetobacter pasteurianus (SFT-13)		0.21 ± 0.01 [fg]	0.29 ± 0.00 [opq]	0.45 ± 0.03 [q]
Acetobacter pasteurianus (SFT-14)		0.21 ± 0.00 [fg]	0.28 ± 0.01 [r]	0.38 ± 0.01 [r]
Acetobacter pasteurianus (SFT-15)		0.13 ± 0.01 [k]	0.32 ± 0.01 [pq]	0.70 ± 0.03 [ijk]
Acetobacter pasteurianus (SFT-16)		0.29 ± 0.02 [a]	0.70 ± 0.02 [bc]	1.28 ± 0.02 [c]
Acetobacter pasteurianus (SFT-17)	0.11 ± 0.00 [1,ns]	0.24 ± 0.02 [c]	0.37 ± 0.02 [lmn]	0.59 ± 0.06 [mno]
Acetobacter pasteurianus (SFT-18)		0.14 ± 0.00 [jk]	0.27 ± 0.02 [r]	0.72 ± 0.06 [ij]
Acetobacter pasteurianus (SFT-19)		0.22 ± 0.00 [ef]	0.36 ± 0.01 [lmno]	0.57 ± 0.01 [mno]
Acetobacter pomorum (SFT-21)		0.18 ± 0.00 [i]	0.30 ± 0.00 [qr]	0.54 ± 0.01 [op]
Acetobacter pomorum (SFT-22)		0.21 ± 0.00 [fg]	0.34 ± 0.00 [nop]	0.54 ± 0.03 [op]
Acetobacter pomorum (SFT-24)		0.22 ± 0.00 [ef]	0.43 ± 0.02 [hi]	0.75 ± 0.06 [hi]
Acetobacter pasteurianus (SFT-25)		0.20 ± 0.00 [gh]	0.38 ± 0.01 [klm]	0.68 ± 0.03 [jk]
Gluconobacter oxydans (SFT-26)		0.15 ± 0.01 [j]	0.20 ± 0.02 [s]	0.45 ± 0.01 [q]
Gluconobacter oxydans (SFT-27)		0.14 ± 0.00 [jk]	0.21 ± 0.00 [s]	0.47 ± 0.02 [q]
Acetobacter pomorum (SFT-28)		0.22 ± 0.00 [ef]	0.39 ± 0.01 [jkl]	0.65 ± 0.01 [kl]
Acetobacter ascendens (SFT-29)		0.23 ± 0.01 [de]	0.41 ± 0.01 [ijk]	0.73 ± 0.01 [hij]
Acetobacter ascendens (SFT-30)		0.27 ± 0.01 [b]	0.69 ± 0.01 [c]	1.36 ± 0.01 [b]
Acetobacter ascendens (SFT-31)		0.13 ± 0.01 [k]	0.35 ± 0.02 [mnop]	0.62 ± 0.01 [lm]
Acetobacter pomorum (SFT-32)		0.14 ± 0.00 [jk]	0.37 ± 0.01 [lmn]	0.70 ± 0.01 [ijk]
Acetobacter malorum (SFT-33)		0.25 ± 0.01 [c]	0.43 ± 0.03 [hi]	0.69 ± 0.01 [jk]
Acetobacter malorum (SFT-36)		0.27 ± 0.00 [b]	0.51 ± 0.00 [f]	0.78 ± 0.04 [gh]
Acetobacter malorum (SFT-37)		0.24 ± 0.01 [cde]	0.33 ± 0.00 [opq]	0.49 ± 0.04 [pq]

[1] All values are mean $\pm$ SD ($n = 3$).; ns, non-significance.; Means with different superscript letters in the same column are significantly different at $p < 0.05$ by Duncan's multiple range test. a > b > c > d > e > f > g > h > I > j > k > l > m > n > o > p > q > r > s.

3.2.2. Acid Resistance

Microbial growth, including that of acetic acid bacteria, is inhibited in an acidic environment; thus, the selection of acid-resistant strains is a critical factor. The results regarding the acid resistance of strains according to the pH concentrations of the culture solution are shown in Table 4. Variation in the resistance confirmed by growth, according to acid concentration, was observed for each strain. However, overall, low growth was observed at pH 8.0 and high growth at pH 5.0 to 7.0. Bang et al. [36] suggested a pH range of 5.5–6.5 as an optimum condition for the growth of acetic acid bacteria. Park et al. [37] reported favorable outcomes for microbial growth at pH 4.0–6.0. These previously reported findings are comparable to the results obtained in this study. Variation in growth rates in the same pH range was observed even in the same genera and species of bacterial strains. The following strains showed a growth rate of 100% or higher in the pH range 4.0–5.0: *A. pasteurianus* SFT-3, 4, 7, 13, 16, 18, and 19; *A. pomorum* SFT-24, 28, and 32; *A. ascendens* SFT-31; and *A. malorum* SFT-36 and 37. *Glu. oxydans* SFT-26 and 27 showed increased growth rates at higher pH, which rose to 144.16% and 251.50% at an optimum growth pH of 6.0. Gupta et al. [32] recommended an optimum pH of 5.5–6.5 to support growth in all strains of *Gluconobacter*, comparable to the outcome of this study.

Table 4. Evaluation of growth rate of acetic acid bacteria according to pH and alcohol concentration.

Treatment / Sample	Acid Resistance (%) pH					Ethanol Tolerance (%) Alcohol Content (%)				
	4.0	5.0	6.0	7.0	8.0	2.0	4.0	6.0	8.0	10.0
Acetobacter pasteurianus (SFT-1)	90.80 ± 0.08 [1],d	112.89 ± 0.16 [a]	98.33 ± 0.09 [c]	100.13 ± 0.09 [b]	84.26 ± 0.06 [e]	95.64 ± 0.07 [a]	86.92 ± 0.14 [e]	94.45 ± 0.07 [b]	88.45 ± 0.21 [c]	87.97 ± 0.07 [d]
Acetobacter pasteurianus (SFT-2)	80.73 ± 0.09 [c]	99.60 ± 0.08 [b]	73.57 ± 0.03 [d]	101.99 ± 0.00 [a]	61.46 ± 0.09 [e]	108.78 ± 0.23 [d]	126.33 ± 0.46 [b]	129.33 ± 0.12 [a]	123.21 ± 0.23 [c]	122.75 ± 0.46 [c]
Acetobacter pasteurianus (SFT-3)	119.84 ± 0.23 [b]	132.10 ± 0.11 [a]	91.19 ± 0.11 [e]	100.66 ± 0.11 [c]	97.10 ± 0.12 [d]	92.76 ± 0.07 [a]	78.27 ± 0.00 [d]	76.56 ± 0.32 [e]	80.61 ± 0.26 [c]	81.37 ± 0.19 [b]
Acetobacter pasteurianus (SFT-4)	100.63 ± 0.19 [b]	118.18 ± 0.019 [a]	95.43 ± 0.10 [c]	100.34 ± 0.29 [b]	87.13 ± 0.29 [d]	97.09 ± 0.08 [a]	91.26 ± 0.14 [b]	80.31 ± 0.22 [c]	77.11 ± 0.08 [d]	67.52 ± 0.36 [e]
Acetobacter pasteurianus (SFT-5)	97.54 ± 0.10 [c]	128.84 ± 0.10 [a]	96.46 ± 0.20 [d]	100.27 ± 0.10 [b]	82.12 ± 0.10 [e]	94.22 ± 0.13 [a]	82.66 ± 0.26 [b]	77.04 ± 0.13 [c]	74.99 ± 0.26 [d]	71.16 ± 0.06 [e]
Acetobacter pasteurianus (SFT-6)	66.05 ± 0.16 [e]	72.01 ± 0.16 [d]	92.97 ± 0.16 [b]	100.51 ± 0.16 [a]	85.35 ± 0.24 [c]	94.01 ± 0.00 [a]	82.02 ± 0.20 [b]	77.98 ± 0.07 [c]	72.26 ± 0.21 [d]	67.61 ± 0.27 [e]
Acetobacter pasteurianus (SFT-7)	111.67 ± 0.12 [b]	136.56 ± 0.45 [a]	102.82 ± 0.23 [c]	99.79 ± 0.00 [d]	81.51 ± 0.11 [e]	88.07 ± 0.24 [a]	64.22 ± 0.12 [e]	80.16 ± 0.18 [c]	74.17 ± 0.06 [d]	83.07 ± 0.12 [b]
Acetobacter pasteurianus (SFT-8)	120.77 ± 0.32 [a]	92.80 ± 0.11 [d]	75.96 ± 0.11 [e]	101.81 ± 0.11 [b]	100.75 ± 0.21 [a]	100.05 ± 0.28 [a]	100.14 ± 0.09 [a]	88.99 ± 0.18 [d]	91.07 ± 0.46 [b]	90.35 ± 0.09 [c]
Acetobacter orientalis (SFT-10)	111.49 ± 0.14 [b]	96.01 ± 0.14 [d]	89.90 ± 0.41 [e]	100.76 ± 0.27 [c]	114.75 ± 0.68 [a]	97.64 ± 0.12 [b]	92.92 ± 0.37 [c]	79.25 ± 0.25 [e]	85.71 ± 0.63 [d]	169.44 ± 0.10 [a]
Acetobacter cibinongensis (SFT-12)	56.19 ± 0.25 [d]	32.96 ± 0.25 [e]	97.43 ± 0.00 [b]	100.19 ± 0.33 [a]	78.19 ± 0.08 [c]	119.63 ± 0.33 [c]	158.89 ± 0.11 [a]	158.24 ± 0.33 [a]	132.97 ± 0.65 [b]	114.32 ± 0.22 [d]
Acetobacter pasteurianus (SFT-13)	115.76 ± 0.11 [b]	144.67 ± 0.11 [a]	107.28 ± 0.11 [c]	99.45 ± 0.22 [d]	98.80 ± 0.11 [e]	99.37 ± 0.08 [c]	98.11 ± 0.34 [d]	96.94 ± 0.25 [e]	102.89 ± 0.09 [b]	106.33 ± 0.09 [a]
Acetobacter pasteurianus (SFT-14)	90.93 ± 0.09 [c]	103.69 ± 0.34 [a]	103.69 ± 0.17 [a]	99.72 ± 0.09 [b]	78.17 ± 0.17 [d]	100.84 ± 0.09 [d]	102.52 ± 0.28 [c]	106.26 ± 0.19 [b]	118.21 ± 0.10 [a]	102.71 ± 0.47 [c]
Acetobacter pasteurianus (SFT-15)	92.04 ± 0.23 [c]	107.68 ± 0.08 [a]	82.95 ± 0.39 [d]	101.28 ± 0.23 [b]	81.41 ± 0.00 [e]	103.47 ± 0.15 [b]	110.41 ± 0.15 [a]	100.27 ± 0.31 [d]	102.10 ± 0.31 [c]	96.30 ± 0.08 [e]
Acetobacter pasteurianus (SFT-16)	108.33 ± 0.10 [b]	117.40 ± 0.30 [a]	101.78 ± 0.10 [c]	99.87 ± 0.10 [d]	80.62 ± 0.10 [e]	86.79 ± 0.05 [a]	60.37 ± 0.11 [b]	60.48 ± 0.22 [b]	59.94 ± 0.06 [c]	55.82 ± 0.22 [d]
Acetobacter pasteurianus (SFT-17)	98.79 ± 0.21 [d]	130.46 ± 0.21 [a]	94.28 ± 0.10 [e]	100.43 ± 0.10 [c]	124.21 ± 0.21 [b]	101.76 ± 0.08 [c]	105.27 ± 0.24 [a]	91.37 ± 0.08 [e]	93.53 ± 0.24 [d]	102.56 ± 0.08 [b]
Acetobacter pasteurianus (SFT-18)	114.56 ± 0.11 [b]	120.88 ± 0.11 [a]	97.88 ± 0.21 [d]	100.16 ± 0.00 [c]	86.07 ± 0.11 [e]	94.82 ± 0.07 [a]	84.46 ± 0.27 [c]	79.41 ± 0.07 [d]	92.16 ± 0.14 [b]	55.76 ± 0.07 [e]
Acetobacter pasteurianus (SFT-19)	108.94 ± 0.10 [b]	119.27 ± 0.20 [a]	100.18 ± 0.10 [c]	99.99 ± 0.20 [c]	83.85 ± 0.59 [d]	93.66 ± 0.13 [a]	80.98 ± 0.06 [b]	74.14 ± 0.13 [c]	72.00 ± 0.00 [d]	66.60 ± 0.07 [e]
Acetobacter pomorum (SFT-21)	100.31 ± 0.26 [a]	99.29 ± 0.09 [b]	93.59 ± 0.09 [c]	100.48 ± 0.09 [a]	89.84 ± 0.09 [d]	92.57 ± 0.17 [a]	77.72 ± 0.25 [b]	76.67 ± 0.09 [c]	71.83 ± 0.32 [d]	68.04 ± 0.08 [e]
Acetobacter pomorum (SFT-22)	93.01 ± 0.21 [d]	120.86 ± 0.07 [a]	91.66 ± 0.14 [e]	100.63 ± 0.14 [b]	93.68 ± 0.27 [c]	94.96 ± 0.06 [a]	84.87 ± 0.11 [b]	83.01 ± 0.06 [c]	61.26 ± 0.12 [d]	55.28 ± 0.23 [e]
Acetobacter pomorum (SFT-24)	102.28 ± 0.07 [c]	114.51 ± 0.07 [b]	123.88 ± 0.50 [a]	98.20 ± 0.22 [d]	80.68 ± 0.22 [e]	95.70 ± 0.13 [a]	87.10 ± 0.13 [c]	87.42 ± 0.07 [b]	57.80 ± 0.07 [e]	65.39 ± 0.00 [d]
Acetobacter pasteurianus (SFT-25)	84.52 ± 0.07 [e]	94.49 ± 0.15 [b]	90.86 ± 0.43 [c]	100.69 ± 0.07 [a]	89.94 ± 0.08 [d]	97.43 ± 0.46 [a]	92.29 ± 0.08 [d]	86.77 ± 0.08 [e]	96.43 ± 0.54 [b]	93.44 ± 0.23 [c]
Gluconobacter oxydans (SFT-26)	43.82 ± 0.23 [e]	60.18 ± 0.11 [d]	144.16 ± 0.23 [b]	96.68 ± 0.12 [c]	185.12 ± 0.35 [a]	137.11 ± 0.19 [c]	211.32 ± 0.19 [a]	94.81 ± 0.56 [e]	130.98 ± 0.19 [d]	172.73 ± 0.56 [b]
Gluconobacter oxydans (SFT-27)	72.19 ± 0.13 [e]	86.73 ± 0.25 [d]	251.50 ± 0.37 [a]	88.60 ± 0.25 [c]	171.23 ± 0.25 [b]	96.48 ± 0.12 [d]	89.44 ± 0.24 [e]	143.57 ± 0.00 [c]	158.50 ± 0.25 [a]	148.18 ± 0.36 [b]
Acetobacter pomorum (SFT-28)	109.16 ± 0.25 [b]	102.64 ± 0.25 [c]	116.01 ± 0.09 [a]	98.79 ± 0.17 [d]	93.19 ± 0.42 [e]	103.35 ± 0.07 [b]	110.05 ± 0.15 [a]	101.33 ± 0.29 [c]	69.72 ± 0.08 [d]	66.33 ± 0.00 [e]
Acetobacter ascendens (SFT-29)	92.54 ± 0.41 [e]	99.01 ± 0.08 [c]	94.89 ± 0.08 [d]	100.38 ± 0.09 [b]	105.23 ± 0.09 [a]	94.46 ± 0.15 [a]	83.37 ± 0.07 [b]	77.61 ± 0.15 [d]	78.79 ± 0.08 [c]	69.70 ± 0.22 [e]
Acetobacter ascendens (SFT-30)	95.06 ± 0.11 [e]	101.59 ± 0.21 [c]	106.45 ± 0.31 [a]	99.51 ± 0.11 [d]	103.97 ± 0.42 [b]	96.04 ± 0.08 [a]	88.12 ± 0.16 [b]	74.21 ± 0.08 [d]	77.33 ± 0.24 [c]	70.37 ± 0.08 [e]
Acetobacter ascendens (SFT-31)	121.31 ± 0.25 [c]	154.97 ± 0.12 [a]	125.67 ± 0.12 [b]	98.07 ± 0.24 [d]	90.80 ± 0.49 [e]	87.24 ± 0.09 [c]	61.72 ± 0.19 [e]	89.13 ± 0.28 [b]	107.37 ± 0.10 [a]	78.36 ± 0.10 [d]
Acetobacter pomorum (SFT-32)	156.12 ± 0.13 [a]	122.40 ± 0.13 [b]	102.64 ± 0.25 [c]	99.80 ± 0.12 [d]	93.75 ± 0.12 [e]	153.29 ± 1.20 [d]	259.88 ± 0.24 [a]	212.69 ± 1.68 [b]	139.64 ± 0.72 [e]	159.28 ± 0.24 [c]
Acetobacter malorum (SFT-33)	117.70 ± 0.08 [a]	89.90 ± 0.29 [d]	75.35 ± 0.07 [e]	101.86 ± 0.22 [c]	117.34 ± 0.15 [b]	72.20 ± 0.16 [a]	16.59 ± 0.07 [e]	23.63 ± 0.00 [d]	43.03 ± 0.07 [b]	29.98 ± 0.04 [c]
Acetobacter malorum (SFT-36)	176.67 ± 0.25 [a]	125.85 ± 0.08 [b]	82.67 ± 0.25 [d]	101.30 ± 0.09 [c]	55.66 ± 0.49 [e]	80.09 ± 0.07 [a]	40.28 ± 0.23 [b]	23.66 ± 0.07 [c]	17.47 ± 0.03 [e]	19.66 ± 0.20 [d]
Acetobacter malorum (SFT-37)	100.68 ± 0.07 [c]	118.63 ± 0.07 [a]	105.50 ± 0.08 [b]	99.59 ± 0.44 [d]	64.57 ± 0.15 [e]	78.76 ± 0.04 [a]	36.27 ± 0.08 [b]	19.86 ± 0.12 [d]	30.14 ± 0.08 [c]	36.3 ± 0.12 [b]

[1] All values are mean ± SD (n = 3); Means with different superscript letters in the same row are significantly different at $p < 0.05$ by Duncan's multiple range test. a > b > c > d > e.

3.2.3. Alcohol Tolerance

Sugars are converted into alcohol by yeast and ethanol is oxidized into acetic acid by acetic acid bacteria during the process of symbiotic fermentation; thus, alcohol concentration is an important factor in microbial growth and acid production [38]. However, a high concentration of ethanol during the initial period can result in a delay of the induction period, leading to deceleration of the growth of acetic acid bacteria along with a reduction in acid productivity [39]. The growth rates of acetic acid bacteria according to ethanol concentrations are shown in Table 4. Most isolated strains of acetic acid bacteria showed reduced growth rates at a concentration of 10% ethanol. This result is consistent with those of an earlier study, which reported lower growth of acetic acid bacteria at an ethanol content of 9% [36]. An increase in growth rates to higher than 100% was observed at ethanol concentrations of 8–10% in *A. pasteurianus* SFT-2, 13, 14, and 17; *A. orientalis* SFT-10; *A. cibinongensis* SFT-12; *A. pomorum* SFT-32; and *Glu. oxydans* SFT-26 and 27, indicating high alcohol tolerance.

An optimum alcohol concentration of 4% for acetic acid fermentation has been reported [40]. According to an earlier study on capacity in the production of acetic acid, the activity of acetic acid production was affected by the characteristics of bacterial strains [41]. Therefore, additional studies are warranted in order to further determine the capacity in the production of acetic acid according to characteristics of the strain. In the selection of strains, this study complied with regulations for food standards and specifications established by the Korean Ministry of Food and Drug Safety for microorganisms approved for use as food materials [5], and assessment of factors impeding the growth and formation of bacteria used in Kombucha fermentation was performed [42–44].

The screening of acetic acid bacteria suitable for use in symbiotic fermentation, including five strains of the genus *Acetobacter* (*A. pasteurianus* SFT-3, 13, 18, and *A. ascendens* SFT-30, 31) and two strains of the genus *Gluconobacter* (*Glu. oxydans* SFT-26, 27) was based on the evaluation of fermentation characteristics (pH and titratable acidity), acid resistance, and alcohol tolerance. An impact evaluation according to fermentation temperature and fermentation period was conducted using the seven selected strains.

3.3. Fermentation Characteristics According to Fermentation Temperature and Time

3.3.1. pH, Titratable Acidity, and Viable Cell Count

The results regarding the optimum temperature for the growth of acetic acid bacteria and determining the incubation period are shown in Table 5. Changes in the overall pH showed a severe decrease from the initial pH between day 0 and day 2, and the lowest pH values were detected in *A. pasteurianus* SFT-18 (pH 3.85) and *Glu. oxydans* SFT-27 (pH 3.56) on the day 3. According to incubation temperature, a large-scale decrease in pH was observed at a temperature range of 30–35 °C. The pH values were reduced by 2.46, compared to the initial pH at an incubation temperature of 35 °C in *A. pasteurianus* SFT-18, and by 2.90 at 30 °C in *Glu. oxydans* SFT-26.

Most strains showed a gradual decrease in titratable acidity. The highest titratable acidity was detected in *A. pasteurianus* SFT-18 (1.68%) and *Glu. oxydans* SFT-27 (0.79%) on the third day. According to incubation temperature, the titratable acidity showed a substantial increase at 30–35 °C. The titratable acidity increased by 1.56 compared with the initial titratable acidity observed at an incubation temperature of 35 °C in *A. pasteurianus* SFT-18 and by 0.67 at 35 °C in *Glu. oxydans* SFT-27.

Table 5. Fermentation characteristics of initially selected acetic acid bacteria strains based on fermentation temperature and period.

Sample	Day	pH 25 °C	pH 30 °C	pH 35 °C	pH 40 °C	Titratable Acidity (%) 25 °C	30 °C	35 °C	40 °C	Microbial Count (logCFU/mL) 25 °C	30 °C	35 °C	40 °C
Acetobacter pasteurianus (SFT-3)	0	6.14 ± 0.01 [1,ns]	6.14 ± 0.00	6.12 ± 0.01	6.13 ± 0.02	0.10 ± 0.01 [bc]	0.12 ± 0.00 [a]	0.11 ± 0.01 [ab]	0.09 ± 0.01 [c]	6.13 ± 0.08 [ns]	6.08 ± 0.15	6.22 ± 0.07	6.22 ± 0.07
	1	5.18 ± 0.01 [b]	4.94 ± 0.01 [c]	4.61 ± 0.00 [d]	5.67 ± 0.00 [a]	0.14 ± 0.01 [c]	0.21 ± 0.00 [b]	0.34 ± 0.03 [a]	0.10 ± 0.02 [d]	6.40 ± 0.52 [ns]	6.36 ± 0.67	6.53 ± 0.25	6.24 ± 0.49
	2	4.61 ± 0.01 [b]	4.60 ± 0.01 [b]	4.22 ± 0.01 [c]	5.07 ± 0.00 [a]	0.35 ± 0.01 [b]	0.38 ± 0.03 [b]	0.81 ± 0.03 [a]	0.18 ± 0.00 [c]	7.15 ± 0.45 [b]	7.17 ± 0.42 [b]	8.05 ± 0.67 [a]	6.95 ± 0.07 [b]
	3	4.35 ± 0.01 [b]	4.32 ± 0.00 [c]	3.97 ± 0.01 [d]	4.50 ± 0.01 [a]	0.57 ± 0.00 [b]	0.60 ± 0.03 [b]	1.32 ± 0.02 [a]	0.41 ± 0.01 [c]	8.81 ± 0.95 [ns]	8.85 ± 0.77	8.92 ± 0.98	8.64 ± 0.81
Acetobacter pasteurianus (SFT-13)	0	6.42 ± 0.00 [a]	6.40 ± 0.01 [b]	6.38 ± 0.01 [c]	6.41 ± 0.00 [ab]	0.09 ± 0.00 [b]	0.10 ± 0.02 [ab]	0.12 ± 0.01 [a]	0.09 ± 0.01 [b]	6.36 ± 0.08 [c]	6.54 ± 0.04 [b]	6.71 ± 0.02 [a]	6.53 ± 0.03 [b]
	1	5.32 ± 0.01 [a]	4.92 ± 0.00 [b]	4.84 ± 0.01 [c]	4.68 ± 0.00 [d]	0.13 ± 0.01 [d]	0.21 ± 0.01 [c]	0.24 ± 0.01 [b]	0.31 ± 0.01 [a]	6.46 ± 0.34 [c]	6.77 ± 0.19 [bc]	6.92 ± 0.19 [b]	7.42 ± 0.12 [a]
	2	4.78 ± 0.01 [a]	4.50 ± 0.00 [b]	4.46 ± 0.01 [c]	4.26 ± 0.00 [d]	0.25 ± 0.01 [d]	0.44 ± 0.00 [c]	0.47 ± 0.02 [b]	0.77 ± 0.02 [a]	7.54 ± 0.34 [ns]	8.18 ± 0.57	8.41 ± 0.66	8.58 ± 0.57
	3	4.50 ± 0.00 [a]	4.24 ± 0.00 [b]	4.11 ± 0.00 [c]	4.00 ± 0.01 [d]	0.42 ± 0.00 [d]	0.71 ± 0.02 [c]	0.92 ± 0.01 [b]	1.29 ± 0.01 [a]	8.70 ± 0.71 [ns]	9.00 ± 0.75	9.05 ± 0.82	8.99 ± 0.86
Acetobacter pasteurianus (SFT-18)	0	6.34 ± 0.00 [b]	6.34 ± 0.01 [b]	6.31 ± 0.02 [c]	6.37 ± 0.01 [a]	0.10 ± 0.01 [b]	0.09 ± 0.01 [b]	0.12 ± 0.01 [a]	0.09 ± 0.00 [b]	5.93 ± 0.13 [ns]	5.93 ± 0.11	5.89 ± 0.10	6.01 ± 0.12
	1	4.89 ± 0.01 [a]	4.87 ± 0.01 [b]	4.49 ± 0.01 [c]	4.89 ± 0.01 [a]	0.22 ± 0.01 [b]	0.22 ± 0.01 [b]	0.43 ± 0.02 [a]	0.22 ± 0.01 [b]	6.67 ± 0.02 [bc]	6.75 ± 0.02 [b]	6.98 ± 0.18 [a]	6.54 ± 0.10 [c]
	2	4.24 ± 0.01 [c]	4.43 ± 0.00 [b]	4.12 ± 0.01 [d]	4.54 ± 0.00 [a]	0.75 ± 0.01 [b]	0.53 ± 0.02 [c]	0.98 ± 0.01 [a]	0.43 ± 0.00 [d]	8.11 ± 0.64 [ns]	7.76 ± 0.35	8.40 ± 0.50	7.43 ± 0.60
	3	3.93 ± 0.00 [c]	4.10 ± 0.01 [b]	3.85 ± 0.00 [d]	4.31 ± 0.01 [a]	1.35 ± 0.01 [b]	0.95 ± 0.02 [c]	1.68 ± 0.02 [a]	0.66 ± 0.01 [d]	9.42 ± 0.81 [ns]	9.23 ± 0.76	9.52 ± 0.49	9.01 ± 0.72
Acetobacter ascendens (SFT-30)	0	6.33 ± 0.01 [ab]	6.31 ± 0.02 [b]	6.33 ± 0.01 [ab]	6.34 ± 0.01 [a]	0.09 ± 0.02 [ns]	0.09 ± 0.01	0.09 ± 0.01	0.11 ± 0.01	6.50 ± 0.03 [c]	6.77 ± 0.04 [a]	6.65 ± 0.02 [b]	6.46 ± 0.06 [c]
	1	5.08 ± 0.00 [b]	4.89 ± 0.02 [d]	4.99 ± 0.00 [c]	5.30 ± 0.01 [a]	0.16 ± 0.02 [b]	0.22 ± 0.01 [a]	0.20 ± 0.01 [a]	0.13 ± 0.01 [c]	7.10 ± 0.23 [b]	7.77 ± 0.02 [a]	6.83 ± 0.07 [c]	6.68 ± 0.15 [c]
	2	4.59 ± 0.00 [c]	4.52 ± 0.00 [d]	4.72 ± 0.01 [b]	4.87 ± 0.00 [a]	0.37 ± 0.01 [b]	0.42 ± 0.02 [a]	0.31 ± 0.01 [c]	0.24 ± 0.01 [d]	8.05 ± 0.89 [ns]	8.51 ± 0.58 [b]	7.51 ± 0.68	7.34 ± 0.58
	3	4.31 ± 0.01 [c]	4.26 ± 0.01 [d]	4.49 ± 0.01 [b]	4.55 ± 0.01 [a]	0.63 ± 0.01 [b]	0.68 ± 0.00 [a]	0.44 ± 0.02 [c]	0.41 ± 0.01 [d]	9.10 ± 0.95 [ns]	9.30 ± 1.19	8.99 ± 0.73	8.91 ± 0.66
Acetobacter ascendens (SFT-31)	0	6.40 ± 0.01 [ab]	6.38 ± 0.02 [bc]	6.37 ± 0.01 [c]	6.41 ± 0.01 [a]	0.08 ± 0.00 [b]	0.11 ± 0.00 [a]	0.12 ± 0.01 [a]	0.08 ± 0.01 [b]	6.36 ± 0.06 [b]	6.48 ± 0.06 [ab]	6.55 ± 0.09 [a]	6.37 ± 0.06 [b]
	1	5.75 ± 0.00 [b]	5.30 ± 0.01 [c]	4.88 ± 0.00 [d]	5.89 ± 0.01 [a]	0.11 ± 0.02 [c]	0.15 ± 0.00 [b]	0.21 ± 0.01 [a]	0.09 ± 0.00 [c]	6.82 ± 0.15 [bc]	7.12 ± 0.08 [ab]	7.41 ± 0.18 [a]	6.39 ± 0.53 [c]
	2	5.04 ± 0.01 [b]	4.65 ± 0.00 [c]	4.57 ± 0.01 [d]	5.71 ± 0.00 [a]	0.19 ± 0.02 [b]	0.38 ± 0.02 [a]	0.39 ± 0.01 [a]	0.12 ± 0.02 [c]	7.52 ± 0.60 [ns]	8.19 ± 0.50	8.40 ± 0.63	7.36 ± 0.51
	3	4.37 ± 0.01 [b]	4.35 ± 0.01 [c]	4.34 ± 0.01 [c]	5.46 ± 0.01 [a]	0.58 ± 0.02 [b]	0.61 ± 0.00 [a]	0.61 ± 0.02 [a]	0.14 ± 0.00 [c]	8.37 ± 0.81 [ns]	8.95 ± 0.75	8.90 ± 0.94	8.04 ± 0.58
Gluconobacter oxydans (SFT-26)	0	6.68 ± 0.01 [a]	6.67 ± 0.00 [a]	6.64 ± 0.01 [b]	6.67 ± 0.00 [a]	0.09 ± 0.01 [ns]	0.09 ± 0.01	0.10 ± 0.00	0.09 ± 0.01	6.11 ± 0.07 [b]	6.20 ± 0.01 [a]	6.27 ± 0.02 [a]	6.23 ± 0.02 [a]
	1	5.06 ± 0.00 [d]	5.24 ± 0.02 [c]	5.76 ± 0.01 [b]	6.12 ± 0.01 [a]	0.13 ± 0.01 [ns]	0.12 ± 0.02	0.11 ± 0.02	0.10 ± 0.01	7.51 ± 0.21 [a]	6.94 ± 0.17 [ab]	6.45 ± 0.38 [b]	6.35 ± 0.76 [b]
	2	4.31 ± 0.01 [d]	4.33 ± 0.01 [c]	4.65 ± 0.00 [b]	6.02 ± 0.00 [a]	0.24 ± 0.02 [a]	0.24 ± 0.02 [a]	0.18 ± 0.00 [b]	0.10 ± 0.00 [c]	8.37 ± 0.79 [ns]	8.35 ± 0.83	8.09 ± 0.84	7.34 ± 0.58
	3	3.80 ± 0.01 [c]	3.77 ± 0.00 [d]	4.46 ± 0.00 [b]	5.89 ± 0.00 [a]	0.52 ± 0.02 [a]	0.52 ± 0.00 [a]	0.21 ± 0.00 [b]	0.11 ± 0.00 [c]	9.54 ± 0.68 [ns]	9.41 ± 0.53	9.37 ± 0.53	8.48 ± 0.63
Gluconobacter oxydans (SFT-27)	0	6.22 ± 0.01 [a]	6.21 ± 0.00 [ab]	6.20 ± 0.00 [b]	6.21 ± 0.01 [ab]	0.11 ± 0.00 [ns]	0.11 ± 0.01	0.12 ± 0.01	0.11 ± 0.00	5.87 ± 0.46 [ns]	5.97 ± 0.69	6.21 ± 0.30	6.11 ± 0.35
	1	5.57 ± 0.01 [b]	5.12 ± 0.01 [d]	5.30 ± 0.00 [c]	5.88 ± 0.00 [a]	0.13 ± 0.01 [a]	0.14 ± 0.01 [a]	0.13 ± 0.00 [a]	0.11 ± 0.01 [b]	7.49 ± 0.29 [ns]	7.28 ± 0.43	7.00 ± 0.47	6.89 ± 0.90
	2	4.68 ± 0.01 [b]	3.84 ± 0.01 [d]	3.90 ± 0.01 [c]	5.83 ± 0.02 [a]	0.16 ± 0.01 [c]	0.45 ± 0.01 [a]	0.38 ± 0.00 [b]	0.11 ± 0.00 [d]	7.86 ± 0.36 [ns]	8.01 ± 0.60	8.04 ± 0.70	7.41 ± 0.61
	3	3.63 ± 0.00 [b]	3.56 ± 0.02 [c]	3.55 ± 0.00 [c]	5.83 ± 0.01 [a]	0.73 ± 0.00 [c]	0.75 ± 0.02 [b]	0.79 ± 0.00 [a]	0.13 ± 0.00 [d]	8.19 ± 0.71 [ns]	8.44 ± 0.87	8.55 ± 0.76	8.09 ± 0.77

[1] All values are mean ± SD ($n = 3$); ns, non-significance; Means with different superscript letters in the same row are significantly different at $p < 0.05$ by Duncan's multiple range test. a > b > c > d.

Regarding changes in viable cell count according to fermentation temperature, all strains showed high growth rates at a temperature of 35 °C or below. High numbers of viable cells were detected according to temperature in *A. pasteurianus* SFT-3 (8.92 logCFU/mL), *A. pasteurianus* SFT-13 (9.05 logCFU/mL), *A. pasteurianus* SFT-18 (9.52 logCFU/mL), and *Glu. oxydans* SFT-27 (8.55 logCFU/mL) at 35 °C; *A. ascendens* SFT-30 (9.30 logCFU/mL) and *A. ascendens* SFT-31 (8.95 logCFU/mL) at 30 °C; and *Glu. oxydans* SFT-26 (9.54 logCFU/mL) at 25 °C. In particular, a wider range of incubation temperatures was observed for growth of *A. pasteurianus* SFT-18 compared with other bacterial strains, with increases in the number of viable cells to 3.49 logCFU/mL at 25 °C, 3.30 logCFU/mL at 30 °C, 3.63 logCFU/mL at 35 °C, and 3.00 logCFU/mL at 40 °C from the initial viable cell count.

Despite variation in acid productivity according to the isolated strain, a significant change in pH and total acidity content was observed as the viable cell count increased in the same strain. Despite an increase in the viable cell count to 1.67–2.25 logCFU/mL in *A. ascendens* SFT-31 and *Glu. oxydans* SFT-26 and 27 at 40 °C compared to the day 0, the change in titratable acidity and pH with the effect of metabolic products was insignificant. According to previous studies reported by Gullo et al. [45] and Sharafi et al. [46], inactivation of acetic acid bacteria may be a result of an irregularity in optimum growth temperature, resulting in a reduction in metabolism caused by injury to membranes. It is believed that these previous findings support the findings of the current study. Therefore, based on its stable fermentation characteristics at a wide range of culture temperatures, *A. pasteurianus* SFT-18 (Accession; CP015168.1, Description; *A. pasteurianus*, Length; 2810721, Start; 874219, End; 875683, Coverage; 0, Bit; 2673, E-Value; 0.0, Match/Total; 1460/1466, Pct. (%); 99.6) can be regarded as the most suitable strain for use in the symbiotic fermentation of Kombucha. The phylogenetic tree is shown in Figure 4.

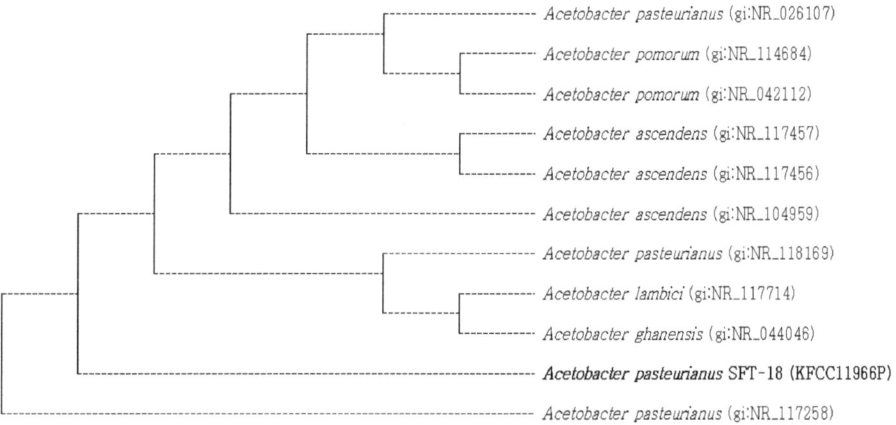

Figure 4. Phylogenetic tree based on 16S rRNA region gene sequences between *Acetobacter pasteurianus* SFT-18 strain and related species.

3.3.2. Content of Gluconate and Glucuronic Acid

The production of functional substances by *Gluconobacter* sp. results from chemical and biological oxidation of glucose into gluconic acid, glucuronic acid, and others. Elimination of many types of toxic substances by glucuronic acid, including exogenous chemicals and excessive steroid hormones, from the human body via the urinary system has been reported [47]. In addition, it can be converted into glucosamine, which is beneficial in the treatment of osteoarthritis and is also known as a precursor of vitamin C biosynthesis [48]. The usefulness of sodium gluconate obtained by the conversion of gluconic acid via microbial fermentation for application in various industries including food and beverage, pharmaceuticals, and others has been reported [49,50].

A culture solution was used at a fermentation temperature of 30 °C based on fermentation characteristics (pH, titratable acidity, and viable cell count) to determine the content of gluconate and glucuronic acid in the isolated sample of *Gluconobacter* sp. The content of gluconate was 25.31 and 25.49 mg/mL, and the content of glucuronic acid was 10.15 and 10.32 mg/mL in *Glu. oxydans* SFT-26 and 27, respectively. Jayabalan et al. [51] reported a glucuronic acid content of 2.33 g/L, and the highest content was detected in Kombucha made from black tea on the 12th day of fermentation. Chen and Liu [52] reported that a glucuronic acid content of approximately 10.0 g/L was detected between the 10th and 20th days and 39.0 g/L was detected on the 60th day after fermentation of Kombucha. A faster rate of glucuronic acid production, as well as a higher overall production amount, was obtained for the two strains (*Glu. oxydans* SFT-26, 27) identified in this study, compared to reports in the existing literature [51,52], both in terms of fermentation time and production rate. However, further verification is required to determine more clearly the impact of Kombucha composition and complex fermentation on changes in content. Therefore, *Glu. oxydans* SFT-27 (Accession; NR_026118.1, Description; *Glu. oxydans*, Length; 1476, Start; 18, End; 1465, Coverage; 98, Bit; 2663, E-Value; 0.0, Match/Total; 1447/1449, Pct. (%); 99.9) can be considered suitable in the production of Kombucha and for enhancing the functionality due to its exceptional capacity for acid resistance and metabolite production. The phylogenetic tree is shown in Figure 5.

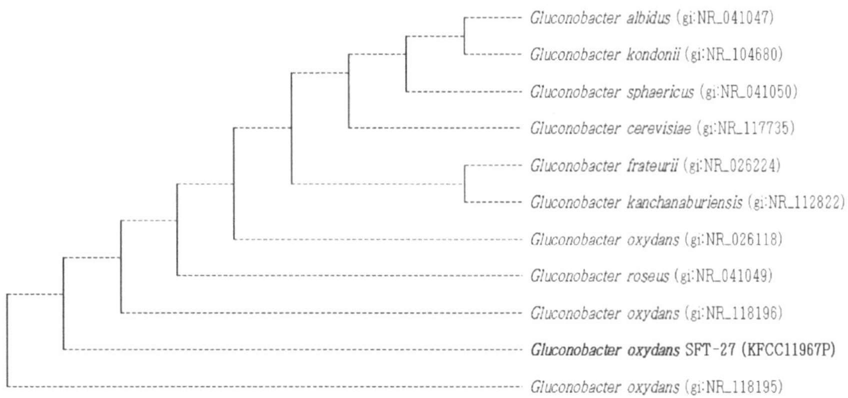

Figure 5. Phylogenetic tree based on 16S rRNA region gene sequences *between Gluconobacter oxydans* SFT-27 strain and related species.

The dominant bacteria in the Kombucha culture belong to the genera *Acetobacter* and *Gluconobacter*, known for producing acetic acid and gluconic acid, respectively [13]. In this study, microbial strains suitable for the complex microbial culture used in the production of Kombucha (SCOBY), including *A. pasteurianus* SFT-18 and *Glu. oxydans* SFT-27, both confirmed for their capacity for producing acetic and gluconic acid, were selected. *A. xylinum, A. pasteurianus, A. acetic, B. gluconicum,* and *Glu. oxydans* were predominantly detected in the currently analyzed Kombucha cultures, and other studies reported similar results [13]. Based on these findings, the two selected strains were frequently detected in Kombucha, indicating their potential for use in the production of Kombucha.

A. pasteurianus, with its high potential in the production of acetic acid [53] and its capacity for producing bacterial cellulose (due to the results of this study), may provide an optimal environment for producing Kombucha [54]. In addition, *Glu. oxydans* [55] may have an important function in conveying various functional and bioactive effects [56] due to the presence of products such as gluconic acid and glucuronic acid (supporting the results of this study).

The main organic acids found in Kombucha include gluconic acid and acetic acid [57], which are known as major compounds contributing to development of the flavor and quality of Kombucha [58]. Gluconic acid is associated with the drink's pleasant sour taste, while acetic acid is responsible for an astringent and acidic off-flavor. Wang et al. [59] reported the potential utilization of *A. pasteurianus* and the enhanced sensory properties of major organic acids in mixed cultures (*Acetobacter* and *Gluconobacter strain*). In addition, mass production of nutritious Kombucha with consistent quality also poses challenges. Obvious differences in the production of major organic acids such as acetic acid and gluconic acid can be observed depending on the type of acetic acid bacteria, and, depending on the complex microbial culture conditions, it can be a major variable affecting the quality of the Kombucha. Therefore, to ensure the manufacture of commercially usable Kombucha, as well as reproducibility, which is the final objective of this study, we plan on building infrastructure for complex microbial culture of the two selected types of acetic acid bacteria, lactic acid bacteria, and yeast.

4. Conclusions

The mass production of nutritious Kombucha with consistent quality presents several challenges. In addition, the type of AAB is known to influence organic acids such as gluconic acid and acetic acid, which are critical factors in determining the quality of Kombucha during the culturing of the microbial complex. Therefore, in this study, two strains determined to be suitable in the production of Kombucha were selected from among the isolated acetic acid bacteria. The results showed that an optimum temperature range of 30–35 °C was suitable in the fermentation of acetic acid bacteria. Regarding the change in pH and titratable acidity, the lowest pH and highest titratable acidity were detected in *Acetobacter pasteurianus* SFT-18 (pH 3.85, 1.68%) and *Gluconobacter oxydans* SFT-27 (pH 3.56, 0.79%). Regarding the change in viable cell count according to fermentation temperature, a high viable cell count was detected at different incubation temperatures in the *Acetobacter pasteurianus* SFT-18 strain. The *Gluconobacter oxydans* SFT-27 strain contained 10.32 mg/mL of glucuronic acid and 25.49 mg/mL of gluconate. This study was conducted to select acetic acid bacteria for use in the production of Kombucha. However, because the production of Kombucha is achieved through a complex fermentation process involving various microorganisms, it must contain all three types of microorganisms: yeast, lactic acid bacteria, and acetic acid bacteria. Thus, a selection study based on the specific quality characteristics of each microorganism, including yeast and lactic acid bacteria, was conducted. The results demonstrated that *Saccharomyces cerevisiae* SFT-71 (microorganism deposit number: KFCC11969P, Korean Culture Center of Microorganisms) and *Leuconostoc mesenteroides* SFT-45 (microorganism deposit number: KFCC11968P, Korean Culture Center of Microorganisms) were the yeast and lactic acid bacteria, respectively, most suitable for use in complex fermentation. Future research on the manufacturing of Kombucha, including studies on the composition of a complex microbial culture matrix in carefully selected strains, is anticipated.

Author Contributions: Conceptualization, D.-H.L., S.-H.K. and C.-K.H.; methodology, D.-H.L., S.-H.K. and C.-K.H.; formal analysis, D.-H.L., C.-Y.L. and E.-H.K.; investigation, D.-H.L., S.-H.K. and B.-K.C.; data curation, D.-H.L., H.-W.J. and W.-H.L.; writing—original draft preparation, D.-H.L. and C.-K.H.; writing—review and editing, D.-H.L., S.-H.K. and C.-K.H.; supervision, C.-K.H.; project administration, S.-H.K. and C.-K.H. All authors have read and agreed to the published version of the manuscript.

Funding: This research was funded by Suncheon city, Jeollanam-do, Republic of Korea, and APC was funded by Sunchon National University.

Institutional Review Board Statement: Not applicable.

Informed Consent Statement: Not applicable.

Data Availability Statement: Data are contained within this article.

Conflicts of Interest: The authors declare no conflicts of interest.

References

1. Saichana, N.; Matsushita, K.; Adachi, O.; Frébort, I.; Frebortova, J. Acetic acid bacteria: A group of bacteria with versatile biotechnological applications. *Biotechnol. Adv.* **2015**, *33*, 1260–1271. [CrossRef] [PubMed]
2. Qin, Z.; Yu, S.; Chen, J.; Zhou, J. Dehydrogenases of acetic acid bacteria. *Biotechnol. Adv.* **2022**, *54*, 107863. [CrossRef] [PubMed]
3. Baek, S.Y.; Kim, J.S.; Mun, J.Y.; Lee, C.H.; Park, Y.K.; Yeo, S.H. Quality characteristics of detoxified *Rhus verniciflua* vinegar fermented using different acetic acid bacteria. *Korean J. Food Preserv.* **2016**, *23*, 347–354. [CrossRef]
4. Gomes, R.J.; De Fatima Borges, M.; De Freitas Rosa, M.; Castro-Gómez, R.J.H.; Spinosa, W.A. Acetic acid bacteria in the food industry: Systematics, characteristics and applications. *Food Technol. Biotechnol.* **2018**, *56*, 139–151. [CrossRef] [PubMed]
5. Deppenmeier, U.; Hoffmeister, M.; Prust, C. Biochemistry and biotechnological applications of *Gluconobacter* strains. *Appl. Microbiol. Biotechnol.* **2002**, *60*, 233–242. [CrossRef] [PubMed]
6. Ministry of Food and Drug Safety. Food Code (No. 2023-13, 2023. 02. 17.). pp. 23–26. Available online: https://various.foodsafetykorea.go.kr/fsd/#/ (accessed on 6 March 2023).
7. Yim, E.J.; Jo, S.W.; Lee, E.S.; Park, H.S.; Ryu, M.S.; Uhm, T.B.; Kim, H.Y.; Cho, S.H. Fermentation characteristics of mulberry (*Cudrania tricuspidata*) fruit vinegar produced by acetic acid bacteria isolated from traditional fermented foods. *Korean J. Food Preserv.* **2015**, *22*, 108–118. [CrossRef]
8. Wang, B.; Rutherfurd-markwick, K.; Zhang, X.X.; Mutukumira, A.N. Kombucha: Production and microbiological research. *Foods* **2022**, *11*, 3456. [CrossRef]
9. Villarreal-soto, S.A.; Beaufort, S.; Bouajila, J.; Souchard, J.P.; Taillandier, P. Understanding kombucha tea fermentation: A review. *J. Food Sci.* **2018**, *83*, 580–588. [CrossRef]
10. Dufresne, C.; Farnworth, E. Tea, Kombucha, and health: A review. *Food Res. Int.* **2000**, *33*, 409–421. [CrossRef]
11. Bishop, P.; Pitts, E.R.; Budner, D.; Thompson-witrick, K.A. Kombucha: Biochemical and microbiological impacts on the chemical and flavor profile. *Food Chem. Adv.* **2022**, *1*, 100025. [CrossRef]
12. Soares, M.G.; Lima, M.D.; Schmidt, V.C.R. Technological aspects of kombucha, its applications and the symbiotic culture (SCOBY), and extraction of compounds of interest: A literature review. *Trends Food Sci. Technol.* **2021**, *110*, 539–550. [CrossRef]
13. Jayabalan, R.; Malbaša, R.V.; Lončar, E.S.; Vitas, J.S.; Sathishkumar, M. A review on kombucha tea—Microbiology, composition, fermentation, beneficial effects, toxicity, and tea fungus. *Compr. Rev. Food Sci. Food Saf.* **2014**, *13*, 538–550. [CrossRef] [PubMed]
14. Laavanya, D.; Shirkole, S.; Balasubramanian, P. Current challenges, applications and future perspectives of SCOBY cellulose of Kombucha fermentation. *J. Clean. Prod.* **2021**, *295*, 126454. [CrossRef]
15. Malbaša, R.V.; Lončar, E.S.; Vitas, J.S.; Čanadanović-Brunet, J.M. Influence of starter cultures on the antioxidant activity of kombucha beverage. *Food Chem.* **2011**, *127*, 1727–1731. [CrossRef]
16. Karabiyikli, S.; Sengun, I. Beneficial effects of acetic acid bacteria and their food products. In *Acetic Acid Bacteria*; CRC Press: Boca Raton, FL, USA, 2017; pp. 321–342. Available online: https://www.taylorfrancis.com/chapters/edit/10.1201/9781315153490-15/ (accessed on 10 November 2023).
17. Villarreal-soto, S.A.; Bouajila, J.; Pace, M.; Leech, J.; Cotter, P.D.; Souchard, J.P.; Taillandier, P.; Beaufort, S. Metabolome-microbiome signatures in the fermented beverage, Kombucha. *Int. J. Food Microbiol.* **2020**, *333*, 108778. [CrossRef] [PubMed]
18. Wang, B.; Rutherfurd-markwick, K.; Zhang, X.X.; Mutukumira, A.N. Isolation and characterisation of dominant acetic acid bacteria and yeast isolated from Kombucha samples at point of sale in New Zealand. *Curr. Res. Food Sci.* **2022**, *5*, 835–844. [CrossRef] [PubMed]
19. Arifuzzaman, M.; Hasan, M.Z.; Rahman, S.M.B.; Pramanik, M.K. Isolation and characterization of *Acetobacter* and *Gluconobacter* spp from sugarcane and rotten fruits. *Res. Rev. Biosci.* **2014**, *8*, 359–365. [CrossRef]
20. Hucker, G.J.; Conn, H.J. *Methods of Gram Staining*; Technical Bulletin of the New York State Agricultural Experiment Station: Ithaca, NY, USA, 1923; pp. 1–23. Available online: http://ecommons.library.cornell.edu/handle/1813/30787 (accessed on 6 March 2023).
21. Karen, R. *Catalase Test Protocol*; American Society of Microbiology: Washington, DC, USA, 2010; pp. 1–9. Available online: https://asm.org/getattachment/72a871fc-ba92-4128-a194-6f1bab5c3ab7/Catalase-Test-Protocol.pdf (accessed on 6 March 2023).
22. Wan, M.; Rosenberg, J.N.; Faruq, J.; Betenbaugh, M.J.; Xia, J. An improved colony PCR procedure for genetic screening of *Chlorella* and related microalgae. *Biotechnol. Lett.* **2011**, *33*, 1615–1619. [CrossRef]
23. Saitou, N.; Nei, M. The neighbor-joining method: A new method for reconstructing phylogenetic trees. *Mol. Biol. Evol.* **1987**, *4*, 406–425. [CrossRef]
24. Guangsen, T.; Xiang, L.; Jiahu, G. Microbial diversity and volatile metabolites of kefir prepared by different milk types. *CYTA J. Food* **2021**, *19*, 339–407. [CrossRef]
25. Kim, M.J.; Kim, B.H.; Han, J.K.; Lee, S.Y.; Kim, K.S. Analysis of Quality Properties and Fermentative Microbial Profiles of Takju and Yakju Brewed with of without Steaming Process. *J. Food Hyg. Saf.* **2011**, *26*, 64–69. Available online: https://koreascience.kr/article/JAKO201117148817199.page (accessed on 6 March 2023).
26. Lee, J.C.; Han, W.C.; Lee, J.H.; Jang, K.H. Quality Evaluation of Vinegar Manufactured using Rice and *Rosa rugosa* Thunb. *Korean J. Food Sci. Technol.* **2012**, *44*, 202–206. [CrossRef]
27. Ansari, F.; Pourjafar, H.; Kangari, A.; Homayouni, A. Evaluation of the glucuronic acid production and antibacterial properties of kombucha black tea. *Curr. Pharm. Biotechnol.* **2019**, *20*, 985–990. [CrossRef] [PubMed]

28. Adnan, A.F.M.; Tan, I.K.P. Isolation of lactic acid bacteria from Malaysian foods and assessment of the isolates for industrial potential. *Bioresour. Technol.* **2007**, *98*, 1380–1385. [CrossRef] [PubMed]

29. Sengun, I.Y.; Karabiyikli, S. Importance of acetic acid bacteria in food industry. *Food Control* **2011**, *22*, 647–656. [CrossRef]

30. Sandu, M.P.; Sidelnikov, V.S.; Geraskin, A.A.; Chernyavskii, A.V.; Kurzina, I.A. Influence of the method of preparation of the Pd-Bi/Al$_2$O$_3$ catalyst on catalytic properties in the reaction of liquid-phase oxidation of glucose into gluconic acid. *Catalysts* **2020**, *10*, 271. [CrossRef]

31. Tomotani, E.J.; Das Neves, L.C.M.; Vitolo, M. Oxidation of glucose to gluconic acid by glucose oxidase in a membrane bioreactor. *Appl. Biochem. Biotechnol.* **2005**, *121*, 149–162. [CrossRef]

32. Gupta, A.; Singh, V.K.; Qazi, G.N.; Kumar, A. *Gluconobacter oxydans*: Its biotechnological applications. *J. Mol. Microbiol. Biotechnol.* **2001**, *3*, 445–456. Available online: https://pubmed.ncbi.nlm.nih.gov/11361077/ (accessed on 6 March 2023).

33. Moonmangmee, D.; Adachi, O.; Ano, Y.; Shinagawa, E.; Toyama, H.; Theeragool, G.; Lotong, N.; Matsushita, K. Isolation and characterization of thermotolerant *Gluconobacter* strains catalyzing oxidative fermentation at higher temperatures. *Biosci. Biotechnol. Biochem.* **2000**, *64*, 2306–2315. [CrossRef]

34. Eom, H.J.; Yoon, H.S.; Kwon, N.R.; Jeong, Y.J.; Kim, Y.H.; Hong, S.T.; Han, N.S. Comparison of the Quality Properties and Identification of Acetic Acid Bacteria for Aronia Vinegar. *J. Korean Soc. Food Sci. Nutr.* **2019**, *48*, 1397–1404. [CrossRef]

35. Shafiei, R.; Delvigne, F.; Babanezhad, M.; Thonart, P. Evaluation of viability and growth of *Acetobacter senegalensis* under different stress conditions. *Int. J. Food Microbiol.* **2013**, *163*, 204–213. [CrossRef] [PubMed]

36. Bang, K.H.; Kim, C.W.; Kim, C.H. Isolation of an Acetic Acid Bacterium *Acetobacter pasteurianus* CK-1 and Its Fermentation Characteristics. *J. Life Sci.* **2022**, *32*, 119–124. [CrossRef]

37. Park, K.S.; Chang, D.S.; Cho, H.R.; Park, U.Y. Investigation of the Cultural Characteristics of High Concentration. *J. Korean Soc. Food Sci. Nutr.* **1994**, *23*, 666–670. Available online: https://koreascience.kr/article/JAKO199411920151954.page (accessed on 6 March 2023).

38. Sim, K.C.; Lee, K.S.; Kim, D.H.; Ryu, I.H.; Lee, J.S. Studies on the Acid Tolerance of *Acetobacter* sp. Isolated from Persimmon Vinegar. *Korean J. Food Sci. Technol.* **2001**, *33*, 574–581. Available online: https://koreascience.kr/article/JAKO200103042136245.page (accessed on 6 March 2023).

39. El-askri, T.; Yatim, M.; Sehli, Y.; Rahou, A.; Belhaj, A.; Castro, R.; Duran-guerrero, E.; Hafidi, M.; Zouhair, R. Screening and characterization of new *Acetobacter fabarum* and *Acetobacter pasteurianus* strains with high ethanol–thermo tolerance and the optimization of acetic acid production. *Microorganisms* **2022**, *10*, 1741. [CrossRef] [PubMed]

40. Kim, Y.D.; Kang, S.K.; Kang, S.H. Studies on the acetic acid fermentation using maesil juice. *J. Korean Soc. Food Sci. Nutr.* **1996**, *25*, 695–700. Available online: https://koreascience.kr/article/JAKO199611921348361.page (accessed on 6 March 2023).

41. Es-sbata, I.; Lakhlifi, T.; Yatim, M.; El-abid, H.; Belhaj, A.; Hafidi, M.; Zouhair, R. Screening and molecular characterization of new thermo- and ethanol-tolerant *Acetobacter malorum* strains isolated from two biomes Moroccan cactus fruits. *Biotechnol. Appl. Biochem.* **2021**, *68*, 476–485. [CrossRef]

42. Antolak, H.; Piechota, D.; Kucharska, A. Kombucha tea—A double power of bioactive compounds from tea and symbiotic culture of bacteria and yeasts(SCOBY). *Antioxidants* **2021**, *10*, 1541. [CrossRef]

43. Filippis, F.D.; Troise, A.D.; Vitaglione, P.; Ercolini, D. Different temperatures select distinctive acetic acid bacteria species and promotes organic acids production during Kombucha tea fermentation. *Food Microbiol.* **2018**, *73*, 11–16. [CrossRef]

44. Magalhaes, K.T.; Pereira, G.V.; Dias, D.R.; Schwan, R.F. Microbial communities and chemical changes during fermentation of sugary Brazilian kefir. *World J. Microbiol. Biotechnol.* **2010**, *26*, 1241–1250. [CrossRef]

45. Gullo, M.; Verzelloni, E.; Canonico, M. Aerobic submerged fermentation by acetic acid bacteria for vinegar production: Process and biotechnological aspects. *Process Biochem.* **2014**, *49*, 1571–1579. [CrossRef]

46. Sharafi, S.M.; Rasooli, I.; Beheshti-maaI, K. Isolation characterization and optimization of indigenous acetic acid bacteria and evaluation of their preservation methods. *Iran. J. Microbiol.* **2010**, *2*, 38–45. Available online: https://pubmed.ncbi.nlm.nih.gov/22347549/ (accessed on 6 March 2023). [PubMed]

47. Vīna, I.; Linde, R.; Patetko, A.; Semjonovs, P. Glucuronic acid containing fermented functional beverages produced by natural yeasts and bacteria associations. *Int. J. Res. Rev. Appl.* **2013**, *14*, 17–25. Available online: https://research.kombuchabrewers.org/wp-content/uploads/kk-research-files/glucuronic-acid-containing-fermented-functional-beverages-produced-by-natural-yeasts-and-bacteria-as.pdf (accessed on 10 November 2023).

48. Merchie, G.; Lavens, P.; Sorgeloos, P. Optimization of dietary vitamin C in fish and crustacean larvae: A review. *Aquaculture* **1997**, *155*, 165–181. [CrossRef]

49. Dai, L.; Lian, Z.; Zhang, R.; Nawaz, A.; Ul-haq, I.; Zhou, X.; Xu, Y. Multi-strategy in production of high titer gluconic acid by the fermentation of concentrated cellulosic hydrolysate with *Gluconobacter oxydans*. *Ind. Crops Prod.* **2022**, *189*, 115748. [CrossRef]

50. Lee, H.C.; Chung, B.W.; Kim, C.Y.; Kim, D.H.; Ra, B.K. The production of sodium gluconate by *Aspergillus niger*. *Korean J. Biotechnol. Bioeng.* **1996**, *11*, 65–70. Available online: https://koreascience.kr/article/JAKO199611919973454.page (accessed on 6 March 2023).

51. Jayabalan, R.; Marimuthu, S.; Swaminathan, K. Changes in content of organic acids and tea polyphenols during kombucha tea fermentation. *Food Chem.* **2007**, *102*, 392–398. [CrossRef]

52. Chen, C.; Liu, B.Y. Changes in major components of tea fungus metabolites during prolonged fermentation. *J. Appl. Microbiol.* **2000**, *89*, 834–839. [CrossRef]

53. Wu, X.; Yao, H.; Liu, Q.; Zheng, Z.; Cao, L.; Mu, D.; Wang, H.; Jiang, S.; Li, X. Producing acetic acid of *Acetobacter pasteurianus* by fermentation characteristics and metabolic flux analysis. *Appl. Biochem. Biotechnol.* **2018**, *186*, 217–232. [CrossRef]
54. Nie, W.; Zheng, X.; Feng, W.; Liu, Y.; Li, Y.; Liang, X. Characterization of bacterial cellulose produced by *Acetobacter pasteurianus* MGC-N8819 utilizing lotus rhizome. *LWT* **2022**, *165*, 113763. [CrossRef]
55. Da silva, G.A.R.; Oliveira, S.S.D.; Lima, S.F.; Nascimento, R.P.; Baptista, A.R.D.; Fiaux, S.B. The industrial versatility of *Gluconobacter oxydans*: Current applications and future perspectives. *World J. Microbiol. Biotechnol.* **2022**, *38*, 134. [CrossRef] [PubMed]
56. Ko, H.M.; Shin, S.S.; Park, S.S. Biological activities of kombucha by stater culture fermentation with *Gluconacetobacter* spp. *J. Korean Soc. Food Sci. Nutr.* **2017**, *46*, 896–902. [CrossRef]
57. Sknepnek, A.; Tomic, S.; Miletic, D.; Levic, S.; Colic, M.; Nedovic, V.; Niksic, M. Fermentation characteristics of novel *Coriolus versicolor* and *Lentinus edodes* kombucha beverages and immunomodulatory potential of their polysaccharide extracts. *Food Chem.* **2021**, *342*, 128344. [CrossRef] [PubMed]
58. Tran, T.; Grandvalet, C.; Verdier, F.; Martin, A.; Alexandre, H.; Tourdot-marechal, R.T. Microbiological and technological parameters impacting the chemical composition and sensory quality of kombucha. *Compr. Rev. Food Sci. Food Saf.* **2020**, *19*, 2050–2070. [CrossRef]
59. Wang, S.; Zhang, L.; Qi, L.; Liang, H.; Lin, X.; Li, S.; Yu, C.; Ji, C. Effect of synthetic microbial community on nutraceutical and sensory qualities of kombucha. *Int. J. Food Sci. Technol.* **2020**, *55*, 3327–3333. [CrossRef]

fermentation

Review

Underexplored Potential of Lactic Acid Bacteria Associated with Artisanal Cheese Making in Brazil: Challenges and Opportunities

Bianca de Oliveira Hosken [1], Gilberto Vinícius Melo Pereira [2,*], Thamylles Thuany Mayrink Lima [1], João Batista Ribeiro [3], Walter Coelho Pereira de Magalhães Júnior [3] and José Guilherme Prado Martin [1,*]

[1]   Microbiology of Fermented Products Laboratory (FERMICRO), Department of Microbiology, Universidade Federal de Viçosa, Viçosa 36570-900, Brazil; bianca.hosken@ufv.br (B.d.O.H.)
[2]   Bioprocess Engineering and Biotechnology Department, Federal University of Paraná, Curitiba 81531-980, Brazil
[3]   Empresa Brasileira de Pesquisa Agropecuária (EMBRAPA), Unidade Embrapa Gado de Leite, Juiz de Fora 36038-330, Brazil; joao-batista.ribeiro@embrapa.br (J.B.R.); walter.magalhaes@embrapa.br (W.C.P.d.M.J.)
*   Correspondence: gilbertovinicius@gmail.com (G.V.M.P.); guilherme.martin@ufv.br (J.G.P.M.)

Abstract: Artisanal cheeses are prepared using traditional methods with territorial, regional and cultural linkages. In Brazil, there is a great diversity of artisanal cheeses (BAC), which have historical, socioeconomic and cultural importance. The diversity of the BAC between producing regions is due to the different compositions of raw milk, the steps involved in the process and the maturation time. The crucial step for cheese differentiation is the non-addition of starter cultures, i.e., spontaneous fermentation, which relies on the indigenous microbiota present in the raw material or from the environment. Therefore, each BAC-producing region has a characteristic endogenous microbiota, composed mainly of lactic acid bacteria (LAB). These bacteria are responsible for the technological, sensory and safety characteristics of the BAC. In this review, the biotechnological applications of the LAB isolated from different BAC were evidenced, including proteolytic, lipolytic, antimicrobial and probiotic activities. In addition, challenges and opportunities in this field are highlighted, because there are knowledge gaps related to artisanal cheese-producing regions, as well as the biotechnological potential. Thus, this review may provide new insights into the biotechnological applications of LAB and guide further research for the cheese-making process.

Keywords: traditional foods; fermentation; bioprospecting; biotechnology

check for **updates**

Citation: Hosken, B.d.O.; Melo Pereira, G.V.; Lima, T.T.M.; Ribeiro, J.B.; Magalhães Júnior, W.C.P.d.; Martin, J.G.P. Underexplored Potential of Lactic Acid Bacteria Associated with Artisanal Cheese Making in Brazil: Challenges and Opportunities. *Fermentation* **2023**, *9*, 409. https://doi.org/10.3390/fermentation9050409

Academic Editors: Roberta Comunian and Luigi Chessa

Received: 28 March 2023
Revised: 18 April 2023
Accepted: 20 April 2023
Published: 25 April 2023

1. Introduction

Lactic acid bacteria (LAB) are a diverse group of Gram-positive bacteria that produce lactic acid as the main fermentation product of the carbohydrate metabolism. The term "LAB" is somewhat ambiguous and is often used to refer to bacteria applied in the production of fermented foods [1]. These include bacteria with high G+C (Bifidobacterium) and low G+C content (Firmicute such as *Lactobacillus*, *Lactococcus* and *Streptococcus*). They are acid-tolerant, meso-aerophilic, not mobile or spore-forming and either rod-shaped (bacilli) or spherical (cocci) [2]. The term LAB has a rather positive connotation, containing bacteria generally considered safe for human consumption, although some strains of enterococci raise concern due to the possible presence of virulence factors and the potential transfer of antibiotic resistance.

LAB are widely spread in the environment and play an important role in fermentation processes. They are employed in the production of pickles, sauerkraut, fermented meats, breads and especially dairy products [3]. Cheese making involves a process of fermentation by LAB. During this process, milk is coagulated by adding rennet or an acid. The acid may be produced by the fermentation of lactose by LAB. Artisanal cheeses are produced by indigenous LAB present in the raw material or from the environment [4]. For all processes, LAB are important for acidification and the ripening process. In addition, they produce key metabolites with antimicrobial activity, including organic acids, ethanol, hydrogen peroxide, diacetyl, CO_2 and bacteriocins [5,6].

In recent years, Brazilian cheeses have been recognized for their quality in several awards, both at national [7] and international [8] levels. In general, the production of Brazilian artisanal cheeses (BAC) involves the use of raw milk and an endogenous ferment consisting of the whey collected the day before, which can be named according to the region, such as "pingo" for Artisanal Minas Cheeses (AMC), the most famous in the country, or "repique" for Porungo cheese, produced in São Paulo state. BAC produced with raw milk must be ripened in accordance with specific legislation in minimum periods in order to guarantee its safety [9,10].

Several studies have demonstrated the diversity of LAB in BAC, with emphasis on *Lactobacillus, Lactococcus, Enterococcus, Weissella, Pediococcus* and *Leuconostoc* genera [11–13]. Different biotechnological applications of LAB isolated from BAC were detected, including probiotic potential [11]; diacetyl [14] and exopolysaccharides (EPS) production [14]; and antimicrobial [5], proteolytic [15] and lipolytic activities [14]. However, there is a lack of knowledge about the biotechnological potential of LAB isolated from BAC. In this review, we present the main gaps detected, indicating the under-investigated artisanal cheese-producing regions, the opportunities for biotechnological exploration, as well as the need to organize a collection of LAB typical of BAC for the purpose of biotechnological research and exploitation. With an ultimate goal, this review provides new insights into the industrial applications of LAB isolated from BAC.

2. Brazilian Artisanal Cheeses (BAC)

In Brazil, there is a great diversity of artisanal cheeses with historical, socioeconomic and cultural importance. In general, cheese production takes place on small farms and includes raw milk and traditional methods, which has been transmitted over hundreds of years by generations of cheesemakers [16,17]. BAC are produced in different geographic regions (Figure 1), such as Marajó cheese in the north; Coalho and Manteiga cheeses in the northeast; Caipira cheese in the central region; Colonial, Serrano, KochKäse and Käschmier in the south; and in the southeast, Artisanal Minas Cheese (AMC), Cabacinha, Parmesan-type cheeses (Alagoa, Vale do Suaçuí and Mantiqueira de Minas), Porungo and Requeijão Moreno [4].

Minas Gerais state is responsible for half of all cheese produced in the country, whose importance is reinforced by the existence of several producing regions. Among them, AMC production is responsible for 50% of the national production [4]. It is produced in the micro-regions of Araxá, Campo das Vertentes, Canastra, Cerrado, Serra do Salitre, Serro, Triângulo Mineiro and, more recently, Serras do Ibitipoca and Entre Serras (Figure 1) [18,19]. The AMC production method has even been recognized as Brazilian intangible heritage. Its production steps consist of milking, filtration, the addition of rennet and endogenous ferment, coagulation, curd cutting, draining, molding, pressing, dry salting and ripening [20]. Its quality has been reinforced by several awards; in 2021, for example, Brazil was one of the leading countries in the ranking of the most famous world cheese contest, winning 57 medals; 4 of the 5 medals in the "super gold" modality were won by cheeses produced in Minas Gerais state [21]. In addition to the socio-cultural relevance of AMC, it has economic importance, representing the main source of income for thousands of rural producer families [22].

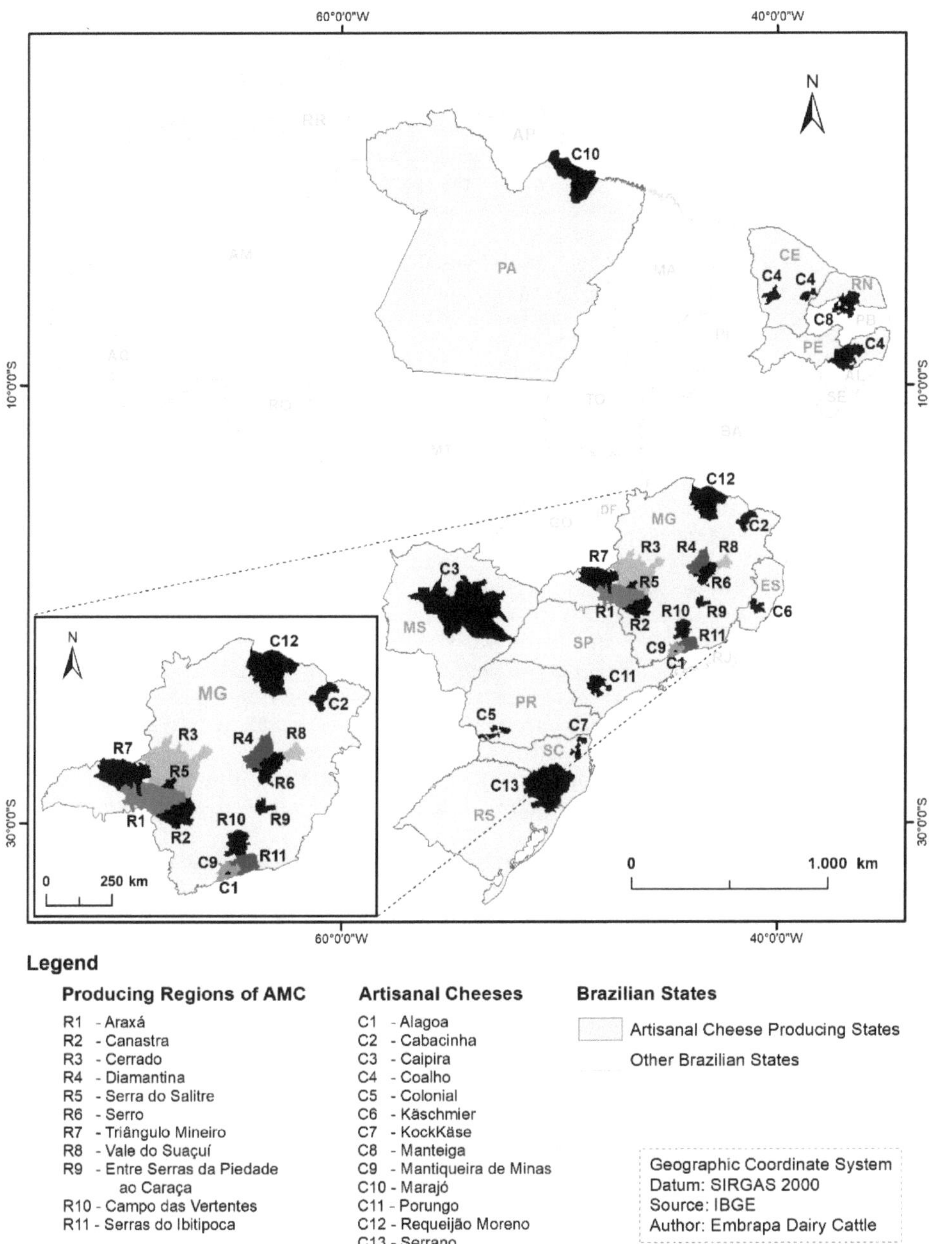

Figure 1. Brazilian artisanal cheeses: types and regions in 2022. In the detail at the bottom left, the producing regions of AMC and the most famous BAC are highlighted.

3. LAB and Food Industry

LAB produce lactic acid as the main fermentation product, generated from two fermentative metabolic pathways: homofermentative and heterofermentative. In cheese making, both LAB metabolisms are reported. Homofermentative LAB includes *Enterococcus, Lactococcus, Pediococcus, Weissella* and *Streptococcus* which produce lactic acid as an end metabolite by Pentose Phosphate or the Embden-Meyerhof-Parnas pathway. Heterofermentative LAB includes *Leuconosctoc* and *Oenococcus* which produce several other products in addition to lactic acid, such as ethanol, acetic acid and CO_2, from the conversion of lactose via the 6-P-gluconate/phosphoketolase pathway. Finally, *Lactobacillus* includes both homofermentative and heterofermentative species [23–25].

Streptococcus thermophilus, Lactococcus lactis and many lactobacilli grow in the presence of a maximum of 2% or eventually 4% of salt, in addition to tolerating environments with a low pH. LAB can also produce several types of glycolytic, lipolytic and proteolytic enzymes. These characteristics reinforce their importance for different applications in the food industry [26,27]. In addition, LAB contribute to the sensory development of various foods, especially flavor (as they produce volatile compounds) and texture (improved by the production of exopolysaccharides). The safety history of LAB contributes to the GRAS (Generally Recognized As Safe) or QPS (Qualified Assumption of Safety) status, enabling its use in food, either as starter cultures or probiotic strains [28]. For fermented foods produced from previously sanitized or pasteurized raw materials, the use of a LAB starter culture is necessary [27]. In addition, there are also non-starter LAB (NSLAB) that are especially important for cheese ripening, for example [26].

In recent years, several studies have explored the potential of LAB to be used as live vectors for in situ synthesis, i.e., the production and delivery of biomolecules at their site of use/application, without removing or transporting them to another site. This is only possible due to the GRAS status of the LAB strains. Another path consists of the direct application of compounds obtained by ex situ synthesis, which means applying the compounds in a place or environment outside their place of use or application [29]. However, in situ synthesis is advantageous as it allows the use of LAB strains instead of purified compounds, enabling the development of polyfunctional cultures, as well as reducing the costs of downstream isolation and purification steps. This strategy may also be better accepted by consumers, because purified compounds are considered food additives [30].

Recent studies have evaluated the use of in situ LAB for the synthesis of gamma-aminobutyric acid (GABA) from L-glutamate—an amino acid released during milk fermentation. This non-essential amino acid plays an important role in the central nervous system as an inhibitory neurotransmitter. Its properties include antidepressant, anxiolytic and antihypertensive activity, as well as the ability to regulate hormone secretion [6]. The production of GABA by LAB appears to be directly related to the acid stress response; thus, LAB strains able to produce GABA could be employed for functional purposes, especially in foods with reduced pH values [31]. Challenges related to the use of LAB in situ include its ability to resist certain types of stress, especially the osmotic pressure resulting from the use of salts by the food industry [32]. In this context, the isolation of LAB from artisanal cheeses aiming at its in situ application is notoriously promising, given its survival in ripened cheeses, which generally have high amounts of salt [11]. Thus, it may represent, in the near future, a promising strategy for the food industry.

Finally, the biotechnological potential of LAB also includes the encapsulation of metabolites produced ex situ by them for the controlled release or application in active packaging. Microencapsulation technology allows food-grade ingredients or bioactive components to be adequately protected and released in a controlled manner over long periods, including at specific sites [33]. The microencapsulation of LAB with probiotic properties for use in livestock, for example, has already been demonstrated [34]. As for the active packaging, antibacterial bioplastic film incorporated with purified bacteriocin from *Lactilactobacillus sakei* was able to reduce the contamination of Coalho cheese by coagulase-positive staphylococci and thermotolerant coliforms [35].

4. LAB Isolated from BAC

In general, BAC are produced from raw milk, which presents a pH close to neutrality, high water activity and significant nutritional value. It also has rich microbiota, mainly composed of LAB [36,37], essential for the fermentation process and, consequently, for the cheese quality and safety [38,39]. The relevant sensory characteristics of artisanal cheeses are provided by the activity of autochthonous LAB, especially related to the production of organic acids, fatty acids and amino acids, as well as peptidases and lipases [40–44].

In BAC, the most frequently reported genera of LAB are *Lactobacillus*, *Lactococcus*, *Enterococcus*, *Pediococcus*, *Leuconostoc*, *Streptococcus* and *Weissella* (Table 1). No data were found regarding the LAB isolated from Cabacinha, Parmesan-type cheeses, Porungo and KochKäse and Käschmier cheeses. It should also be noted that LAB also correspond to the majority group in an endogenous ferment used in the production of various types of BAC, in addition to the milking and production environment; therefore, the LAB diversity of BAC is influenced by the geographic location, climatic conditions and processing steps [15,43,44].

Table 1. LAB isolated from BAC produced in several producing regions.

BAC	LAB *	References
Caipira	*Enterococcus* sp., *E. faecium*, *E. durans*, *E. faecalis*, *E. hermanniensis*, *Lactococcus*, *Lb. plantarum* subsp. *plantarum*, *Lb. paracasei* subsp. *paracasei*, *Lb. casei.*	[12,39]
Coalho	*Enterococcus* sp., *E. faecium*, *E. casseliflavus*, *E. durans*, *E. faecalis*, *E. gallinarum*, *E. italicus*, *E. hermanniensis*, *Lactobacillus* sp., *Lb. acidophilus*, *Lb. curvatus*, *Lb. fermentum*, *Lb. paracasei* subsp. *paracasei*, *Lb. plantarum* subsp. *plantarum*, *Lb. rhamnosus* , *Lactococcus* sp., *Lc. lactis*, *Lc. lactis* subsp. *lactis*, *Lc. garvieae*, *Leuconostoc* sp., *Lc. mesenteroides* subsp. *mesenteroides*, *Streptococcus* sp., *S. infantarius*, subsp. *infantarius*, *S. lutetiensis*, *S. macedonicus*, *S. waiu*, *Weisella* sp., *W. paramesenteroides*	[12,39,45–47]
Colonial	*E. faecium*, *E. durans*, *E. faecalis,, E. hermanniensis*, *Lactococcus* sp., *Lc. lactis*, *Lc. piscium*, *Lc. raffinolactis* group, *Lactobacillus* sp., *Lb. brevis*, *Lb. casei-paracasei*, *Leuconostoc* sp., *S. equinus-lutetiensis*, *S. parauberis*, *S. porcorum/sanguinis*	[12,39,48]
Manteiga	*E. faecium*, *E. durans*, *E. faecalis*, *E. hermanniensis*, *Lactobacillus* sp., *Lactococcus* sp., *Leuconostoc* sp., *Streptococcus* sp	[12,39]
Marajó	*E. durans*, *E. faecium*, *E. faecalis*, *E. gilvus*, *E. hermanniensis*, *Lactobacillus* sp., *Lactococcus* sp., *Leuconostoc* sp., *Streptococcus* sp	[12,39,49]
Artisanal Minas	*Enterococcus* spp., *E. durans*, *E. faecalis*, *E. faecium*, *E. gilvus*, *E. hermanniensis*, *E. raffinosus*, *E. rivorum*, *Lactobacillus* sp., *Lb. casei*, *Lb. paracasei* subsp. *paracasei*, *Lb. plantarum* subsp. *plantarum*, *Lb. paraplantarum*, *Lb. rhamnosus*, *Lb. hilgardii*, *Lb. brevis*, *Lb. buchneri* subsp. *buchneri*, *Lb. parabuchneri*, *Lb. acidipiscis*, *Lactococcus* spp., *Lc. lactis*, *Lc. garvieae*, *Leuconostoc* sp., *Ln. mesenteroides*, *Pediococcus* sp., *P. acidilactici*, *Streptococcus* sp., *S. agalactiae*, *S. macedonicus*, *S. porcorum/sanguinis*, *S. thermophilus*, *S. infantarius*, *W. paramesenteroides*	[39,40,44,48,50–55]
Serrano	*Enterococcus* sp., *E. faecium*, *E. durans*, *E. faecalis*, *E. hermanniensis.* *Lactobacillus* sp., *Lb. casei*, *Lb. plantarum* subsp. *plantarum*, *Lb. paracasei* subsp. *paracasei*, *Lb. rhamnosus*, *Lb. acidophilus*, *Lb. curvatus*, *Lb. fermentum*, *Lactococcus* sp., *Lc. lactis*, *Lc. piscium*, *Lc. raffinolactis*, *Leuconostoc* sp., *Ln. mesenteroides*, *Streptococcus* sp., *S. equinus-lutetiensis-infantarius*, *S. parauberis*, *S. porcorum/sanguinis.*	[39,56–58]

* *Lactobacillus* species updated according to the reclassification [59]. *E.* = *Enterococcus*, *Lc.* = *Lactococcus*, *Lb.* = *Lactobacillus*, *Ln* = *Leuconostoc*, *S.* = *Streptococcus*, *P.* = *Pediococcus*, *W.* = *Weissella*.

The importance of LAB in cheese production is due to the presence of starter cultures and NSLAB. Starter cultures, mainly *Lc. lactis* and *S. thermophilus*, are responsible for converting lactose into lactic acid at a controlled rate. This process results in a gradual decrease in pH, which has a significant impact on various aspects of cheese production and ultimately determines the cheese's composition and quality. During the early stages of cheese ripening, *Lb. delbrueckii* and *Lb. helveticus* play a critical role, breaking down proteins, metabolizing lactose, producing aromatic compounds and providing substrates that can be further consumed by other microbial groups, such as NSLAB [60]. NSLAB mainly include the facultative heterofermentative *Lactobacillus* genus, followed by *Pediococcus pentosaceus* [61]. They can impact the cheese flavor and texture due to the production of compounds from the catabolism of amino acids, mainly methionine, aromatic amino acids and branched-chain amino acids, in addition to the synthesis of EPS [39,61–63]. In addition, bacteriocins, hydrogen peroxide, diacetyl and CO_2 are also produced by NSLAB, acting as biopreservatives and contributing to the cheese safety [5,11,64,65].

5. Biotechnological Potential of LAB Isolated from BAC

The self-sufficiency in inputs, the increasing demand for clean-label products and food production in the bioeconomy context have stimulated the development of research for bioprospecting microbial and bioactive compounds from different types of products, especially fermented foods [66,67]. Among them, dairy products stand out due to their recognized microbial diversity, especially LAB. In this context, artisanal cheeses have proved to be an important source for the isolation of microorganisms with biotechnological purposes [11].

Recent studies have demonstrated the potential for the industrial application of LAB, such as the production of enzymes, diacetyl, EPS, antimicrobial compounds, probiotic and prebiotic effects, among others, aimed mainly at improving food quality and safety [32,68,69]. In Brazil, research has been carried out to discover novel LAB strains isolated from BAC for industrial exploitation (Table 2). In the next sections, the main biotechnological applications of LAB identified in BAC by different studies published in recent years are discussed.

Table 2. Biotechnological potential of LAB isolated from BAC.

BAC	Biotechnological Potential	References
Marajó	Antimicrobial activity against *L. monocytogenes*, *St. aureus* and *Es. coli*, lipolytic activity, proteolytic activity, acidification capacity, diacetyl production	[11,14,49,70,71]
Manteiga	Antimicrobial activity against *L. monocytogenes* and *St. aureus*, lipolytic activity, proteolytic activity, acidification capacity, diacetyl production, probiotic potential	[11,14,39,72]
Coalho	Antimicrobial activity against *Listeria* sp., *B. cereus*, *B. subtilis*, *E. faecalis*, *St. aureus*, *Es. coli*, *K. pneumoniae* and *P. aeruginosa*, lipolytic activity, proteolytic activity, acidification capacity, probiotic potential, β-galactosidase synthesis	[11,14,39,47,72–74]
Serrano	Antimicrobial activity against *L. monocytogenes*, *St.aureus*, *Es. coli*, *S. enterica* and *Penicillium*, lipolytic activity, proteolytic activity, acidification capacity, diacetyl production, probiotic potential	[11,14,39,72,75]
Caipira	Antimicrobial activity against *L. monocytogenes* and *St.aureus*, lipolytic activity, proteolytic activity, acidification capacity, diacetyl production, probiotic potential	[11,14,39,72]
AMC	Antimicrobial activity against *Listeria* sp., *Enterococcus* sp., *St. aureus*, *S.* Typhimurium and *S.* Enteritidis, lipolytic activity, proteolytic activity, acidification capacity, diacetyl production, probiotic potential, EPS production	[11,14,39,43,72,76–83]
Colonial	Antimicrobial activity against *L. monocytogenes* and *St. aureus*, lipolytic activity, proteolytic activity, acidification capacity, diacetyl production, probiotic potential	[11,14,39,72]

L. = Listeria, St. = Staphylococcus, Es. = Escherichia, K. = Klebsiella, P. = Pseudomonas, B. = Bacillus, S. = Salmonella.

5.1. Bacteriocin Production

Bacteriocins are proteins or peptides ribosomally synthesized by Gram-positive and -negative bacteria, with recognized antimicrobial activity (bacteriostatic, bactericidal or bacteriolytic) against taxonomically related or unrelated microorganisms [84–86]. They can be broad spectrum, inhibiting a wide variety of bacteria, or narrow spectrum, inhibiting taxonomically close bacteria [86,87]. In general, they are cationic and exhibit amphipathic properties, with the cell membrane being, in most cases, the target of their activity [88]. The first studies about the antimicrobial activity of LAB date back to the 1920s, with the discovery of colicin V; the discovery of nisin, in 1969, intensified the search for bioactive peptides synthesized by LAB, more specifically bacteriocins. Its use by the food and medical industries represents an alternative to the use of chemical additives and antibiotics, respectively, which has stimulated the interest in novel research in the area [31,89–91].

The industrial application of bacteriocins has several advantages, such as the activity against pathogens and spoilage microorganisms in foods, relative stability in different pH and temperature values, possibility of use as natural preservatives in foods and selective toxicity and inactivation by digestive proteases, with little influence on gut microbiota. Furthermore, a genetic determinant is usually encoded by plasmids, which allows facilitated genetic manipulation [92,93]. In addition, bacteriocins produced by LAB are considered GRAS, which favors their industrial application. However, the only bacteriocin approved by the Food and Drug Administration (FDA) for use as a preservative in foods is nisin, produced by *Lactococcus lactis* and commercially available as Nisaplin® [94]. Nisin can also be applied in veterinary practice, for example, in the treatment of mastitis as an alternative to conventional antibiotics [95,96]. However, the low stability and solubility of nisin at neutral pH, the hydrophobic nature and the selection of resistant bacteria reinforce the importance of studies focused on the discovery of new bacteriocins [97,98].

In this context, artisanal cheeses consist of an important source of bacteriocins [39]. A recent evaluation of the phylogenetic distribution of the LAB bacteriocin repertoire associated with artisanal cheeses reported bacteriocins not yet characterized, for example, two novel putative glycocins and one lasso peptide in the genome of some strains belonging to the *E. faecalis* species, reinforcing their relevance as a potential source [84]. Pediocins produced by four different strains of *Pediococcus pentosaceus* isolated from AMC were able to inhibit the growth of *Listeria monocytogenes*, a relevant foodborne pathogen [98]. The *Pediococcus* and *Lactobacillus* strains isolated from sheep cheese produced in southern Brazil and artisanal cheese produced in Minas Gerais state have also been identified as producing bacteriocins with anti-listeria activity [99,100]. In addition to this pathogen, *Bacillus cereus*, one of the most important causes of food poisoning, and *Pseudomonas fluorescens*, common spoilage bacteria, were inhibited by bacteriocins (not identified yet) produced by the LAB isolated from Colonial cheese produced in southern Brazil [101].

5.2. Acidification Capacity

The acidification capacity is a widely studied aspect in LAB isolated from artisanal cheeses and can vary significantly depending on the strain and substrate. LAB are mainly responsible for the acidification of the raw milk, resulting in the pH decreasing and, consequently, affecting the activity of the rennet. Acidification also contributes to the solubilization of calcium phosphate, impacting the cheese texture, as well as the syneresis process, with reflections on its centesimal composition. Finally, acidification plays an important role in the microbial succession during cheese ripening, favoring the enzymatic activity of NSLAB, with desirable effects on the cheese flavor and texture [62,102,103].

Furthermore, the decreasing pH resulting from the production of organic acids can inhibit the growth of spoilage and pathogenic microorganisms. The release of short-chain weak organic acids, especially lactic, acetic, sorbic and propionic, during the fermentation process corresponds to one of the main mechanisms of biopreservation in fermented foods [104]. The increase in the lipid solubility of organic acids under conditions of high acidity interferes with the cell membrane potential, impairing the metabolic functions of undesirable microorganisms [105].

The acidification capacity of LAB isolated from BAC varied according to the microbial species and producing region; *Lacticaseibacillus paracasei* and *Levilactobacillus brevis* were more efficient in acidifying the substrate under the LAB isolated from AMC, Coalho and Caipira cheeses and presented a high acidification capacity [11], which was attributed to the type of herd feeding, differences in the cheese pressing stage, as well as the higher proportion of carbohydrate in the cheese. A low acidification capacity was observed for *Weissella* spp. isolated from BAC [71], reinforcing that acidification depends on the LAB species. It is also worth mentioning that the acidification capacity may vary according to the culture medium used for isolation; LAB isolated from M17 agar showed a greater acidification capacity than those isolated from MRS agar, which makes it difficult to compare the results of LAB isolated from different culture media [43].

5.3. Probiotic Potential

According to the Food and Agriculture Organization [106], probiotics are live microorganisms that, when administered in adequate amounts, confer health benefits on the host. The term prebiotic refers to substrates that, when metabolized by the host's gut microbiota, result in health benefits. However, prebiotics can also be found in other sources, such as food, where they can stimulate the growth or activity of beneficial microorganisms [107]. The consumption of probiotics and/or prebiotics corresponds to one of the most efficient ways to maintain the balance of the intestinal microbiota (eubiosis) [108].

A probiotic strain must present some requirements, such as the ability to resist the acidic conditions, adhere to the gut environment, inhibit pathogens, modulate the immune system and confer benefits on the host's health; in addition, it does not present virulence factors. Several LAB strains meet these requirements, which make them even more relevant for application by the food industry [109]. Regarding probiotic food, it must comply with legal rules, demonstrating that viable microorganisms confer health benefits and are in a sufficient minimum number until the expiration date. If the food does not meet all these requirements, it only contains probiotics but is not considered as a probiotic food. This is mainly applied for artisanal fermented foods, in which the microbial species present, as well as their quantities, are generally not known [110].

Several probiotic LAB strains are widely used by the food industry, especially in the production of functional foods. Recent studies have demonstrated different types of benefits of probiotic LAB and their respective functional applications [111–113]. Many of these properties are related to the increasing values of proteins, minerals and vitamins in foods. In addition, the releasing of products from microbial metabolism, such as peptides, GABA, conjugated linoleic acids (CLA) and EPS, can contribute to health promotion [114]. Other benefits of probiotic LAB include the prevention of cardiovascular diseases, diarrhea, allergies, certain types of cancer and immunomodulation, among others [115].

The probiotic potential of LAB isolated from BAC has been demonstrated by different studies. Strains isolated from Colonial cheese showed high resistance to gastric acidity, with significant potential for use as a probiotic [116]. In vitro and in vivo probiotic potential was demonstrated for a *Lb. plantarum* strain isolated from AMC produced in the Canastra region, Minas Gerais state [117]. *Lb. plantarum* and *Lb. rhamnosus* isolated from the same type of cheese have already been evaluated as probiotic cultures in fermented milk [76].

Regarding the prebiotic property of compounds produced by LAB, it is generally related to the production of EPS (as will be discussed further in the next section), because it can favor the growth of probiotic strains. In cheeses, the supplementation with prebiotics can increase the populations of viable probiotic microorganisms; for example, the use of galactooligosaccharides (GOS), fructooligosaccharides (FOS) and inulin as nutraceuticals has stimulated the growth, survival and activity of probiotic strains in cheeses [118]. In artisanal cheeses, lactulose promoted the growth of lactobacilli and induced the production of short-chain fatty acids (SCFA) in Portuguese Serpa cheese [29]. SCFA contribute to health benefits, such as the regulation of energy metabolism, protection against colorectal cancer and inflammatory bowel disorders and obesity prevention [119]. At this moment, the prebiotic potential of LAB isolated from artisanal cheeses still remains unexplored in Brazil.

5.4. Exopolysaccharide (EPS) Production

Exopolysaccharides are biopolymers produced by microorganisms, whose composition and production yield are strain-dependent, both impacted by fermentation conditions [120,121]. *Xanthomonas campestris* and *Acetobacter xylinum* are recognized as excellent EPS-producing species; however, for industrial use, it is preferable that the producing microorganisms are GRAS, which reduces costs with purification processes. Furthermore, the application of purified EPS results in different effects on food when compared to EPS produced in situ, with better results [120].

The production of EPS by LAB has already been reported for *Lactobacillus*, *Lactococcus*, *Leuconostoc* and *Streptococcus* genera [122–124]. *Lb. rhamnosus* and *Lactobacillus kefiranofaciens* are even recognized as excellent EPS-producing species [109]. *Leuconostoc mesenteroides* and *Streptococcus salivarius* subsp. *thermophilus*, for example, have already been identified as producing species of dextran and fructan homopolysaccharides, respectively [120]. In this sense, the EPS production by LAB is especially important for the food industry, mainly for obtaining viscosity, stabilizing, emulsifying or gelling agents [109,125].

In cheeses, the production of EPS by NSLAB results in curd strengthening and the reduction of syneresis as a result of its binding with water molecules in the casein network [126]; thus, it contributes especially for the improvement in appearance and texture attributes in cheeses. In addition, EPS can minimize the harmful effects of bacteriophages during the fermentation process of dairy products, as they make the virus adsorption on the surface of the microbial cell difficult [127]. In BAC, the potential for EPS production by LAB has been little explored, with the first results indicating the AMC from Canastra, Campo das Vertentes, Serro and Cerrado, as well as Serrano cheese, as a source of LAB for this purpose [39]. These authors reinforce that obtaining EPS from the LAB of BAC constitutes a cheap, natural and sustainable strategy, with lower exploration costs, aiming at application in dairies.

In addition to the technological properties of EPS, its prebiotic effect stands out, reinforcing the relevance of the LAB (Table 2). The EPS produced by LAB favor the tolerance of probiotic strains to gut stress conditions, resulting in increased viability [109,121]. Because they can be metabolized in the gut, EPS constitute a substrate for the growth of probiotic strains, favoring the health benefits already demonstrated for this microbial group. For example, EPS produced by *Lb. plantarum* favored the growth of probiotic bacteria [128], and it also showed a bifidogenic effect, reducing the damages related to putrefactive bacteria [109].

5.5. Diacetyl Production

Diacetyl (2,3-butanedione) is a volatile compound produced by some LAB species during the conversion of citrate to pyruvate in the fermentation process, although it is not an exclusive feature of LAB [74,129,130]. The presence of diacetyl in certain foods is desirable, contributing to the buttery aroma and flavor [39,131]. It also presents antimicrobial activity against food pathogens, for which the mechanism of action consists of blocking the binding site of the microbial enzyme responsible for the use of arginine, affecting protein synthesis [129,132].

The production of diacetyl by *Lc. lactis* subsp. *lactis* biovar *diacetylactis* isolated from raw goat milk has already been reported [105]. In BAC, the LAB isolated from Marajó, Manteiga and AMC cheeses were able to produce diacetyl (Table 2). The *Leuconostoc* and *Streptococcus* strains isolated from Coalho cheese also showed this ability [74]. Finally, it was found that strains of *Weissella cibaria* and *Weissella paramesenteroides* isolated from cheeses produced in several regions of Brazil can produce diacetyl; *W. paramesenteroides* also stood out as an excellent producer of protease and had a high acidification capacity, desirable characteristics for cultures used in the dairy industry [71].

Interestingly, the occurrence of diacetyl-producing LAB may vary in BAC depending on the type of endogenous ferment used in the cheesemaking process. [78] evaluated the diacetyl production capacity in LAB isolated from "pingo" (the endogenous ferment used in the production of AMC) and from "rala", a kind of alternative inoculant consisting of portions of grated cheese; 66% of the "rala" isolates were able to produce diacetyl, much higher compared to the "pingo" isolates (25%). This difference can be explained by the predominance of NSLAB in the "rala" (a group that includes the main producers of diacetyl), because it is obtained from cheeses ripened for 3 to 5 days, unlike the "pingo" which consists of the whey collected from the still-fresh cheese produced on the previous day.

5.6. Proteolytic and Lipolytic Activities

Microbial cultures presenting proteolytic activity are widely used in the food industry, such as in the production of several types of dairy products, including cheeses and fermented milks; in the meat industry, to improve its texture, aroma and color; in the bakery industry, to break down the gluten net, improving the bread texture; in the alcoholic and non-alcoholic beverage industry, to reduce turbidity; and even in the production of animal feed [133].

In cheeses, proteolytic LAB play important roles for their quality, especially in ripened cheeses; therefore, the use of proteolytic cultures or purified enzymes is of great relevance for the cheese industry [134,135]. Proteolytic LAB strains can be used as adjunct cultures, acting on the peptide bonds of the matrix with the consequent release of amino acids and improvement in the cheese aroma, flavor and texture [70]. In addition, they can be used in the elaboration of dairy products with lower allergenic potential, reducing the risks for consumers with greater sensitivity to milk proteins [32].

Pediococcus acidilactici and *Weissella viridescens* proteolytic strains were isolated from ripened BAC [136], in addition to *Enterococcus* spp. isolates from AMC produced in the Campo das Vertentes, Serro and Cerrado regions, and from Coalho, Colonial, Serrano and Caipira cheeses [14]. The cheese-producing region can influence the microbial diversity of the product and, consequently, the occurrence of LAB with proteolytic activity. The LAB isolated from the AMC produced in the Campo das Vertentes region showed greater proteolytic activity than the LAB from cheeses produced in the Canastra region [43].

The contribution of LAB to the lipolysis processes in BAC is secondary, being more relevant for certain types of cheeses, such as blue cheeses (Gorgonzola and Roquefort) and cheddar [62]. However, lipases play an important role in the releasing of free fatty acids, precursors of volatile aromatic compounds that improve the sensory quality of the product [32,130]. It has been shown that BAC are good sources for the isolation of LAB with lipolytic activity, especially LAB isolated from AMC produced in the Araxá, Canastra and Serro regions, as well as from Colonial and Serrano cheeses, in addition to *Pediococcus acidilactici* isolated from Marajó cheese [11] and *Enterococcus* spp. isolated from AMC produced in the Araxá, Campo das Vertentes and Cerrado regions [14].

5.7. β-Galactosidase Activity

β-galactosidases are widely used for the hydrolysis of lactose by the food industry, with the aim of reducing its content in dairy products. This enzyme prevents crystallization and increased sweetness, flavor and solubility in several types of dairy products. In addition, the hydrolysis of lactose into D-glucose and D-galactose enables the development of lactose-free products, suitable for intolerant consumers, who correspond to about 70% of the world's adult population [137]. β-galactosidases are also able to catalyze transgalactosylation reactions, being successfully applied in the synthesis of lactose-based prebiotics, such as GOS, lactulose and lactosaccharose [138].

Another application of β-galactosidases that has been evaluated in recent years is the increase in safety due to the reduction in pH during the fermentation process. The glucose released from its activity can be consumed by the microbiota with the consequent production of lactic acid, increasing acidification rates, and thus contributing to the inhibition of pathogens [73]. Furthermore, β-galactosidases have also been used for the treatment of whey. Its inadequate disposal has been shown to be a serious environmental problem, especially regarding the eutrophication of rivers and water courses. In this context, the application of β-galactosidases can help to mitigate the damage resulting from the disposal of whey, in addition to allowing its reuse for the production of ingredients to be used in confectionery and bakery products [139].

To the best of our knowledge, there are no reports of the isolation of β-galactosidase-producing LAB from BAC. The production of β-galactosidase by strains of *Lacticaseibacillus casei* and *Limosilactobacillus fermentum* isolated from buffalo mozzarella has been demonstrated [140]; a strain of *Leuconostoc mesenteroides* subsp. *mesenteroides* with β–galactosidase activity, also from buffalo mozzarella, has been reported [141]. In BAC, only one study demonstrated the production of β-galactosidases in the strains of *E. durans* and *E. faecium* isolated from Coalho cheese [73].

6. Underexplored Biopotential and Opportunities for LAB from BAC

Brazil is one of the largest economies in the world, but it still depends on the import of inputs widely used in different industries, such as food, pharmaceuticals and biofuels, among others [142]. It is a paradox, given that the country has the greatest biodiversity on the planet and, therefore, a practically inexhaustible source for prospecting microorganisms with biotechnological potential. In this context, Brazilian fermented foods represent a relevant source of bacteria and fungi aimed at industrial exploitation; among these, BAC has stood out in recent years [11,66].

For this review, studies about LAB with biotechnological potential isolated from BAC were evaluated. Despite considerable progress in recent years, reinforced by the promising results presented here, there is still a gap to be filled by further studies. Most of the research carried out has focused mainly on the evaluation of antimicrobial activity, acidification capacity and enzyme and diacetyl production by LAB (Table 2). A few studies demonstrated the EPS production in different LAB isolated from BAC [39,71,78]. A similar situation was observed for the β-galactosidase synthesis, more specifically by LAB strains isolated from Coalho cheese [47,73]. As for prebiotics, there are no studies, so far, that have demonstrated their potential for use in LAB isolated from BAC.

In addition, most of the studies have been carried out in traditional and nationally recognized cheese-making regions, especially those involved in the production of AMC (Figure 2). Therefore, some types of cheese still lack information about their microbial diversity; for example, there are no studies of the isolation and identification of LAB isolated from Cabacinha cheeses, Parmesan-type cheeses, Porungo and KochKäse and Käschmier cheeses, produced in the south by German immigrants. It is, therefore, a niche opportunity for exploring the biotechnological potential of LAB; new insights into the genetic heritage of these traditional products can be provided from studies with cheesemakers in these regions.

Finally, it is worth to emphasize the urgent necessity to create and maintain a Brazilian collection of LAB that includes researchers from different regions in the country. Considering the continental dimension of Brazil, it is a complex and onerous effort. However, the articulation of researchers from universities and research institutions with public agents is essential to obtain human and financial resources aiming at the establishment of a national collection of LAB with scientific legitimacy and that becomes a reliable source of microorganisms for future research. We believe that this collection will have the potential to become a world reference in the cataloguing of LAB strains isolated from cheeses, with inestimable biotechnological value.

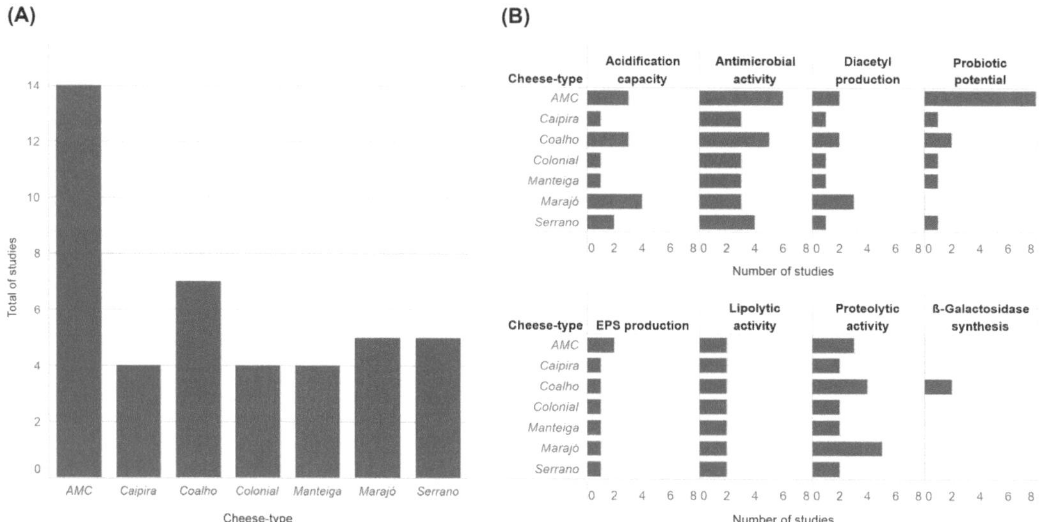

Figure 2. Studies about biotechnological potential from LAB isolated from BAC. (**A**) Number of studies about LAB in BAC published last 10 years; (**B**) Distribution of the studies according to biotechnological potential of LAB.

7. Conclusions

This review demonstrated the challenges and opportunities little explored for the application of LAB isolated from BAC. The discovery and characterization of new LAB strains isolated from BAC allow to increase the knowledge of the variety of compounds and enzymes produced by these bacteria and, consequently, expand the opportunities of applications. The use of producer strains or even isolated substances can be used for the elaboration of new functional foods, with improved sensorial and rheological characteristics, and also with greater microbiological safety.

Author Contributions: Conceptualization, B.d.O.H.; Investigation and Writing—Original Draft Preparation, B.d.O.H. and T.T.M.L.; Writing—Reviewing and Editing, G.V.M.P., J.B.R. and J.G.P.M.; Writing—Original Draft Preparation and Visualization, W.C.P.d.M.J. All authors have read and agreed to the published version of the manuscript.

Funding: The current study was funded by Coordenação de Aperfeiçoamento de Pessoal de Nível Superior 357—Brasil (CAPES) Finance Code 001.

Institutional Review Board Statement: Not applicable.

Informed Consent Statement: Not applicable.

Data Availability Statement: Not applicable.

Conflicts of Interest: The authors declare no conflict of interest.

References

1. Ferchichi, M.; Sebei, K.; Boukerb, A.M.; Karray-Bouraoui, N.; Chevalier, S.; Feuilloley, M.G.J.; Connil, N.; Zommiti, M. Enterococcus spp.: Is It a Bad Choice for a Good Use—A Conundrum to Solve? *Microorganisms* **2021**, *9*, 2222. [CrossRef] [PubMed]
2. Gopal, P.K. *Bacteria, Beneficial: Probiotic Lactic Acid Bacteria: An Overview*, 3rd ed.; Elsevier: Amsterdam, The Netherlands, 2021; Volume 4, ISBN 9780128187661.
3. Lindner, J.D.D.; Pereira, G.V.M.; Miotto, M. Culturas microbianas para aplicação em alimentos fermentados. In *Microbiologia de Alimentos Fermentados*, 1st ed.; Blucher: São Paulo, Brazil, 2022; Volume 1, pp. 137–162. ISBN 9786555061321.
4. Lima, T.T.M.; Hosken, B.O.; Venturim, B.C.; Lopes, I.L.; Martin, J.G.P. Traditional Brazilian fermented foods: Cultural and technological aspects. *J. Ethn. Foods* **2022**, *9*, 35. [CrossRef]

5. Cavicchioli, V.Q.; Camargo, A.C.; Todorov, S.D.; Nero, L.A. Novel bacteriocinogenic Enterococcus hirae and Pediococcus pentosaceus strains with antilisterial activity isolated from Brazilian artisanal cheese. *J. Dairy Sci.* **2017**, *100*, 2526–2535. [CrossRef]

6. Fidan, H.; Esatbeyoglu, T.; Simat, V.; Trif, M.; Tabanelli, G.; Kostka, T.; Montanari, C.; Ibrahim, S.A.; Ozogul, F. Recent developments of lactic acid bacteria and their metabolites on foodborne pathogens and spoilage bacteria: Facts and gaps. *Food Biosci.* **2022**, *47*, 101741. [CrossRef]

7. Prêmio Queijo Brasil. Resultado V Prêmio Queijo Brasil. 2021. Available online: http://www.premioqueijobrasil.com.br (accessed on 15 July 2022).

8. Mondial du Fromage. Concours International Produits. 2021. Available online: https://www.mondialdufromage.com/concours-produits.php/ (accessed on 12 August 2022).

9. Dores, M.T.; Nobrega, J.E.; Ferreira, C.L.L.F. Room temperature aging to guarantee microbiological safety of brazilian artisan Canastra cheese. *Food Sci. Technol.* **2013**, *33*, 180–185. [CrossRef]

10. Martins, J.M.; Galinari, É.; Pimentel-Filho, N.J.; Ribeiro, J.I.; Furtado, M.M.; Ferreira, C.L.L. Determining the minimum ripening time of artisanal Minas cheese, a traditional Brazilian cheese. *Braz. J. Microbiol.* **2015**, *46*, 219–230. [CrossRef]

11. Margalho, L.P.; Kamimura, B.A.; Brexó, R.P.; Alvarenga, V.O.; Cebeci, A.S.; Janssen, P.W.M.; Dijkstra, A.; Starrenburg, M.J.C.; Sheombarsing, R.S.; Cruz, A.G.; et al. High throughput screening of technological and biopreservation traits of a large set of wild lactic acid bacteria from Brazilian artisanal cheeses. *Food Microbiol.* **2021**, *100*, 103872. [CrossRef]

12. Kamimura, B.A.; Filippis, F.; Sant'Anna, A.S.; Ercolini, D. Large-scale mapping of microbial diversity in artisanal Brazilian cheeses. *Food Microbiol.* **2019**, *80*, 40–49. [CrossRef] [PubMed]

13. Domingos-Lopes, M.F.P.; Santon, C.; Ross, P.R.; Dapkevicius, M.L.E.; Silva, C.G.C. Genetic diversity, safety and technological characterization of lactic acid bacteria isolated from artisanal Pico cheese. *Food Microbiol.* **2017**, *63*, 178–190. [CrossRef]

14. Margalho, L.P.; Schalkwijk, S.; Bachmann, H.; Sant'Anna, A.S. Enterococcus spp. in Brazilian artisanal cheeses: Occurrence and assessment of phenotypic and safety properties of a large set of strains through the use of high throughput tools combined with multivariate statistics. *Food Control* **2020**, *118*, 107425. [CrossRef]

15. Carneiro, J.O.; Chaves, A.C.S.D.; Stephan, M.P.; Boari, C.A.; Koblitz, M.G.B. Artisan minas cheese of Serro: Proteolysis during ripening. *Heliyon* **2020**, *6*, e04446. [CrossRef]

16. Monteiro, R.P.; Matta, V.M. Queijo Minas Artesanal: Valorizando a Agroindústria Familiar. 2018. Available online: https://ainfo.cnptia.embrapa.br/digital/bitstream/item/199625/1/Livro-Queijo-Minas-Artesanal-Ainfo.pdf (accessed on 19 August 2022).

17. Roldan, B.B.; Revillion, J.P.P. Convenções de qualidade em queijos artesanais no Brasil, Espanha e Itália. *Rev. Inst. Laticínios Cândido Tostes* **2019**, *74*, 108–122. [CrossRef]

18. Instituto Mineiro de Agropecuária (IMA). Ordinance n° 2016. November 26th. Identifies the Serras da Ibitipoca Region as Producer of Minas Artesanal Cheese and Revokes IMA Ordinance No. 1834, of 4 July 2018. 2020. Available online: http://www.ima.mg.gov.br/institucional/portarias#ano-2020 (accessed on 10 May 2021).

19. Instituto Mineiro de Agropecuária (IMA). Ordinance n° 2.141, of 19 April 2022. Identifies the Entre Serras da Piedade ao Caraça Region as Producer of Minas Artisanal Cheese. Available online: http://ima.mg.gov.br/index.php?preview=1&option=com_dropfiles&format=&task=frontfile.download&catid=1829&id=19230&Itemid=1000000000000 (accessed on 10 May 2021).

20. Instituto do Patrimônio Histórico e Artístico Nacional (IPHAN). Certidão do Registro do Modo Artesanal de Fazer Queijo de Minas, nas Regiões do Serro e nas Serras da Canastra e do Salitre. 2008. Available online: http://portal.iphan.gov.br/uploads/ckfinder/arquivos/Certid%C3%A3o%20(Modo%20artesanal%20de%20fazer%20Queijo%20de%20Minas).pdf (accessed on 11 October 2022).

21. Giannini, A. Queijos Brasileiros Conquistam 57 Medalhas em Concurso Mundial na França. 2021. Available online: https://veja.abril.com.br/gastronomia/queijos-brasileiros-conquistam-57-medalhas-em-concurso-mundial-na-franca/ (accessed on 3 September 2022).

22. Pires, M.F.A. Queijo Artesanal Ganha Status, Internet e o Mundo. In *Anuário do Leite 2019*; Texto Comunicação Corporativa: São Paulo, Brazil, 2019. Available online: https://www.infoteca.cnptia.embrapa.br/handle/doc/1109959 (accessed on 5 September 2022).

23. Abdel-Rahman, M.L.; Sonomoto, K. Opportunities to overcome the current limitations and challenges for efficient microbial production of optically pure lactic acid. *J. Biotechnol.* **2016**, *236*, 176–192. [CrossRef] [PubMed]

24. König, H.; Unden, G.; Fröhlich, J. Lactic acid bacteria. In *Biology of Microorganisms on Grapes, in Must and in Wine*; Spring: Berlin, Germany, 2017; pp. 3–41. [CrossRef]

25. Liu, W.; Pang, H.; Zhang, H.; Cai, Y. Biodiversity of lactic acid bacteria. In *Lactic Acid Bacteria: Fundamentals and Practice*; Springer: Dordrecht, The Netherlands, 2014; ISBN 978-94-017-8840-3.

26. Settanni, L.; Moschetti, G. Non-starter lactic acid bacteria used to improve cheese quality and provide health benefits. *Food Microbiol.* **2010**, *27*, 691–697. [CrossRef]

27. Widyastuti, Y.; Rohmatussolihat, Y.; Febrisiantosa, A. The role of lactic acid bacteria in milk fermentation. *Food Nutr. Sci.* **2014**, *5*, 435–442. [CrossRef]

28. Diana, M.; Tres, A.; Quílez, J.; Llombart, M.; Rafecas, M. Spanish cheese screening and selection of lactic acid bacteria with high gamma-aminobutyric acid production. *LWT-Food Sci. Technol.* **2014**, *56*, 351–355. [CrossRef]

29. Ruiz-Moyano, S.; Santos, M.T.P.G.; Galván, A.I.; Merchán, A.V.; González, E.; Córdoba, M.G.; Benito, M.J. Screening of autochthonous lactic acid bacteria strains from artisanal soft cheese: Probiotic characteristics and prebiotic metabolism. *LWT-Food Sci. Technol.* **2019**, *114*, 108388. [CrossRef]

30. O'Connor, P.M.; Kuniyoshi, T.M.; Oliveira, R.P.S.; Hill, C.; Ross, R.P.; Cotter, P.D. Antimicrobials for food and feed; a bacteriocin perspective. *Curr.Opin. Biotechnol.* **2020**, *61*, 160–167. [CrossRef]

31. Mayo, B.; Rodríguez, J.; Vázquez, L.; Flórez, A.B. Microbial Interactions within the Cheese Ecosystem and Their Application to Improve Quality and Safety. *Foods* **2021**, *10*, 602. [CrossRef]

32. Meng, Z.; Zhang, L.; Xin, L.; Lin, K.; Yi, H.; Han, X. Technological characterization of Lactobacillus in semihard artisanal goat cheeses from different Mediterranean areas for potential use as nonstarter lactic acid bacteria. *J. Dairy Sci.* **2018**, *101*, 2887–2896. [CrossRef]

33. Chawda, P.J.; Shi, J.; Xue, S.; Quek, S.Y. Co-encapsulation of bioactives for food applications. *Food Qual. Saf.* **2017**, *1*, 302–309. [CrossRef]

34. Pupa, P.; Apiwatsiri, P.; Sirichokchatchawan, W.; Pirarat, N.; Muangsin, N.; Shah, A.A.; Prapasarakul, N. The efficacy of three double-microencapsulation methods for preservation of probiotic bacteria. *Sci. Rep.* **2021**, *11*, 13753. [CrossRef] [PubMed]

35. Contessa, C.R.; Souza, N.B.; Gonçalo, G.B.; Moura, C.M.; Rosa, G.S.; Moraes, C.C. Development of active packaging based on agar-agar incorporated with bacteriocin of Lactobacillus sakei. *Biomolecules* **2021**, *11*, 1869. [CrossRef]

36. Pellegrino, M.S.; Frola, I.D.; Natanael, B.; Gobelli, D.; Nader-Macias, M.E.F.; Bogni, C.I. In vitro characterization of lactic acid bacteria isolated from bovine milk as potential probiotic strains to prevent bovine mastitis. *Probiotics Antimicrob. Proteins* **2019**, *11*, 74–84. [CrossRef] [PubMed]

37. Vanniyasingam, J.; Kapilan, R.; Vasantharuba, S. Isolation and characterization of potential probiotic lactic acid bacteria isolated from cow milk and milk products. *J. Agric. Sci.* **2019**, *13*, 32–43. [CrossRef]

38. Djadouni, F.; Kihal, M. Antimicrobial activity of lactic acid bacteria and the spectrum of their biopeptides against spoiling germs in foods. *Braz. Arch. Biol. Technol.* **2012**, *55*, 435–444. [CrossRef]

39. Margalho, L.P.; Feliciano, M.D.E.; Silva, C.E.; Abreu, J.S.; Piran, M.V.F.; Sant'Anna, A.S. Brazilian artisanal cheeses are rich and diverse sources of nonstarter lactic acid bacteria regarding technological, biopreservative, and safety properties—Insights through multivariate analysis. *J. Dairy Sci.* **2020**, *103*, 7908–7926. [CrossRef]

40. Castro, R.D.; Oliveira, L.G.; Sant'Anna, F.M.; Luiz, L.M.P.; Sandes, S.H.C.; Silva, C.I.F.; Silva, A.M.; Nunes, A.C.; Penna, C.F.A.M.; Souza, M.R. Lactic acid microbiota identification in water, raw milk, endogenous starter culture, and fresh Minas artisanal cheese from the Campo das Vertentes region of Brazil during the dry and rainy seasons. *J. Dairy Sci.* **2016**, *99*, 6086–6096. [CrossRef]

41. Matera, J.; Luna, A.S.; Batista, D.B.; Pimentel, T.C.; Moraes, J.; Kamimura, B.A.; Ferreira, M.V.S.; Silva, H.L.A.; Mathias, S.P.; Esmerino, E.A.; et al. Brazilian cheeses: A survey covering physicochemical characteristics, mineral content, fatty acid profile and volatile compounds. *Food Res. Int.* **2018**, *108*, 18–26. [CrossRef] [PubMed]

42. Wilkinson, M.G.; Lapointe, G. Invited review: Starter lactic acid bacteria survival in cheese: New perspectives on cheese microbiology. *J. Dairy Sci.* **2020**, *103*, 10963–10985. [CrossRef]

43. Campagnollo, F.B.; Margalho, L.P.; Kamimura, B.A.; Feliciano, M.D.; Freire, L.; Lopes, L.S.; Alvarenga, V.O.; Cadavez, V.A.P.; Gonzales-Barron, U.; Schaffner, D.W.; et al. Selection of indigenous lactic acid bacteria presenting anti-listerial activity, and their role in reducing the maturation period and assuring the safety of traditional Brazilian cheeses. *Food Microbiol.* **2018**, *73*, 288–297. [CrossRef]

44. Perin, L.M.; Sardaro, M.L.S.; Nero, L.A.; Neviani, E.; Gatti, M. Bacterial ecology of artisanal Minas cheeses assessed by culture-dependent and -independent methods. *Food Microbiol.* **2017**, *65*, 160–169. [CrossRef] [PubMed]

45. Bruno, L.M.; Marcó, M.B.; Capra, M.L.; Carvalho, J.D.G.; Meinardi, C.; Quiberoni, A. Wild Lactobacillus strains: Technological characterisation and design of Coalho cheese lactic culture. *Int. J. Dairy Technol.* **2017**, *70*, 572–582. [CrossRef]

46. Medeiros, N.C.; Abrantes, M.R.; Medeiros, J.M.S.; Campêlo, M.C.S.; Rebouças, M.O.; Costa, M.G.A.; Silva, J.B.A. Quality of milk used in informal artisanal production of coalho and butter cheeses. *Semin. Ciências Agrárias* **2017**, *38*, 1955–1962. [CrossRef]

47. Lima, C.P.; Dias, G.M.P.; Soares, M.T.C.V.; Bruno, L.M.; Porto, A.L.F. Queijo coalho como fonte de bactérias ácido láticas probióticas. *Res. Soc. Dev.* **2020**, *9*, e266984958. [CrossRef]

48. Kothe, C.I.; Mohellibi, N.; Renault, P. Revealing the microbial heritage of traditional Brazilian cheeses through metagenomics. *Food Res. Int.* **2022**, *157*, 111265. [CrossRef]

49. Figueiredo, E.L.; Andrade, N.J.; Pires, A.C.S.; Peña, W.E.L.; Figueiredo, H.M. Caracterização do potencial tecnológico e identificação genética de bactérias ácido láticas isoladas de queijo do marajó, tipo creme, de leite de búfala. *Rev. Bras. Prod. Agroind.* **2016**, *18*, 293–303. [CrossRef]

50. Almeida, T.T.; Andretta, M.; Ferreira, L.R.; Carvalho, A.F.; Nero, L.A. The complex microbiota of artisanal cheeses interferes in the performance of enumeration protocols for lactic acid bacteria and staphylococci. *Int. Dairy J.* **2020**, *109*, 104791. [CrossRef]

51. Camargo, A.C.; Costa, E.A.; Fusieger, A.; Freitas, R.; Nero, L.A.; de Carvalho, A.F. Microbial shifts through the ripening of the "Entre Serras" Minas artisanal cheese monitored by high-throughput sequencing. *Food Res. Int.* **2020**, *139*, 109803. [CrossRef] [PubMed]

52. Lima, C.D.L.C.; Lima, L.A.; Cerqueira, M.M.O.P.; Ferreira, E.G.; Rosa, C.A. Bactérias do ácido lático e leveduras associadas com o queijo de minas artesanal produzido na região da Serra do Salitre, Minas Gerais. *Arq. Bras. Med. Vet. Zootec.* **2009**, *61*, 266–272. [CrossRef]

53. Luiz, L.M.P.; Castro, R.D.; Sandes, S.H.C.; Silva, J.G.; Oliveira, L.G.; Sales, G.A.; Nunes, A.C.; Souza, M.R. Aislamiento e identificación de las bacterias ácido-lácticas del queso brasileño Minas artesanal. *CYTA—J. Food* **2017**, *15*, 125–128. [CrossRef]
54. Resende, M.F.S.; Costa, H.H.S.; Andrade, E.H.P.; Acúrcio, L.B.; Drummond, A.F.; Cunha, A.F.; Nunes, A.C.; Moreira, J.L.S.; Penna, C.F.A.M.; Souza, M.R. Queijo de minas artesanal da Serra da Canastra: Influência da altitude das queijarias nas populações de bactérias ácido lácticas. *Arq. Bras. Med. Vet. Zootec.* **2011**, *63*, 1567–1573. [CrossRef]
55. Sant'Anna, F.M.; Wetzels, S.U.; Cicco, S.H.S.; Figueiredo, R.C.; Sales, G.A.; Figueiredo, N.C.; Nunes, C.A.; Schmitz-Esser, S.; Mann, E.; Wagner, M.; et al. Microbial shifts in Minas artisanal cheeses from the Serra do Salitre region of Minas Gerais, Brazil throughout ripening time. *Food Microbiol.* **2019**, *82*, 349–362. [CrossRef] [PubMed]
56. Delamare, A.P.L.; Andrade, C.C.P.; Mandelli, F.; Almeida, R.C.; Echeverrigaray, S. Microbiological, physico-chemical and sensorial characteristics of Serrano, an artisanal Brazilian cheese. *Food Nutr. Sci.* **2012**, *3*, 1068–1075. [CrossRef]
57. Rosa, T.D.; Wassermann, G.E.; Souza, C.F.V.; Caron, D.; Carlini, C.R.; Ayub, M.A.Z. Microbiological and physicochemical characteristics and aminopeptidase activities during ripening of Serrano cheese. *Int. J. Dairy Technol.* **2008**, *61*, 70–79. [CrossRef]
58. Souza, C.F.V.; Rosa, T.D.; Ayub, M.A.Z. Changes in the microbiological and physicochemical characteristics of Serrano cheese during manufacture and ripening. *Braz. J. Microbiol.* **2003**, *34*, 260–266. [CrossRef]
59. Zheng, J.; Wittouck, S.; Salvetti, E.; Franz, C.M.A.P.; Harris, H.M.B.; Mattarelli, P.; O'toole, P.W.; Pot, B.; Vandamme, P.; Walter, J.; et al. A Taxonomic Note on the Genus Lactobacillus: Description of 23 Novel Genera, Emended Description of the Genus Lactobacillus Beijerinck 1901, and Union of Lactobacillaceae and Leuconostocaceae. *Int. J. Syst. Evol. Microbiol.* **2020**, *70*, 2782–2858. [CrossRef]
60. Powell, I.B.; Broome, M.C.; Limsowtin, G.K.Y. Cheese: Starter Cultures: General Aspects. In *Encyclopedia of Dairy Sciences*, 2nd ed.; Academic Press: Cambridge, MA, USA, 2011; pp. 552–558. [CrossRef]
61. Broadbent, J.R.; Budinich, M.F.; Steele, J.L. Cheese: Non-Starter lactic acid bacteria. In *Encyclopedia of Dairy Sciences*, 2nd ed.; Academic Press: Cambridge, MA, USA, 2011; pp. 639–644. [CrossRef]
62. Blaya, J.; Barzideh, Z.; LaPointe, G. Symposium review: Interaction of starter cultures and nonstarter lactic acid bacteria in the cheese environment. *J. Dairy Sci.* **2018**, *101*, 3611–3629. [CrossRef] [PubMed]
63. Gobbetti, M.; Di Cagno, R.; Calasso, M.; Neviani, E.; Fox, P.F.; De Angelis, M. Drivers that establish and assembly the lactic acid bacteria biota in cheeses. *Trends Food Sci. Technol.* **2018**, *78*, 244–254. [CrossRef]
64. Campagnollo, F.B.U.; Gonzales-Barron, V.A.; Pilão, C.A.S.; Sant'Ana, D.W.S. Quantitative risk assessment of Listeria monocytogenes in traditional Minas cheeses: The cases of artisanal semi-hard and fresh soft cheeses. *Food Control* **2018**, *92*, 370–379. [CrossRef]
65. Melini, F.; Melini, V.; Luziatelli, F.; Ruzzi, M. Raw and heat-treated milk: From public health risks to nutritional quality. *Beverages* **2017**, *3*, 54. [CrossRef]
66. Bortolomedi, B.M.; Paglarini, C.S.; Brod, F.C.A. Bioactive compounds in kombucha: A review of substrate effect and fermentation conditions. *Food Chem.* **2022**, *385*, 132719. [CrossRef] [PubMed]
67. Kocabaş, D.S.; Lyne, J.; Ustunol, Z. Hydrolytic enzymes in the dairy industry: Applications, market and future perspectives. *Trends in Food Sci. Technol.* **2022**, *119*, 467–475. [CrossRef]
68. Ağagündüz, D.; Şahin, T.Ö.; Ayten, Ş.; Yılmaz, B.; Güneşliol, B.E.; Russo, P.; Spano, G.; Özogul, F. Lactic acid bacteria as pro-technological, bioprotective and health-promoting cultures in the dairy food industry. *Food Biosci.* **2022**, *47*, 101617. [CrossRef]
69. Sanlibaba, P.; Çakmak, G.A. Exopolyssacharide production by lactic acid bacteria. *App. Microbiol.* **2016**, *2*, 1000115. [CrossRef]
70. Martins, M.C.F. Diversidade de bactérias láticas e identificação molecular de Lactococcus isolados de ambientes lácteos e não lácteos. Ph.D. Thesis, Universidade Federal de Viçosa, Viçosa, Brasil, 2018.
71. Teixeira, C.G.; Fusieger, A.; Martins, E.; Freitas, R.; Vakarelova, M.; Nero, L.A.; de Carvalho, A.F. Biodiversity and technological features of Weissella isolates obtained from Brazilian artisanal cheese-producing regions. *LWT-Food Sci. Technol.* **2021**, *147*, 111474. [CrossRef]
72. Margalho, L.P.; Jorge, G.P.; Noleto, D.A.P.; Silva, C.E.; Abreu, J.S.; Piran, M.V.F.; Brocchi, M.; Sant'Anna, A.S. Biopreservation and probiotic potential of a large set of lactic acid bacteria isolated from Brazilian artisanal cheeses: From screening to in product approach. *Microbiol. Res.* **2021**, *242*, 126622. [CrossRef]
73. Brito, L.P.; Silva, E.C.; Calaça, P.R.A.; Medeiros, R.S.; Soares, M.T.C.V.; Porto, A. Bactérias ácido láticas isoladas de queijo de Coalho do nordeste brasileiro na produção de laticínios: Uma triagem para aplicação tecnológica. *Res. Soc. Dev.* **2020**, *9*, e5249108457. [CrossRef]
74. Dias, G.M.P.; Silva, A.B.; Granja, N.M.C.; Silva, T.N.; Lima, G.V.M.; Cavalcanti, M.T.H.; Porto, A.L.F. Can Coalho cheese lactic microbiota be used in dairy fermentation to reduce foodborne pathogens? *Sci. Plena* **2019**, *15*, 1–9. [CrossRef]
75. Mareze, J.; Ramos-Pereira, J.; Santos, J.A.; Beloti, V.; López-Díaz, T.M. Identification and characterisation of lactobacilli isolated from an artisanal cheese with antifungal and antibacterial activity against cheese spoilage and mycotoxigenic Penicillium spp. *Int. Dairy J.* **2022**, *130*, 105367. [CrossRef]
76. Acurcio, L.B.; Sandes, S.H.C.; Bastos, R.W.; Sant'Anna, F.M.; Pedroso, S.H.S.P.; Reis, D.C.; Nunes, Á.C.; Cassali, G.D.; Souza, M.R.; Nicoli, J.R. Milk fermented by Lactobacillus species from Brazilian artisanal cheese protect germ-free-mice against Salmonella Typhimurium infections. *Beneficial Microbes* **2017**, *8*, 579–588. [CrossRef]

77. Andrade, C.R.G.; Souza, M.R.; Penna, C.F.A.M.; Acurcio, L.B.; Sant'Anna, F.M.; Castro, R.D.; Oliveira, D.L.S. Propriedades probióticas in vitro de Lactobacillus spp. isolados de queijos minas artesanais da Serra da Canastra—MG. *Arq. Bras. Med. Vet. Zootec.* **2014**, *66*, 1592–1600. [CrossRef]

78. Brumano, E.C.C. Impacto do tipo de fermento endógeno na qualidade e tempo de maturação de queijo Minas artesanal produzido em propriedades cadastradas pelo IMA (Instituto Mineiro de Agropecuária) na região do Serro—MG. Ph.D. Thesis, Universidade Federal de Viçosa, Viçosa, Brasil, 2016.

79. Costa, H.H.S.; Souza, M.R.; Acúrcio, L.B.; Cunha, A.F.; Resende, M.F.S.; Nunes, Á.C. Potencial probiótico in vitro de bactérias ácido-láticas isoladas de queijo-de-minas artesanal da Serra da Canastra, MG. *Arq. Bras. Med. Vet. Zootec.* **2013**, *65*, 1858–1866. [CrossRef]

80. Sant'Anna, F.M.; Acúrcio, L.B.; Alvim, L.B.; Castro, R.D.; Oliveira, L.G.; Silva, A.M.; Nunes, A.C.; Nicoli, J.R.; Souza, M.R. Assessment of the probiotic potential of lactic acid bacteria isolated from Minas artisanal cheese produced in the Campo das Vertentes region, Brazil. *Int. J. Dairy Technol.* **2017**, *70*, 592–601. [CrossRef]

81. Todorov, S.D.; Holzapfel, W.; Nero, L.A. Safety evaluation and bacteriocinogenic potential of Pediococcus acidilactici strains isolated from artisanal cheeses. *LWT-Food Sci. Technol.* **2020**, *139*, 110550. [CrossRef]

82. Tulini, F.L.; Winkelströter, L.K.; Martinis, E.C.P. Identification and evaluation of the probiotic potential of Lactobacillus paraplantarum FT259, a bacteriocinogenic strain isolated from Brazilian semi-hard artisanal cheese. *Anaerobe* **2013**, *22*, 57–63. [CrossRef]

83. Valente, G.L.C.; Acúrcio, L.B.; Freitas, L.P.V.; Nicoli, J.R.; Silva, A.M.; Souza, M.R.; Penna, C.F.A.M. Short communication: In vitro and in vivo probiotic potential of Lactobacillus plantarum B7 and Lactobacillus rhamnosus D1 isolated from Minas artisanal cheese. *J. Dairy Sci.* **2019**, *102*, 5957–5961. [CrossRef]

84. Gontijo, M.T.P.; Silva, J.S.; Vidigal, M.P.; Martin, J.G.P. Phylogenetic distribution of the bacteriocin repertoire of lactic acid bacteria species associated with artisanal cheese. *Food Res. Int.* **2020**, *128*, 108783. [CrossRef] [PubMed]

85. Perez, R.H.; Zendo, T.; Sonomoto, K. Circular and leaderless bacteriocins: Biosynthesis, mode of action, applications and prospects. *Front. Microbiol.* **2018**, *9*, 2085. [CrossRef]

86. Silva, C.C.G.; Silva, S.P.M.; Ribeiro, S.C. Application of bacteriocins and protective cultures in dairy food preservation. *Front. Microbiol.* **2018**, *9*, 594. [CrossRef]

87. Mills, S.; Serrano, L.; Griffin, C.; O'Connor, P.M.; Schaad, G.; Bruining, C.; Hills, C.; Ross, R.; Meijer, W.C. Inhibitory activity of Lactiplantibacillus plantarum subsp. plantarum LMG P-26358 against Listeria innocua when used as an adjunct starter in the manufacture of cheese. *Microb. Cell Fact.* **2011**, *10*, S7. [CrossRef]

88. Ogaki, M.B.; Furlaneto, M.C.; Maia, L.F. Revisão: Aspectos gerais das bacteriocinas. *Braz. J. Food Technol.* **2015**, *18*, 267–276. [CrossRef]

89. Balciunas, E.M.; Martinez, F.A.C.; Todorov, S.D.; Franco, B.D.G.M.; Converti, A.; Oliveira, R.P.S. Novel biotechnological applications of bacteriocins: A review. *Food Control* **2013**, *32*, 134–142. [CrossRef]

90. Coelho, M.C.; Silva, C.C.G.; Ribeiro, S.C.; Dapkevicius, M.L.N.E.; Rosa, H.J.D. Control of Listeria monocytogenes in fresh cheese using protective lactic acid bacteria. *Int. J. Food Microbiol.* **2014**, *191*, 53–59. [CrossRef] [PubMed]

91. Montel, M.C.; Buchin, S.; Mallet, A.; Delbes-Paus, C.; Vuitton, D.A.; Desmasures, N.; Berthier, F. Traditional cheeses: Rich and diverse microbiota with associated benefits. *Int. J. Food Microbiol.* **2014**, *177*, 136–154. [CrossRef]

92. Galvéz, A.; Abriquel, H.; López, R.L.; Omar, N.B. Bacteriocin-based strategies for food biopreservation. *Int. J. Food Microbiol.* **2007**, *120*, 51–70. [CrossRef]

93. Heredia-Castro, P.Y.; Hérnández-Mendoza, A.; González-Córdova, A.; Vallejo-Cordoba, B. Bacteriocinas de bacterias ácido lácticas: Mecanismos de acción y actividad antimicrobiana contra patógenos en quesos. *Interciencia* **2017**, *42*, 340–346. Available online: https://www.redalyc.org/journal/339/33951621002/html/ (accessed on 21 May 2021).

94. Yang, S.C.; Lin, C.H.; Sung, C.T.; Fang, J.Y. Antibacterial activities of bacteriocins: Application in foods and pharmaceuticals. *Front. Microbiol.* **2014**, *5*, 241. [CrossRef]

95. Ahmad, V.; Khan, M.S.; Jamal, Q.M.S.; Alzohairy, M.A.; Karaawi, M.A.A.; Siddiqui, M.U. Antimicrobial potential of bacteriocins: In therapy, agriculture and food preservation. *Int. J. Antimicrob. Agents* **2017**, *49*, 1–11. [CrossRef]

96. Godoy-Santos, F.; Pinto, M.S.; Barbosa, A.A.T.; Brito, M.A.V.P.; Mantovani, H.C. Efficacy of a ruminal bacteriocin against pure and mixed cultures of bovine mastitis pathogens. *Indian J. Microbiol.* **2019**, *59*, 304–312. [CrossRef] [PubMed]

97. Ibarra-Sánchez, L.A.; El-Haddad, N.; Mahmoud, D.; Miller, M.J.; Karam, L. Invited review: Advances in nisin use for preservation of dairy products. *J. Dairy Sci.* **2020**, *103*, 2041–2052. [CrossRef] [PubMed]

98. Gutiérrez-Cortés, C.; Suarez, H.; Buitrago, G.; Nero, L.A.; Todorov, S.D. Characterization of bacteriocins produced by strains of Pediococcus pentosaceus isolated from Minas cheese. *Ann. Microbiol.* **2018**, *68*, 383–398. [CrossRef]

99. Nespolo, C.R.; Brandelli, A. Production of bacteriocin-like substances by lactic acid bacteria isolated from regional ovine cheese. *Braz. J. Microbiol.* **2010**, *41*, 1009–1018. [CrossRef] [PubMed]

100. Winkelströter, L.K.; Tulini, F.L.; De Martinis, E.C.P. Identification of the bacteriocin produced by cheese isolate Lactobacillus paraplantarum FT259 and its potential influence on Listeria monocytogenes biofilm formation. *LWT-Food Sci. Technol.* **2015**, *64*, 586–592. [CrossRef]

101. Pegoraro, K.; Sereno, M.J.; Cavicchioli, V.Q.; Viana, C.; Nero, L.A.; Bersot, L.S. Bacteriocinogenic potential of lactic acid bacteria isolated from artisanal colonial type—Cheese. *Arch. Vet. Sci.* **2020**, *25*. [CrossRef]

102. Paula, J.C.J.; Carvalho, A.F.; Furtado, M.M. Princípios básicos de fabricação de queijo: Do histórico à salga. *Rev. Inst. Latícinios Cândido Tostes* **2009**, *64*, 19–25. Available online: https://www.revistadoilct.com.br/rilct/article/view/76/82 (accessed on 17 May 2021).

103. Pontarolo, G.H.; Melo, F.D.; Martini, C.L.; Wildemann, P.; Alessio, D.R.M.; Sfaciotte, R.; Neto, A.T.; Vaz, E.K.; Ferraz, S.M. Quality and safety of artisan cheese produced in the serrana region of Santa Catarina. *Semina Ciências Agrárias* **2017**, *38*, 739–747. [CrossRef]

104. Martin, J.G.P.; Lindner, J.D. Alimentos fermentados: Passado, presente e futuro. In *Microbiologia de Alimentos Fermentados*, 1st ed.; Blucher: São Paulo, Brazil, 2022; Volume 1, pp. 29–62. ISBN 9786555061321.

105. Souza, J.V.; Dias, F.S. Protective, technological, and functional properties of select autochthonous lactic acid bacteria from goat dairy products. *Curr. Opin. Food Sci.* **2017**, *13*, 1–9. [CrossRef]

106. Food and agriculture organization of united nations (FAO); World health organization (WHO). Evaluation of health and nutritional properties of probiotics in food including powder milk with live lactic acid bacteria. In *Probiotics in Food*; FAO: Rome, Italy, 2001; ISSN 0254-4725.

107. Davani-Davari, D.; Negahdaripour, M.; Karimzadeh, I.; Seifan, M.; Mohkam, M.; Masoumi, S.J.; Berenjian, A.; Ghasemi, Y. Prebiotics: Definition, Types, Sources, Mechanisms, and Clinical Applications. *Foods* **2019**, *8*, 92. [CrossRef]

108. Ferreira, C.L.L.F.; Carvalho, M.M.; Píccolo, M.P. Alimentos probióticos. In *Microbiologia de Alimentos Fermentados*, 1st ed.; Blucher: São Paulo, Brazil, 2022; Volume 1, pp. 163–221. ISBN 9786555061321.

109. Caggianiello, G.; Kleerebezem, M.; Spano, G. Exopolysaccharides produced by lactic acid bacteria: From health-promoting benefits to stress tolerance mechanisms. *Appl. Microbiol. Biotechnol.* **2016**, *100*, 3877–3886. [CrossRef]

110. Marco, M.L.; Sanders, M.E.; Gänzle, M.; Arrieta, M.C.; Cotter, P.D.; De Vuyst, L.; Hill, C.; Holzapfel, W.; Leeber, S.; Merenstein, D.; et al. The International Scientific Association for Probiotics and Prebiotics (ISAPP) consensus statement on fermented foods. *Nat. Rev. Gastroenterol. Hepatol.* **2021**, *18*, 196–208. [CrossRef]

111. Li, J.; Feng, S.; Yu, L.; Zhao, J.; Tian, F.; Chen, W.; Zhai, Q. Capsular polysaccharides of probiotics and their immunomodulatory roles. *Food Sci. Hum. Wellness* **2022**, *11*, 1111–1120. [CrossRef]

112. Marquez, A.; Andrada, E.; Russo, M.; Bolondi, M.L.; Fabersani, E.; Medina, R.; Gauffin-Cano, P. Characterization of autochthonous lactobacilli from goat dairy products with probiotic potential for metabolic diseases. *Heliyon* **2022**, *8*, e10462. [CrossRef]

113. Mohd-Zubri, N.S.; Ramasamy, K.; Abdul-Rahman, N.Z. Characterization and potential oral probiotic properties of Lactobacillus plantarum FT 12 and Lactobacillus brevis FT 6 isolated from Malaysian fermented food. *Arch Oral Biol.* **2022**, *143*, 105515. [CrossRef]

114. Evivie, S.E.; Huo, G.C.; Igene, J.O.; Bian, X. Some current applications, limitations and future perspectives of lactic acid bacteria as probiotics. *Food Nutr. Res.* **2017**, *61*, 1318034. [CrossRef] [PubMed]

115. Tarique, M.; Abdalla, A.; Masad, R.; Al-Sbiei, A.; Kizhakkayil, J.; Osaili, T.; Olaimat, A.; Liu, S.-Q.; Fernandez-Cabezudo, M.; Ramadi, B.; et al. Potential probiotics and postbiotic characteristics including immunomodulatory effects of lactic acid bacteria isolated from traditional yogurt-like products. *LWT-Food Sci. Technol.* **2022**, *159*, 113207. [CrossRef]

116. Agostini, C.; Eckert, C.; Vincenzi, A.; Machado, B.L.; Jordon, B.C.; Kipper, J.P.; Dullius, A.; Dullius, C.H.; Lehn, D.N.; Sperotto, R.A.; et al. Characterization of technological and probiotic properties of indigenous Lactobacillus spp. from south Brazil. *3 Biotech.* **2018**, *8*, 451. [CrossRef]

117. Borges, L.; Bastos, R.; Sandes, S.; Guimarães, A.C.C.; Alves, A.G.; Reis, D.C.; Wuyts, S.; Nunes, A.C.; Cassali, G.; Lebeer, S.; et al. Protective effects of milk fermented by Lactiplantibacillus plantarum subsp. plantarum B7 from Brazilian artisanal cheese on a Salmonella enterica serovar Typhimurium infection in BALB/c mice. *J. Funct. Foods* **2017**, *33*, 436–445. [CrossRef]

118. Chourasia, R.; Abedin, M.M.; Phukon, L.C.; Sahoo, D.; Singh, S.P.; Rai, A.K. Biotechnological approaches for the production of designer cheese with improved functionality. *Compr. Rev. Food Sci. Food Saf.* **2021**, *20*, 960–979. [CrossRef] [PubMed]

119. Chambers, E.S.; Preston, T.; Frost, G.; Morrison, D.J. Role of gut microbiota-generated short-chain fatty acids in metabolic and cardiovascular health. *Curr. Nutr. Rep.* **2018**, *7*, 198–206. [CrossRef]

120. Behare, P.V.; Singh, R.; Kumar, M.; Prajapati, J.B.; Singh, R.P. Exopolysaccharides of lactic acid bacteria: A review. *J. Food Sci. Technol.* **2009**, *46*, 1–11. Available online: https://www.researchgate.net/publication/261474901 (accessed on 10 May 2021).

121. Zhou, Y.; Cui, Y.; Qu, X. Exopolysaccharides of lactic acid bacteria: Structure, bioactivity and associations: A review. *Carbohydr. Polym.* **2019**, *207*, 317–332. [CrossRef] [PubMed]

122. Angmo, K.; Kumari, A.; Savitri; Bhalla, T.C. Probiotic characterisation of lactic acid bacteria isolated from fermented foods and beverage of Ladakh. *LWT-Food Sci. Technol.* **2016**, *66*, 428–435. [CrossRef]

123. Breyer, G.M.; Arechavaleta, N.N.; Siqueira, F.M.; Motta, A.S. Characterization of lactic acid bacteria in raw buffalo milk: A screening for novel probiotic candidates and their transcriptional response to acid stress. *Probiotics Antimicro. Proteins* **2021**, *13*, 468–483. [CrossRef]

124. Kumari, A.; Angmo, K.; Monika; Bhalla, T.C. Probiotic attributes of indigenous Lactobacillus spp. isolated from traditional fermented foods and beverages of north-western Himalayas using in vitro screening and principal component analysis. *J. Food Sci. Technol.* **2016**, *53*, 2463–2475. [CrossRef]

125. Lynch, K.M.; Zanini, E.; Coffey, A.; Arendt, E.K. Lactic acid bacteria exopolysaccharides in foods and beverages: Isolation, properties, characterization and health benefits. *Annu Rev. Food Sci. Technol.* **2018**, *9*, 155–176. [CrossRef]

126. Zhang, L.; Li, X.; Ren, H.; Liu, L.; Ma, L.; Li, M.; Bi, W. Impact of using exopolysaccharides (EPS)- Producing strain on qualities of half-fat cheddar cheese on qualities of half-fat cheddar cheese. *Int. J. Food Prop.* **2015**, *18*, 1546–1559. [CrossRef]
127. Badel, S.; Bernardi, T.; Michaud, P. New perspectives for Lactobacilli exopolysaccharides. *Biotechnol. Adv.* **2011**, *29*, 54–66. [CrossRef]
128. Das, D.; Baruah, R.; Goyal, A. A food additive with prebiotic properties of an α-d-glucan from Lactobacillus plantarum DM5. *Int. J. Biol. Macromol.* **2014**, *69*, 20–26. [CrossRef] [PubMed]
129. Antonio, M.B.; Borelli, B.M. A importância de bactérias láticas na segurança e qualidade dos queijos Minas artesanais. *Rev. Inst. Laticínios Cândido Tostes* **2020**, *75*, 204–221. [CrossRef]
130. Bintsis, T. Lactic acid bacteria as starter cultures: An update in their metabolism and genetics. *AIMS Microbiol.* **2018**, *4*, 665–684. [CrossRef]
131. Hernandez-Valdes, J.A.; Solopova, A.; Kuipers, O.P. Development of *Lactococcus lactis* Biosensors for Detection of Diacetyl. *Front. Microbiol.* **2020**, *11*, 1032. (accessed on 1 September 2021). [CrossRef] [PubMed]
132. Kamarinou, C.S.; Papadopoulou, O.S.; Doulgeraki, A.I.; Tassou, C.C.; Galanis, A.; Chorianopoulos, N.G.; Argyri, A.A. Mapping the key technological and functional characteristics of indigenous lactic acid bacteria isolated from greek traditional dairy products. *Microorganisms* **2022**, *10*, 246. [CrossRef] [PubMed]
133. Kieliszek, M.; Pobiega, K.; Piwowarek, K.; Kot, A.M. Characteristics of the Proteolytic Enzymes Produced by Lactic Acid Bacteria. *Molecules* **2021**, *26*, 1858. [CrossRef]
134. Bachmann, H.P.; Fröhlich-Wyder, M.T.; Jakob, E.; Roth, E.; Wechsler, D.; Beuvier, E.; Buchin, S. Cheese: Raw milk cheeses. In *Encyclopedia of Dairy Sciences*, 2nd ed.; Academic Press: Cambridge, MA, USA, 2011; pp. 652–660. [CrossRef]
135. Coolbear, T.; Weimer, B.; Wilkinson, M.G. Lactic acid bacteria: Lactic acid bacteria in flavor development. In *Encyclopedia of Dairy Sciences*, 2nd ed.; Academic Press: Cambridge, MA, USA, 2011; pp. 160–165. [CrossRef]
136. Biscola, V.; Choiset, Y.; Rabesona, H.; Chobert, J.-M.; Haertlé, T.; Franco, B.D.G.M. Brazilian artisanal ripened cheeses as sources of proteolytic lactic acid bacteria capable of reducing cow milk allergy. *J. Appl. Microbiol.* **2018**, *125*, 564–574. [CrossRef]
137. Barbosa, N.E.A.; Ferreira, N.C.J.; Vieira, T.L.E.; Brito, A.P.S.O.; Garcia, H.C.R. Intolerância à lactose: Revisão sistemática. *Para Res. Med. J.* **2020**, *4*, e33. [CrossRef]
138. Silvério, S.C.; Macedo, E.A.; Teixeira, J.A.; Rodrigues, L.R. New β- galactosidase producers with potential for prebiotic synthesis. *Bioresour. Technol.* **2018**, *250*, 131–139. [CrossRef]
139. Saqib, S.; Akram, A.; Halim, S.A.; Tassaduq, R. Sources of β-galactosidase and its applications in food industry. *Biotechnol* **2017**, *7*, 79. [CrossRef]
140. Souza, B.M.S.; Borgonovi, T.F.; Casarotti, S.N.; Todorov, S.D.; Penna, A.L.B. Lactobacillus casei and Lactobacillus fermentum Strains Isolated from Mozzarella Cheese: Probiotic Potential, Safety, Acidifying Kinetic Parameters and Viability under Gastrointestinal Tract Conditions. *Probiotics Antimicrob. Proteins* **2019**, *11*, 382–396. [CrossRef] [PubMed]
141. Paula, A.T.; Jeronymo-Ceneviva, A.B.; Silva, L.F.; Todorov, S.D.; Franco, B.D.G.M.; Penna, A.L.B. Leuconostoc Mesenteroides SJRP55: A Potential Probiotic Strain Isolated from Brazilian Water Buffalo Mozzarella Cheese. *Ann. Microbiol.* **2015**, *65*, 899–910. [CrossRef]
142. Confederação Nacional da Indústria (CNI). Mercado de Insumos e Matérias Primas. 2021. Available online: https://static.portaldaindustria.com.br/portaldaindustria/noticias/media/filer_public/58/3b/583b1a48-b74e-4c31-8600-9dbad69f802f/sondagem_especial_-_insumos_e_materias-primas.pdf (accessed on 22 August 2022).

 *fermentation*

Article

Functional Characterization of *Saccharomyces* Yeasts from Cider Produced in Hardanger

Urban Česnik [1], Mitja Martelanc [1], Ingunn Øvsthus [2], Tatjana Radovanović Vukajlović [1], Ahmad Hosseini [3], Branka Mozetič Vodopivec [1] and Lorena Butinar [1,*]

[1] Wine Research Centre, University of Nova Gorica, Glavni trg 8, 5271-SI Vipava, Slovenia; urban.cesnik@ung.si (U.Č.); mitja.martelanc@ung.si (M.M.); tatjana.radovanovic@ung.si (T.R.V.); branka.mozetic@ung.si (B.M.V.)

[2] NIBIO, Norwegian Institute of Bioeconomy Research, P.O. Box 115, NO-1431 Ås, Norway; ingunn.ovsthus@nibio.no

[3] Center for Information Technologies and Applied Mathematics, University of Nova Gorica, Glavni trg 8, 5271-SI Vipava, Slovenia; ahmad.hosseini@ung.si

* Correspondence: lorena.butinar@ung.si; Tel.: +386-5-90-99-702

Abstract: *Saccharomyces cerevisiae* is commonly used for the production of alcoholic beverages, including cider. In this study, we examined indigenous *S. cerevisiae* and *S. uvarum* strains, both species commonly found in cider from Hardanger (Norway), for their strain-specific abilities to produce volatile and non-volatile compounds. Small-scale fermentation of apple juice with 20 *Saccharomyces* strains was performed to evaluate their aroma-producing potential as a function of amino acids (AAs) and other physicochemical parameters under the same experimental conditions. After fermentation, sugars, organic acids, AAs, and biogenic amines (BAs) were quantified using the HPLC–UV/RI system. A new analytical method was developed for the simultaneous determination of nineteen AAs and four BAs in a single run using HPLC–UV with prior sample derivatization. Volatile compounds were determined using HS-SPME-GC-MS. Based on 54 parameters and after the removal of outliers, the nineteen strains were classified into four groups. In addition, we used PLS regression to establish a relationship between aroma compounds and predictor variables (AAs, BAs, organic acids, sugars, hydrogen sulfide (H_2S) production, CO_2 release) of all 19 strains tested. The results of the VIP show that the main predictor variables affecting the aroma compounds produced by the selected yeasts are 16, belonging mainly to AAs.

Keywords: *Saccharomyces*; Hardanger; characterization; fermentation; cider; non-volatile compounds; volatile organic compounds; partial least squares (PLS) regression

Citation: Česnik, U.; Martelanc, M.; Øvsthus, I.; Radovanović Vukajlović, T.; Hosseini, A.; Mozetič Vodopivec, B.; Butinar, L. Functional Characterization of *Saccharomyces* Yeasts from Cider Produced in Hardanger. *Fermentation* **2023**, *9*, 824. https://doi.org/10.3390/fermentation9090824

Academic Editor: Roberta Comunian

Received: 30 July 2023
Revised: 25 August 2023
Accepted: 31 August 2023
Published: 8 September 2023

1. Introduction

Norwegian cider is becoming more and more popular in Norway in recent years among producers and consumers. Especially in the Southwest part of Norway, in the Hardanger region, there is a long tradition of producing ciders. Available data show that traditional cider from Hardanger is very different from French, English, or Spanish ciders in terms of sensory characteristics, apple cultivars, and fermentation process. A recent comparison of the aromatic component composition of different French and Norwegian ciders, including ciders from Hardanger, has confirmed that Norwegian ciders contain more aromas, which are behind fruity and fresh sensory sensations [1]. Cider from Hardanger is mostly made from desert apples, which have different chemical compositions in comparison to cider apples [2,3]. Ciders in Hardanger were traditionally produced by employing long spontaneous fermentation, even over winter, at low temperatures, often with the addition of sucrose and nothing else to increase the alcoholic strength [2]. Nowadays, spontaneous fermentation is more and more replaced by inoculated commercial yeasts. In our previous study of yeast ecology from ciders produced in the Hardanger area, we had seen that in ciders, apart from the presence of non-*Saccharomyces* species at the early stages of fermentation, the most predominant species isolated during fermentation were *S. cerevisiae* and *S. uvarum* [4].

S. uvarum is a cryotolerant yeast and belongs to the *Saccharomyces sensu stricto* clade, being the furthest relative from *S. cerevisiae*, and is now recognized as a pure species, distinct from *S. bayanus*, which is a hybrid of *S. uvarum* and *S. eubayanus* [5,6]. Due to its problematic taxonomy and incorrect identification, it is difficult to trace it in the scientific literature and obtain data about its origin, diversity, and potential usage in cider and winemaking. *S. uvarum* is not only related to cider fermentations, but it is also known in white wine production from colder grape-growing regions [7].

S. uvarum in grape juice, in comparison to *S. cerevisiae*, produces less acetic acid, acetaldehyde, and ethanol but more glycerol, succinic acid, and malic acid [7]. It is also capable of producing substantially more 2-phenyl ethanol (rose note), isoamyl alcohol (whisky), iso-butanol (solvent, bitter), and ethyl acetate (pineapple) [8,9]. There are also reports about the use of *S. uvarum* in cider production [10,11]. The phenotypic differences between *S. uvarum* and *S. cerevisiae*, the primary yeast species used worldwide for wine and cider making, are associated with pronounced proteomic differences [12].

S. cerevisiae gives ciders consistent aroma and taste and less risk of spoilage, however, resulting in less complex ciders [13]. Thus, to modulate the profile of ciders by enhancing microbial diversity seems a rational approach and was shown in recent studies [14,15]. To better evaluate the potential of different natural *Saccharomyces* strains, we need to assess the impact of nitrogen sources on fermentative behavior and possible undesirable production of metabolites such as acetic acid, H_2S, and BAs as well. In spontaneous fermentations, yeasts generally use naturally present amino acids, which results in higher aromatic complexity but also higher production of BAs.

The impact of *Saccharomyces* strains on the aroma profile related to different fermentation conditions is already well described in wine and beer production [7]; however, in recent years, cider production has been supported by research mainly based on non-*Saccharomyces* yeast strains [14–17], and not so much on potential alternative *Saccharomyces* species, such as *S. uvarum*.

S. cerevisiae is known to be more controllable in its fermentation output and performance in wine and beer production [7,18]. Therefore, it is very important to study them in natural media for cider production, namely apple juice. Most of the yeast characterization studies are still performed in synthetic must.

This study aimed to characterize traits of isolated *S. cerevisiae* and *S. uvarum* strains from ciders produced in Hardanger [4], an important area for cider production. We aimed to take a deeper look at their amino acid and sugar consumption in typical Hardanger apple juice (*Malus domestica* cv. 'Aroma'), the conversion of present sugars into ethanol, the yeast production potential of characteristic volatile compounds, and possible production of undesirable compounds, such as BAs, acetic acid and H_2S. To sum up, a comprehensive study of 20 yeast metabolite phenotypes has been used to classify the yeasts into groups with similar properties with the help of statistical methods, and partial least squares (PLS) regression has been used to reveal the correlation between amino acids and other primary metabolites with a synthesis of five different chemical groups of volatile compounds in apple ciders.

2. Materials and Methods

2.1. Yeast Strains, Media, and Culture Conditions

A list of the 20 strains from the *Saccharomyces* genus used in this study is provided in Table S1 (in Supplementary Materials). The yeast strains were isolated during the biodiversity study on cider yeasts in cider from Hardanger (in preparation for MDPI Foods) [4] and kept as cryo-cultures at -80 °C in 15% glycerol in the in-house culture collection at NIBIO Ullensvang (Lofthus, Norway) and the Wine Research Centre at the University of Nova Gorica (Vipava, Slovenia).

2.2. Screening for Sulfite Reductase Activity Using BiGGY Agar

Strains were tested for H_2S production on Bismuth Sulfite Glucose Glycine Yeast agar (BiGGY) [19]. BiGGY plates were spot-inoculated with a one-day-old liquid culture pregrown in Yeast Extract-Peptone-Dextrose (YPD) medium (10 g/L yeast extract, 20 g/L peptone, 20 g/L glucose) at 25 °C and 150 rpm. BiGGY plates were inoculated at 25 °C, and colony color was assessed after 5 days. The assay was performed in triplicate.

2.3. Micro-Fermentations

Monoculture fermentations were performed with selected *Saccharomyces* strains listed in Table S1. For the fermentation experiment, apples from the apple cultivar Aroma (*M. domestica* cv. 'Aroma'), grown in the Hardanger area, were milled and pressed with a belt press; apple juice was immediately frozen at −20 °C till the experimental set-up. Apples and the obtained juice were not pre-treated with enzymes or other enological additives during processing.

Pre-cultures were prepared by inoculating single colonies of Wallerstein Laboratory Nutrient Agar (WL) (VWR) plate cultures in 3 mL YPD medium in 15 mL tubes. After incubation for 24 h at 25 °C and 150 rpm, the pre-cultures were centrifuged (2000 rpm, 10 min, room temperature (RT)) and washed with 0.85% NaCl solution. Finally, the yeast pellets were resuspended in diluted sterile apple juice (1:1 with sterile water). The optical density at 600 nm (OD 600 nm) of the yeast suspensions was adjusted to 1.0 and left at RT for 30 min for adaptation.

The apple juice was unfrozen and sterile filtered using a vacuum filtration system (500 mL Polyethersulfone (PES), 0.2 μm membrane filter (VWR)). Then, 20 mL of the apple juice was placed in 40 mL glass vials and inoculated with the pre-culture to achieve a final OD of 600 nm 0.1 AU. Fermentations were prepared in triplicates and conducted at 15 °C for 26 days.

During fermentation, mass loss was monitored and H_2S was quantified using 120SF gas detector tubes (Komyo Kitagawa, Kawasaki-City, Japan), as described by Ugliano and Henschke [20]. The detector tubes were inserted into the vials through a hole in the PTFE/silicone partition of the lids.

At the end of fermentation, samples were centrifuged (20 min, 6000 rpm) and stored at −20 °C before chemical analysis.

2.4. Determination of Sugars, Acids, and Ethanol

Reagents, Materials, and Standards for HPLC–UV/RI Analyses

Chemicals: we used glucose (99%) (Acros Organics, Fair Lawn, NJ, USA), fructose (99%) (Acros Organics, Fair Lawn, NJ, USA), sucrose (99.9%) (Acros Organics, NJ, USA), tartaric acid (Alfa Aesar, Karlsruhe, Germany), lactic acid (30%) (Sigma, Steinheim, Germany), D-L malic acid (99%) (Aldrich, Steinheim, Germany), and citric acid (99.9%) (Sigma, Steinheim, Germany). Concentrated sulfuric acid (VI) was purchased from VWR Chemicals (Leuven, Belgium).

Three in-house developed HPLC–UV/RI methods were used for the determination of sugars, organic acids, and ethanol in cider samples, respectively. An Agilent 1100 series HPLC system (Agilent Technologies©, Palo Alto, CA, USA) was equipped with Agilent OpenLab CDS ChemStation 2.3.54 software, a UV detector (G1314A VWD) for the analysis of organic acids (detection at 210 nm), and a refractive index detector (model G7162A) for the analysis of sugars (fructose and glucose) and ethanol. Samples were filtered using Polytetrafluoroethylene (PTFE) 0.45 μm syringe filters (VWR® International, Radnor, PA, USA). For the determination of glucose and fructose, 4 μL of each sample was injected onto a Phenomenex Luna® Omega Sugar HPLC column (150 mm long and ø of 4.6 mm, particle size of 3 μm) with a precolumn (5 mm long and ø of 4.6 mm, particle size of 3 μm) using a mobile phase of acetonitrile/water = 75:25 (*v/v*) at a flow rate of 0.9 mL/min and a run time of 15 min [21]. Separation of the organic acids was performed on two HPLC columns, which were coupled sequentially: Phenomenex C18 Kinetex F5 (dimensions

150 × 4.6 mm with a particle size of 2.6 μm) with a precolumn (5 mm long and ø of 4.6 mm, particle size of 3 μm) and a Phenomenex C18 Kinetex EVO (dimensions 250 × 4.6 mm with a particle size of 5 μm) kept at 30 °C during analyses. The injection volume was 4 μL, the mobile phase consisted of 5 mM H_2SO_4, the flow rate was 0.7 mL/min, and the run time was 20 min [22]. Ethanol was evaluated on a multimodal ROA Organic Acid H+ (8%) column (Phenomenex) with a size of 300 × 7.8 mm and a sample injection volume of 5 μL. Isocratic elution was performed using 5 mM H_2SO_4 as the mobile phase with a flow rate of 0.9 mL/min and a run time of 25 min [23]. All HPLC columns were kept at 30 °C during chromatographic analyses.

Method validation data are summarized in Table S2 (in Supplementary Materials).

2.5. HPLC–UV Determination of Amino Acids and Biogenic Amines

2.5.1. Reagents, Materials, and Standards for HPLC–UV Analyses

Amino acids: L-tyrosine disodium salt hydrate (98%) (Sigma, Steinheim, Germany), L-aspartic acid (99,5%) (AppliChem, Darmstadt, Germany), L-serine (99%) (Sigma, Steinheim, Germany), L-leucine (98%) (Sigma, Steinheim, Germany), L-cystein (99%) (Merck, Darmstadt, Germany), isoleucine (99.5%) (AppliChem, Darmstadt, Germany), L-phenylalanine (99.5%) (Sigma, Steinheim, Germany), L-asparagine (98%) (Sigma, Steinheim, Germany), L-lysine monohydrochloride (98%) (Sigma, Steinheim, Germany), L-glycine (99%) (Sigma, Steinheim, Germany), L-glutamine (99%) (Sigma, Steinheim, Germany), L-thryptophan (98%) (Sigma, Steinheim, Germany), L-arginine monohydrochloride (99.5%) (Sigma, Steinheim, Germany), L-alanine (99.5%) (AppliChem, Darmstadt, Germany), L-lysine monohydrochloride (99%) (Acros Organics, NJ, USA), L-proline (99.5%) (AppliChem, Darmstadt, Germany), L-glutamic acid (99%) (Sigma, Steinheim, Germany), L-valine (98%) (Sigma, St. Louis, MO, USA), L-methionine (99.5%) (Fisher Scientific, Geel, Belgium), threonine (99%) (Fisher Scientific, Geel, Belgium), L-hystidine monohydrochloride monohydrate (99%) (VWR Chemicals, Leuven, Belgium).

Biogenic amines: putrescine dihydrochloride (98%) (AppliChem, Darmstadt, Germany), cadaverine dihydrochloride (98%) (Sigma, Steinheim, Germany), histamine dihydrochloride (98%) (Alfa Aesar, Karlsruhe, Germany), tyramine hydrochloride (98%) (Alfa Aesar, Karlsruhe, Germany).

Other reagents and chemicals: hypochloric acid (Gram-Mol, Zagreb, Croatia), sodium hydroxide, sodium hydrogen carbonate and ammonia were purchased from (Sigma, Steinheim, Germany), ethanol, acetonitrile (HPLC grade) (J.T. Baker, Gliwice, Poland), sodium acetate (Carl Roth, Karlsruhe, Germany). Derivatization reagent dansyl chloride was supplied from Sigma (Steinheim, Germany).

Chemicals were prepared with ultrapure water, which was prepared using a Milli-Q water purification system Purelab Option-Q system (ELGA LabWater, High Wycombe, UK) to a specific resistance of >18.0 MΩ cm^{-1} at 25 °C.

A mix stock standard solution of 19 amino acids and a mix stock solution of 4 biogenic amines (1000 mg/L) were prepared in 0.1 M HCl. The solutions were stirred in an ultrasonic bath for 5 min.

2.5.2. Derivatization Procedure for Determining Amino Acids and Biogenic Amines

Derivatization was performed according to the procedure described by Topić Božič et al. [24] with modifications. 250 μL of the standard (amino acids or biogenic amines) and cider samples were mixed with 70 μL of saturated $NaHCO_3$, 75 μL of 0.1 M NaOH, and then with 1.5 mL of dansyl chloride derivatization reagent (0.2% in acetonitrile). The mixture was then shaken and incubated in an oven at 40 °C for 45 min. Then, 100 μL of ammonia was added to the reaction mixture and the mixture was incubated at RT for 30 min. Samples were filtered using 0.45 μm PTFE syringe filters before HPLC analysis. Calibration curves for amino acids and biogenic amines were generated in the range of 1–100 mg/L by diluting the standard stock solution (1000 mg/L).

2.5.3. HPLC–UV Analysis of AAs and BAs

Separation and quantification of 19 amino acids and 4 biogenic amines was performed on Agilent's HPLC–UV system (described in Section 2.4) using Kinetex® 2.6 μm EVO C18 RP column (150 mm long and ø of 4.6 mm, Phenomenex) with a pre-column (Kinetex® 2.6 μm EVO C18 RP, 5 mm long and ø of 4.6 mm). The separation was done at 35 °C with gradient elution at a flow rate of 0.8 mL/min. The total run time was 65 min.

The mobile phase was prepared from 20 mM sodium acetate. The pH of the mobile phase was pH adjusted to 6.5 with 0.8 M acetic acid. An injection volume of 3 μL was used. Derivatized amino acids and biogenic amines were detected at 246 nm. The gradient profile is described in Table S3 (in Supplementary Materials).

2.5.4. Method Validation

Method validation was performed by testing linearity, repeatability, the limit of detection (LOD), the limit of quantification (LOQ) and recovery (presented in Supplementary Materials in Table S4). The derivatization step was included in the validation procedure.

2.6. Determination of Volatile Compounds Using HS-SPME-GC-MS

Selected aroma compounds were determined in the ciders using an automated robotic system for Solid Phase Micro-Extraction (SPME) in head space (HD) and injected on a gas chromatograph coupled with a mass spectrometric detector (GC-MS). Esters, C6 alcohols, and volatile phenols were analyzed using a method adapted from a previously published protocol [25]. Samples were extracted by headspace Solid Phase Micro-Extraction (HS–SPME) using an SPME fiber assembly (50/30 μm DVB/CAR/PDMS, Stableflex, 24 Ga, Autosampler, Gray (Supelco, St. Louis, MO, USA)). To a 20 mL SPME vial, 3 mL of the cider sample was added with 2 g NaCl and 3 mL deionized water, and 20 μL solution of internal deuterated standards (ethyl butyrate-4,4,4 d3, ethyl d5 hexanoate, ethyl octanoate d15, ethyl trans-cinnamate d5) was added. The solution was then homogenized using a vortex mixer and the samples were loaded into a Gerstel MPS Robotic Autosampler. The program consisted of introducing the fiber into the SPME Arrow Conditioning Module for 2 min at 270 °C. The fiber was then introduced into the headspace of a sample vial for 30 min at 40 °C while simultaneously vortexing the sample with the agitator at 250 rpm. The fiber was then transferred to the injector for desorption at 250 °C for 15 min. The time for sample injection into the GC column was set to 30 s, followed by cleaning of the fibers in the SPME Arrow Conditioning Module at 270 °C for 10 min.

2.7. Multivariate Data and Statistical Analysis

Data were presented as means ± standard deviation (SD) from three repetitions. ANOVA and Tukey's method were employed using IBM SPSS Statistics 27 to compare the variances among the means of various groups. A significance level of $\alpha = 5\%$ was chosen to determine statistical significance.

Data analysis was performed by a multivariate approach. Principal components analysis (PCA) and the partial least squares (PLS) regression analysis were carried out to explore the differences among ciders produced from apple juice by different *Saccharomyces* strains. To deal with non-detectable values, the data matrix was pre-processed, and non-detectable values were replaced with LLOQ/2.

All computational efforts and multivariate data analysis were implemented in IBM SPSS Statistics 27, Minitab 21, GraphPad Prism 9.5.1, and XLSTAT 2023 on a Lenovo PC with Intel(R) Core (TM) i7-6600U CPU @ 2.60 GHz and 16 GB of RAM, Microsoft Windows 10 OS. Boxplots for AAs utilization were prepared by R v. 4.1.2 for macOS.

3. Results and Discussion

3.1. H_2S Production

According to the results obtained on BiGGY plates, seven strains were classified as non-H_2S producers (white colony color), two as moderate (white, light brown edge), and the rest as strong H_2S producers (brown color) (Table 1). Using detector tubes inserted into a hole in the septum of the vial cap, we quantitatively assessed H_2S formation by the strains tested under fermentative conditions. The H_2S formation potential determined by color staining on BiGGY agar did not agree with the results obtained with detector tubes. Five *S. uvarum* strains produced H_2S; in strain 2176, we detected an average of 58.3 ppm H_2S, followed by strains 2402 and 2128 with 48.3 and 36.7 ppm, respectively (Table 1). The lowest H_2S production was detected in *S. uvarum* 2401 (average 13.3 ppm) (Table 1).

Table 1. Results of color staining on BiGGY agar and measured H_2S production during fermentation for 20 *Saccharomyces* strains tested.

Yeast Species and Strain Code	Colony Color on BiGGy Agar [1]	H_2S Detector Tubes [2] (ppm)
S. uvarum 2046	white	0 ± 0.00 E
S. uvarum 2071	white	35 ± 8.7 BC
S. uvarum 2120	brown	0 ± 0.00 E
S. uvarum 2186	brown	0 ± 0.00 E
S. uvarum 2401	white	13.3 ± 5.8 DE
S. cerevisiae 2003	brown	10 ± 0.00 DE
S. cerevisiae 2095	brown	0 ± 0.00 E
S. cerevisiae 2265	white, light brown edge	0 ± 0.00 E
S. cerevisiae 2273	white	0 ± 0.00 E
S. cerevisiae 2303	brown	0 ± 0.00 E
S. cerevisiae 2349	brown	0 ± 0.00 E
S. uvarum 2128	brown	36.7 ± 10.4 BC
S. uvarum 2204	brown	0 ± 0.00 E
S. uvarum 2216	white	23.3 ± 2.9 CD
S. uvarum 2376	white	0 ± 0.00 E
S. uvarum 2083	white, light brown edge	0 ± 0.00 E
S. uvarum 2104	brown	0 ± 0.00 E
S. uvarum 2176	white	58.3 ± 18.9 A
S. uvarum 2402	brown	48.3 ± 2.9 AB
S. uvarum 2061	brown	0 ± 0.00 E

[1] White colony color = no H_2S production; White colony color with light brown edge = moderate H_2S production; Brown colony color = strong H_2S production. [2] H_2S was measured with gas detector tubes (120SF; Komyo Kitagawa, Kawasaki-City, Japan). Values are reported as mean ± SD of three replicates. Values not connected by the same letter are significantly different (ANOVA, Tukey's method).

3.2. HPLC–UV Determination of Amino Acids and Biogenic Amines

Determination of AAs and BAs is challenging when using the HPLC–UV system since both AAs and BAs lack chromophores for detection. Therefore, if HPLC coupled with mass spectrometric detection (MS) is not used, a sample derivatization step requiring HPLC coupled with a fluorescence detector (FLD) is necessary before analysis, especially if quantitation below 1 mg/L is required [26]. The HPLC–FLD system has more than a 10-fold higher sensitivity for such analyses compared to HPLC–UV [27]. All in all, only two methods for simultaneous determination of AAs and BAs have been published to the best of our knowledge, one based on the HPLC–MS system [28] and the second based on ultra-performance liquid chromatography (UPLC) coupled with a diode array detector (DAD) [29]. Here, we present a newly developed method for the simultaneous determination of nineteen AAs and four BAs based on the HPLC–UV system using dansyl chloride as a derivatizing agent for AAs and BAs. The method was validated and applied for the analysis of fermented beverages based on apple juice (cider) and can also be used for wine samples or, as in our study, for the in-depth characterization of AA utilization and BA production of a larger number of yeasts. The separation system is based on the C18 reverse phase (RP) system with 3 μm particles, which allows better resolution as well as higher sensitivity for the analyzed compounds (Figure 1). Together with the previous derivatization with dansyl chloride, we were able to obtain adequate LODs and LOQs for the determination of AAs and BAs in the analyzed samples. These LODs and LOQs are comparable to the published method [29], although the UPLC system with 1.8 μm particles was used. The linearity range of the method for each compound is between 0.5 mg/L and 200 mg/L, which corresponds to the actual range of occurrence of the analyzed compounds in juices and fermented beverages (the detection limits are between 0.03 and 0.3 mg/L). All validated parameters indicate that the method presented here can be a relevant and useful tool for the quality control of cider by monitoring fermentation, especially due to the fact that the HPLC–UV system is an easily accessible tool in analytical/research laboratories.

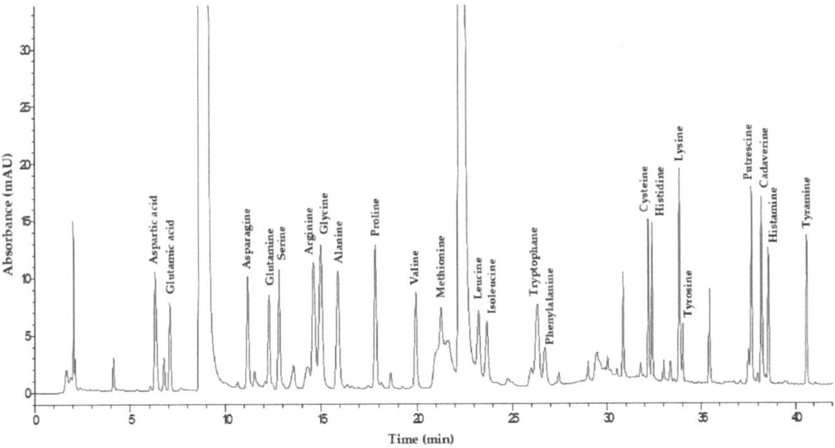

Figure 1. HPLC–UV chromatogram of simultaneously separated 19 AAs and 4 BAs (each at 25 mg/L con-centration level) with prior derivatization using dansyl chloride agent.

3.3. Behavior of Saccharomyces Strains under Fermentative Conditions

The total AA content in apple juice from the cultivar Aroma averaged 303.97 mg/L. The evaluation of the AA content revealed that asparagine (42.2%) was the most important AA in juice, followed by tyrosine (13.1%), aspartic acid (8.2%), arginine (6.7%), and glutamine (5.8%) (Table S5 in Supplementary Materials).

The AA utilization profile of both species showed a similar utilization pattern of AAs present in apple juice, and the pattern of AAs utilized also reflected the strain effect (Figure 2, Table S6 in Supplementary Materials). The major AA source for all strains tested was asparagine, which provided an average of 57.7% of the total AAs utilized. Note that the initial asparagine concentration in apple juice was 3–65 times higher than that of the other AAs (Table S5). Aspartic acid, arginine, and glutamine were the next most utilized AAs, followed by glutamic acid, serine, proline, and then valine and tyrosine (Figure 2, Table S6). The remaining AAs were present in very low concentrations in apple juice and were mostly utilized during fermentation. For methionine, histidine, and alanine, some residual amounts were still detected at the end of fermentation (Figure 2, Table S6).

The good utilization of asparagine and glutamine by the *Saccharomyces* strains tested in this study is consistent with what has been reported in the literature, as these two AAs, along with ammonium, have often been reported as preferred nitrogen sources for *S. cerevisiae* [26,27]. Aspartic acid, arginine, glutamic acid, and serine were also included among the preferred nitrogen sources [26,27], which we also observed in our study.

Although tyrosine was the second most abundant AA in apple juice, less than a quarter of it was utilized by the strains, which is consistent with the literature where tyrosine is considered a non-preferred nitrogen source [26,27].

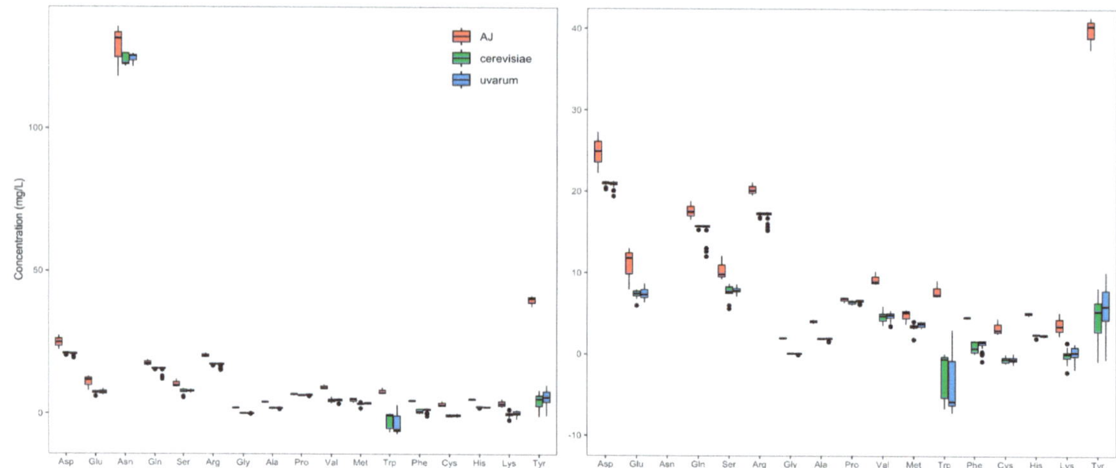

Figure 2. Boxplots showing the utilization pattern for AAs of 6 *S. cerevisiae* (cerevisiae, green box) and 14 *S. uvarum* strains (uvarum, blue box), compared to the initial concentration in apple juice (AJ, red box). On the left side, a pattern for all amino acids is shown, while on the right side, an enlarged plot for lower concentrations is presented. Asp-aspartic acid, Glu-glutamic acid, Asn-asparagine, Gln-glutamine, Ser-serine, Arg-arginine, Gly-glycine, Ala-Alanine, Pro-proline, Val-valine, Met-methionine, Trp-tryptophane, Phe-phenylalanine, Cys-cysteine, His-histidine, Lys-lysine, Tyr-tyrosine.

Cysteine, lysine, and especially tryptophan were found in higher concentrations at the end of fermentation compared to what was present in the apple juice. This could be explained by the release of these AAs into the medium during fermentation (Tables S5 and S6). Tryptophane and methionine were also previously classified as non-preferred nitrogen sources [26,27]. However, the data in the literature on the classification of nitrogen sources were mostly based on their ability to support yeast growth of laboratory strains under non-fermentative conditions.

The HPLC–UV method enabled us to determine AAs and BAs simultaneously. Thus, we were able to detect strains with BA-producing ability using this method (Table S6). At the end of fermentation, strains of both species formed two BAs, putrescine and tyramine. Putrescine was detected at relatively low concentrations in all strains tested, whereas the highest concentrations for tyramine averaged 3.9 mg/L in the high-producing strains 2104, 2186, and 2402 (Table S6). The BA production capability is another characteristic that is important for the selection of starter yeasts for cider production, especially when substrates for the production of BA are present in apple juice, such as tyrosine, which can be further decarboxylated to tyramine.

The initial sugar content of apple juice was 101.6 g/L, and the main sugar was fructose (52.9%), followed by sucrose (35.2%) and glucose (11.9%), as shown in Table S7 (in Supplementary Materials). At the end of fermentation, most of the sugar was consumed (98.6% on average), and the residual sugar consisted mainly of remaining glucose, which varied in very low concentrations from 1.3 to 1.6 g/L (Table S7). Hence, studies showed that *Saccharomyces* yeasts display a clear preference for glucose over fructose [29–31]. If we assume that hydrolysis of sucrose was a limiting step during our fermentation, the glucophilic nature of *Saccharomyces* yeasts would still leave more fructose at the end. However, there are few systematic studies on the preference for glucose and fructose of cider yeast strains, especially in mixed sugar media, such as apple juice [17]. Therefore, it would be of immense importance to focus on the utilization of glucose, fructose, and sucrose by cider yeasts under different fermentation conditions in the near future. Nevertheless, such studies could provide tools for the evaluation and selection of yeast strains for cider production, especially yeasts with fructophilic character, since fructose is the main sugar in apple juice and may pose a problem for stuck fermentation. In addition, sucrose is usually added during fermentation in the production of cider from Hardanger to increase the alcoholic strength. Moreover, from this point of view, it would be important to study the consumption of fructose, glucose, and sucrose during the fermentation process.

Ethanol content averaged between 3.99 and 4.82% (*v*/*v*) and did not differ statistically among the *Saccharomyces* strains tested (Table S7).

In terms of organic acids in apple juice, malic acid was the most abundant (97.0% of total organic acids), which is consistent with previous studies [2,32]. Citric acid and tartaric acid were also detected, and their contents were much lower than those of malic acid. In cider, malic acid averaged between 5.2 g/L and 6.6 g/L, being least affected by the degradation of *S. uvarum* 2046 and most affected by strain *S. uvarum* 2176 (Table S7). Acetic acid was within acceptable levels [13], with the highest concentration determined in the ferments of *S. cerevisiae* 2349; otherwise, the levels were less than 0.09 g/L in the other yeast strains (Table S7). Citric acid and tartaric acid were also detected in very low concentrations in the finished ciders, ranging from 0.46 to 1.31 g/L and from 0.04 to 0.06 g/L, respectively (Table S7).

3.4. Volatile Compound Production Profiles of the Saccharomyces Ferments

The concentrations of the volatile compounds are listed in Table 2. We measured a total of twenty-six aroma compounds, eighteen esters (seven ethyl esters of fatty acids, two ethyl esters of branched acids, and nine acetate esters), three C6 alcohols, and five volatile phenols in experimental ciders.

The main group of aroma compounds in our experiment were C6 alcohols, ethyl esters of fatty acids, and volatile phenols. When we sum all aroma compounds, we see a large variability between strains, ranging from 2503 µg/L (*S. cerevisiae* 2095) to 8654 µg/L (*S. uvarum* 2376).

The major aroma compound in all ciders was hexanol, which varied from 1222 µg/L (*S. uvarum* 2083) to 2539 µg/L (*S. cerevisiae* 2349). According to Waterhouse et al. [33], the olfactory threshold value for 1-hexanol is 8000 µg/L, which means that its contribution to the aroma profile is most likely negligible. In the study of Scandinavian and British ciders, 1-hexanol was the major C6 alcohol [34], ranging from 32 to 6541 µg/L. In our recent study, Norwegian ciders, on average, contained 5137 µg/L of 1-hexanol, and French ones 6555 µg/L [1].

Higher alcohols are known to be the most abundant group of aroma compounds in cider and apple juice [35,36], but they are mainly important as precursors of esters, which are known for their fruity and sweet aroma [37]. The concentration of higher alcohols generally decreases or disappears during cider fermentation [16,34], but in some cases, it also increases or remains unchanged [36].

Seven different ethyl esters of fatty acids were quantified, namely ethyl propanoate, ethyl butyrate, ethyl hexanoate, ethyl octanoate, ethyl decanoate, ethyl valerate, and ethyl dodecanoate. The amounts varied greatly between strains, as shown in Table 2. Ethyl esters are known to be an important component of the Norwegian cider flavor (Øvsthus et al., 2023, in press [1]) and also for other ciders [34,38]. Although some esters may be originally present in apple juice before fermentation, most esters in the ciders are formed by the esterification of alcohols with carboxylic acids during fermentation and aging [36,39].

The yeast with the lowest total concentration of fatty acid ethyl esters (average 269.3 µg/L) (*S. cerevisiae* 2273) produced almost no ethyl decanoate and ethyl dodecanoate (0 and 15.5 µg/L, respectively). Whereas in the cider with the highest total ethyl ester of fatty acids contents among all yeast strains (*S. uvarum* 2376) (5123.67 µg/L), these two esters were among the two most abundant ones (1905.93 and 1899.66 µg/L, respectively). None of the *Saccharomyces* yeasts produced ethyl valerate in higher concentrations (on average, less than 1% of all ethyl esters of fatty acids).

Table 2. Volatile compounds (μg/L) of ciders produced from apple juice fermentation carried out by 19 *Saccharomyces* strains. Values are reported as mean ± standard deviation of three replicates. Groups are indicated by letters. The values (i.e., the means) on the same column are significantly different according to Tukey's method (and 95% confidence) if they do not share a letter.

Yeast Species	Strain no.	Acetate Esters(AE)								
		Propyl Acetate	Isobutyl Acetate	Butyl Acetate	Isoamyl Acetate	Z-3-Hexenyl Acetate	E-2-Hexenyl Acetate	Ethyl Phenyl Acetate	Hexyl Acetate	OctylAcetate
S. uvarum	2104	7.72 ± 0.47 ABC	7.98 ± 1.19 AB	145.97 ± 3.54 ABC	69.99 ± 12.52 ABC	3.71 ± 0.13 AB	0.35 ± 0.07 ABCD	1.45 ± 0.75 ABCD	29.77 ± 0.89 ABC	2.39 ± 0.29 BC
S. uvarum	2120	10.90 ± 0.58 A	11.11 ± 4.64 A	220.86 ± 41.45 A	103.57 ± 52.16 AB	4.68 ± 0.98 AB	0.51 ± 0.16 A	1.03 ± 0.65 BCD	38.47 ± 9.16 A	2.76 ± 0.14 AB
S. uvarum	2128	9.29 ± 1.32 ABC	10.01 ± 1.20 ABC	191.05 ± 25.9 AB	91.34 ± 11.65 ABC	4.09 ± 0.10 AB	0.42 ± 0.07 AB	1.17 ± 0.07 ABCD	33.84 ± 1.79 AB	3.18 ± 0.74 A
S. uvarum	2176	9.46 ± 0.43 ABC	5.28 ± 0.31 ABC	196.73 ± 5.9 AB	56.26 ± 9.23 ABC	3.41 ± 0.02 ABC	0.35 ± 0.04 ABCD	2.48 ± 1.02 A	29.86 ± 1.21 ABC	2.10 ± 0.09 BCD
S. uvarum	2186	9.62 ± 0.99 ABC	10.07 ± 2.36 AB	207.93 ± 31.8 AB	80.29 ± 27.62 ABC	3.94 ± 0.45 AB	0.42 ± 0.08 AB	1.22 ± 0.07 ABCD	34.65 ± 3.52 AB	2.15 ± 0.50 BC
S. uvarum	2204	10.38 ± 0.63 AB	9.283 ± 1.86 AB	213.61 ± 22.2 AB	90.32 ± 29.43 ABC	3.84 ± 0.50 AB	0.40 ± 0.05 ABC	1.02 ± 0.07 BCD	31.96 ± 5.61 AB	2.07 ± 0.38 BCD
S. uvarum	2216	9.29 ± 1.02 ABC	4.84 ± 0.60 ABC	194.17 ± 48.7 AB	39.08 ± 8.34 ABC	3.14 ± 0.60 ABCD	0.36 ± 0.04 ABCD	1.23 ± 0.06 ABCD	23.87 ± 6.13 ABCD	1.94 ± 0.37 CDE
S. cerevisiae	2349	8.94 ± 1.36 ABC	4.40 ± 0.61 BC	161.28 ± 18.41 ABC	99.41 ± 23.76 AB	3.52 ± 0.41 AB	0.27 ± 0.04 ABCD	0.48 ± 0.05 CD	33.91 ± 4.70 AB	1.01 ± 0.0.3 F
S. uvarum	2376	9.10 ± 2.46 ABC	6.91 ± 2.59 ABC	158.67 ± 61.76 ABC	49.23 ± 24.20 ABC	2.81 ± 0.84 ABCD	0.32 ± 0.08 ABCD	1.36 ± 1.00 ABCD	22.91 ± 8.88 ABCD	0.99 ± 0.0.6 F
S. uvarum	2402	8.41 ± 1.47 ABC	9.25 ± 2.37 AB	166.11 ± 49.54 ABC	65.48 ± 20.68 ABC	3.09 ± 0.81 ABCD	0.31 ± 0.11 ABCD	1.81 ± 0.42 ABC	24.66 ± 8.59 ABCD	1.31 ± 0.5 DEF
S. uvarum	2401	9.00 ± 1.94 ABC	10.20 ± 3.26 ABC	159.13 ± 43.61 ABC	86.14 ± 28.94 ABC	3.08 ± 0.92 ABCD	0.28 ± 0.07 ABCD	1.27 ± 0.24 ABCD	22.25 ± 9.79 ABCD	1.17 ± 0.08 EF
S. uvarum	2071	5.82 ± 1.36 BC	4.98 ± 1.71 ABC	95.61 ± 39.4 BC	29.48 ± 18.71 BC	1.36 ± 0.69 D	0.15 ± 0.07 D	0.74 ± 0.16 CD	8.48 ± 4.83 D	0.87 ± 0.08 F
S. uvarum	2061	5.63 ± 0.99 C	9.60 ± 3.30 AB	63.46 ± 53.23 C	45.96 ± 20.88 ABC	1.42 ± 0.68 CD	0.16 ± 0.09 CD	2.24 ± 0.56 AB	9.25 ± 6.02 CD	1.17 ± 0.09 EF
S. uvarum	2046	7.29 ± 2.01 ABC	6.17 ± 2.77 ABC	124.13 ± 35.94 ABC	50.56 ± 26.21 ABC	2.17 ± 0.76 BCD	0.20 ± 0.09 BCD	1.21 ± 0.30 ABCD	17.46 ± 8.16 BCD	0.90 ± 0.09 F
S. cerevisiae	2265	10.43 ± 2.05 AB	5.82 ± 1.32 ABC	170.95 ± 51.36 ABC	88.59 ± 18.02 ABC	3.66 ± 0.67 AB	0.33 ± 0.10 ABCD	1.15 ± 0.06 ABCD	35.88 ± 8.79 AB	1.11 ± 0.05 F
S. cerevisiae	2273	6.73 ± 1.40 ABC	2.36 ± 0.60 C	105.34 ± 32.26 ABC	12.01 ± 1.02 C	1.23 ± 0.36 D	0.21 ± 0.11 BCD	0.29 ± 0.07 D	8.98 ± 4.03 D	1.12 ± 0.03 F
S. cerevisiae	2003	9.12 ± 1.88 ABC	6.94 ± 1.97 ABC	171.20 ± 50.52 ABC	118.73 ± 62.59 A	3.89 ± 0.24 AB	0.32 ± 0.08 ABCD	0.68 ± 0.14 CD	39.72 ± 12.89 A	0.95 ± 0.01 F
S. cerevisiae	2095	7.65 ± 1.14 ABC	4.93 ± 0.56 ABC	119.12 ± 21.09 ABC	59.46 ± 3.74 ABC	2.54 ± 0.29 BCD	0.25 ± 0.03 ABCD	0.66 ± 0.03 CD	24.27 ± 2.46 ABCD	1.06 ± 0.16 F
S. uvarum	2083	5.99 ± 0.74 BC	4.93 ± 1.12 ABC	111.12 ± 30.50 ABC	35.92 ± 13.09 ABC	2.17 ± 0.46 ABCD	0.26 ± 0.08 ABCD	1.20 ± 0.47 ABCD	19.74 ± 5.42 ABCD	0.95 ± 0.05 F

Yeast Species	Strain no.	Ethyl Esters from Fatty Acids(EEFA)						
		Ethyl propanoate	Ethyl butyrate	Ethyl hexanoate	Ethyl octanoate	Ethyl decanoate	Ethyl valerate	Ethyl dodecanoate
S. uvarum	2104	57.36 ± 4.95 CDE	45.16 ± 1.23 FG	86.53 ± 3.80 EFGH	291.56 ± 45.08 EFGHI	247.30 ± 33.52 DEFG	0.42 ± 0.03 CD	258.68 ± 32.96 DEFG
S. uvarum	2120	70.66 ± 16.71 BCD	57.37 ± 7.25 CDEFG	121.80 ± 21.89 DEFG	373.86 ± 67.47 CDEFG	392.82 ± 90.60 DEF	0.59 ± 0.13 BCD	401.77 ± 89.09 DEF
S. uvarum	2128	53.38 ± 9.49 DE	52.17 ± 3.14 EFG	120.43 ± 7.44 DEFG	425.35 ± 76.54 BCDEF	488.00 ± 63.42 CD	0.52 ± 0.7 BCD	495.37 ± 62.37 CD
S. uvarum	2176	158.94 ± 10.53 A	93.79 ± 5.46 AB	191.53 ± 16.28 BCDEF	606.22 ± 49.90 B	887.72 ± 87.30 B	1.39 ± 0.08 A	888.43 ± 85.84 B
S. uvarum	2186	69.13 ± 11.88 CD	55.23 ± 5.87 DEFG	123.36 ± 14.0 CDEFG	359.55 ± 38.35 CDEFGH	468.80 ± 103.79 CD	0.82 ± 0.17 B	476.49 ± 102.06 CD
S. uvarum	2204	100.91 ± 9.56 B	59.85 ± 3.87 CDEFG	110.91 ± 6.62 DEFGH	452.85 ± 25.5 BCDE	741.37 ± 50.63 BC	0.82 ± 0.05 B	744.51 ± 49.76 BC
S. uvarum	2216	172.37 ± 10.62 A	74.33 ± 3.9 BCDE	133.56 ± 8.04 BCDEF	548.31 ± 6.95 BC	463.66 ± 16.62 D	1.18 ± 0.02 A	471.43 ± 16.35 D
S. cerevisiae	2349	41.54 ± 6.13 DEF	104.21 ± 17.0 A	256.10 ± 49.27 A	339.71 ± 77.20 DEFGH	141.68 ± 27.57 FG	0.61 ± 0.12 BCD	154.82 ± 27.11 FG
S. uvarum	2376	88.12 ± 19.10 BC	80.15 ± 16.9 ABCD	185.19 ± 30.43 BC	973.86 ± 40.27 A	1905.93 ± 189.17 A	0.73 ± 0.20 BC	1889.66 ± 186.02 A
S. uvarum	2402	45.20 ± 8.91 DEF	49.15 ± 5.69 EFG	106.44 ± 11.39 DEFGH	293.59 ± 93.67 DEFGHI	147.69 ± 57.34 FG	0.50 ± 0.05 BCD	160.72 ± 56.39 FG
S. uvarum	2401	70.76 ± 12.89 BCD	55.83 ± 7.83 CDEFG	83.51 ± 18.44 FGH	485.70 ± 60.40 BCD	911.34 ± 145.23 B	0.57 ± 0.07 BCD	911.66 ± 142.80 B
S. uvarum	2071	30.65 ± 7.57 EF	40.60 ± 5.69 G	56.81 ± 16.0 H	320.02 ± 29.51 DEFGH	434.32 ± 136.17 DE	0.39 ± 0.03 D	442.58 ± 133.90 DE
S. uvarum	2061	31.84 ± 1.64 EF	40.55 ± 6.28 G	101.68 ± 25.6 DEFGH	402.24 ± 151.09 CDEFG	246.06 ± 90.58 DEFG	0.45 ± 0.08 CD	257.46 ± 89.07 DEFG
S. uvarum	2046	43.61 ± 15.19 DEF	47.99 ± 8.93 FG	83.97 ± 21.60 FGH	459.05 ± 92.84 BCDE	813.25 ± 137.77 B	0.46 ± 0.10 CD	815.20 ± 135.47 B
S. cerevisiae	2265	40.92 ± 5.91 DEF	80.59 ± 8.62 ABC	147.85 ± 16.34 BCDE	247.14 ± 17.84 FGHI	67.17 ± 13.19 FG	0.69 ± 0.06 BCD	81.55 ± 12.97 G
S. cerevisiae	2273	30.44 ± 7.57 EF	45.83 ± 7.78 FG	63.54 ± 15.44 GH	113.50 ± 23.76 I	0.009 ± 0.01 G	0.45 ± 0.08 CD	15.50 ± 0.01 G
S. cerevisiae	2003	21.63 ± 5.0 F	68.59 ± 11.4 BCDEF	153.89 ± 26.90 BCD	345.91 ± 39.18 DEFGH	168.46 ± 38.26 EFG	0.73 ± 0.23 BC	181.15 ± 37.63 EFG
S. cerevisiae	2095	43.44 ± 10.08 DEF	51.09 ± 4.62 EFG	95.96 ± 10.40 DEFGH	177.46 ± 18.36 HI	21.61 ± 19.33 G	0.61 ± 0.10 BCD	36.75 ± 19.01 G
S. uvarum	2083	25.70 ± 1.83 EF	39.03 ± 4.13 G	70.54 ± 13.79 GH	223.80 ± 52.53 GHI	269.30 ± 78.80 DEFG	0.36 ± 0.07 D	280.31 ± 77.48 DEFG

Table 2. *Cont.*

Yeast Species	Strain no.	Ethyl Esters from Branched Acids(EEBA)				C6-Alcohols(C6-OH)	
		Ethyl isobutyrate	Ethyl 2-methylbutyrate	Z-3-Hexenol	E-3-Hexenol	Hexanol	
S. uvarum	2104	2.56 ± 0.34 CDEF	0.46 ± 0.04 A	140.19 ± 6.51 ABCD	4.39 ± 0.46 A	1731.77 ± 89.15 BCDE	
S. uvarum	2120	4.92 ± 1.69 ABC	0.64 ± 0.20 A	176.86 ± 25.80 ABC	5.00 ± 0.52 A	2171.38 ± 229.22 ABC	
S. uvarum	2128	4.93 ± 0.76 ABC	0.64 ± 0.09 A	151.29 ± 15.1 ABCD	3.94 ± 0.41 A	1797.3 ± 218.7 BCDE	
S. uvarum	2176	6.69 ± 0.60 A	0.74 ± 0.03 A	148.12 ± 17.15 ABCDE	4.58 ± 0.38 A	1869.75 ± 83.91 ABCDE	
S. uvarum	2186	4.65 ± 0.58 ABCD	0.57 ± 0.09 A	154.85 ± 2.82 ABCD	4.65 ± 0.36 A	1967.90 ± 44.01 ABCD	
S. uvarum	2204	4.31 ± 0.21 ABCDE	0.56 ± 0.05 A	168.81 ± 12.2 ABC	5.42 ± 0.40 A	2208.67 ± 161.80 ABC	
S. uvarum	2216	5.05 ± 0.34 ABC	0.66 ± 0.06 A	181.20 ± 10.05 AB	5.53 ± 0.92 A	2234.25 ± 185.44 ABC	
S. cerevisiae	2349	2.81 ± 0.48 CDEF	0.36 ± 0.09 A	191.59 ± 7.24 A	5.69 ± 0.30 A	2539.14 ± 96.89 A	
S. uvarum	2376	4.28 ± 0.83 ABCDE	0.54 ± 0.14 A	158.60 ± 10.66 ABCD	5.08 ± 0.45 A	1803.84 ± 127.14 BCDE	
S. uvarum	2402	4.67 ± 1.13 ABCD	0.52 ± 0.12 A	187.92 ± 10.84 A	5.86 ± 0.72 A	2417.87 ± 172.85 AB	
S. uvarum	2401	6.58 ± 1.62 A	0.62 ± 0.14 A	181.73 ± 11.85 AB	5.16 ± 0.94 A	2242.75 ± 140.58 ABC	
S. uvarum	2071	5.53 ± 1.41 AB	0.37 ± 0.12 A	121.63 ± 22.1 CDE	3.67 ± 0.54 A	1626.81 ± 308.32 CDE	
S. uvarum	2061	5.14 ± 1.10 ABC	0.44 ± 0.09 A	148.37 ± 29.6 ABCD	3.87 ± 0.44 A	1841.51 ± 434.03 ABCDE	
S. uvarum	2046	4.09 ± 1.01 ABCDE	0.43 ± 0.15 A	108.39 ± 14.9 DE	3.75 ± 0.78 A	1353.55 ± 238.15 DE	
S. cerevisiae	2265	2.13 ± 0.24 DEF	2.28 ± 0.38 A	147.23 ± 9.61 ABCDE	4.26 ± 0.97 A	1883.32 ± 143.94 ABC	
S. cerevisiae	2273	1.20 ± 0.46 F	2.17 ± 0.46 A	129.62 ± 10.8 BCDE	3.90 ± 0.35 A	1749.91 ± 106.43 BCDE	
S. cerevisiae	2003	2.24 ± 0.60 DEF	2.35 ± 0.31 A	129.60 ± 28.17 BCDE	4.02 ± 1.21 A	1622.19 ± 377.31 CDE	
S. cerevisiae	2095	1.97 ± 0.51 EF	2.28 ± 0.35 A	128.68 ± 36.2 BCDE	3.96 ± 0.28 A	1655.67 ± 447.66 CDE	
S. uvarum	2083	3.23 ± 0.20 BCDEF	2.26 ± 0.32 A	92.40 ± 20.39 E	6.43 ± 4.27 A	1222.57 ± 292.76 E	

Yeast Species	Strain no.	VolatilePhenols(VP)					
		4-Ethyl phenol	4-Ethyl guaiacol	4-Vinyl guaiacol	4-Vinyl phenol	Guaiacol	
S. uvarum	2104	5.74 ± 0.93 BCD	0.36 ± 0.15 AB	609.05 ± 116.3 ABCDE	212.40 ± 37.09 ABCDE	1.10 ± 0.09 A	
S. uvarum	2120	7.35 ± 1.12 ABCD	0.29 ± 0.09 AB	1087.30 ± 325.63 E	29.27 ± 19.12 E	1.45 ± 0.72 A	
S. uvarum	2128	6.83 ± 0.75 BCD	0.29 ± 0.01 AB	1456.3 ± 17.87 BCDE	153.97 ± 198.2 BCDE	2.40 ± 2.10 A	
S. uvarum	2176	12.91 ± 0.31 A	0.30 ± 0.02 AB	2001.80 ± 161.37 E	39.46 ± 7.96 E	3.42 ± 4.89 A	
S. uvarum	2186	5.75 ± 1.31 BCD	0.23 ± 0.03 AB	1175.05 ± 273.51 E	39.72 ± 11.56 E	2.09 ± 2.22 A	
S. uvarum	2204	7.29 ± 0.60 ABCD	0.26 ± 0.02 AB	1216.44 ± 149.23 DE	47.71 ± 8.01 DE	0.91 ± 0.24 A	
S. uvarum	2216	6.15 ± 0.30 BCD	0.32 ± 0.01 AB	2121.76 ± 107.63 ABCD	413.98 ± 324 ABCD	1.80 ± 1.32 A	
S. cerevisiae	2349	4.10 ± 0.96 CD	0.29 ± 0.05 AB	1632.24 ± 134.28 A	561.04 ± 26.01 A	0.89 ± 0.81 A	
S. uvarum	2376	12.92 ± 2.33 A	0.29 ± 0.03 AB	1199.19 ± 156.97 CDE	81.33 ± 14.71 CDE	11.91 ± 17.74 A	
S. uvarum	2402	5.07 ± 0.78 BCD	0.34 ± 0.08 AB	1599.26 ± 121.19 AB	473.78 ± 33.05 AB	0.60 ± 0.40 A	
S. uvarum	2401	7.56 ± 1.09 ABCD	0.31 ± 0.03 AB	1793.76 ± 158.55 BCDE	187.33 ± 259.49 BCDE	3.38 ± 0.13 A	
S. uvarum	2071	2.66 ± 0.90 D	0.23 ± 0.04 AB	884.52 ± 148.71 ABCDE	327.35 ± 47.5 ABCDE	8.99 ± 14.11 A	
S. uvarum	2061	9.06 ± 5.41 ABC	0.22 ± 0.03 AB	1085.67 ± 366.57 ABCDE	274.44 ± 206 ABCDE	3.35 ± 0.13 A	
S. uvarum	2046	10.82 ± 2.52 AB	0.18 ± 0.03 B	653.55 ± 113.66 CDE	82.09 ± 35.37 CDE	4.82 ± 4.44 A	
S. cerevisiae	2265	6.57 ± 1.83 BCD	0.40 ± 0.04 A	1608.44 ± 145.22 ABC	418.23 ± 29.5 ABC	1.08 ± 0.17 A	
S. cerevisiae	2273	6.46 ± 0.95 BCD	0.17 ± 0.04 B	610.93 ± 127.20 ABCDE	228.56 ± 43.8 ABCDE	0.93 ± 0.18 A	
S. cerevisiae	2003	8.12 ± 1.82 ABCD	0.24 ± 0.07 AB	680.06 ± 196.04 ABCDE	246.31 ± 78.99 ABCDE	1.09 ± 0.48 A	
S. cerevisiae	2095	8.33 ± 2.80 ABCD	0.28 ± 0.18 AB	36.01 ± 10.53 E	18.18 ± 2.88 E	0.75 ± 0.02 A	
S. uvarum	2083	9.58 ± 1.53 ABC	0.21 ± 0.06 AB	357.71 ± 34.07 E	26.92 ± 2.24 E	6.30 ± 1.32 A	

When comparing the relative values in the ethyl fatty acid ester group, all yeasts produced from 19–40% of ethyl octanoate. *S. uvarum* strains produced, in general, more ethyl esters of decanoic and dodecanoic acids (18–36% of each), while ethyl esters of propanoic, butanoic, and hexanoic acids were less abundant in this group of ciders (from 2–10% of all ethyl esters). The later esters were more abundant in *S. cerevisiae* strains and less in the contribution of decanoate or dodecanoate ethyl esters to the total ethyl ester of fatty acids fingerprint.

There was no or low significant difference in the concentration of the acetate esters between the yeast strains. The group of acetate esters was represented by propyl acetate, isobutyl acetate, butyl acetate, isoamyl acetate, Z-3-hexenyl acetate, E-2-hexenyl acetate, ethyl phenyl acetate, hexyl acetate, and octyl acetate. The major acetate ester was butyl acetate, which varied between 63.5–220.8 µg/L among yeasts but was generally quite comparable among samples, followed by isoamyl acetate (12.0–118.73 µg/L) and hexyl acetate (8–39 µg/L).

We determined five different volatile phenols in our ciders, namely 4-ethylphenol, 4-ethylguaiacol, 4-vinylguaiacol, 4-vinylphenol, and guaiacol. These five volatile phenols together accounted, on average, for between 1% (*S. cerevisiae* 2303) and 40% (*S. cerevisiae* 2365) of the measured volatile fingerprint of the ciders.

The major volatile phenolics were 4-vinylguaiacol and 4-vinylphenol, the presence of which varied among samples due to differences in yeast metabolic characteristics. Moreover, 4-vinylguaiacol varied from 36 to 2121 µg/L and 4-vinylphenol from 18 to 473 µg/L.

Ethyl phenols varied in low concentrations, below the odor threshold (OT). The OT determined in water/10% ethanol solution at pH 3.2 for 4-ethylguaiacol and 4-ethylphenol was 33 and 440 µg/L, respectively [40]. Their presence imparts equine, peasant, smoky, and medicinal aromatic odors when present above their OTs concentrations.

3.5. Correlations of AAs and Physico-Chemical Parameters with Aroma Compound Formation
Data Analysis and Data Configuration

As described in previous sections, we determined 54 different compounds (Table 1 (measured H_2S production), Table 2, Tables S6 and S7) in the resulting ciders. Strain *S. cerevisiae* 2303 was identified as an outlier and omitted in further statistical analyses (see Appendix A). All 54 continuous variables (summarized in Table 1 (measured H_2S production), Table 2, Tables S6 and S7) associated with 19 different yeast strains were selected to draw a heat map (Figure 3). In Figure 3, the rows represent the measured compounds (and the corresponding clusters), and the columns represent the different yeast strains (and the resulting clusters) used for the single-strain fermentations in the fermentation experiment.

Before generating the heat map, the data were standardized to a value between 0 and 100 using the following equation:

$$\text{new score} = (\text{score} - \min(x))/(\max(x) - \min(x)) * 100, \tag{1}$$

The dendrograms on the top and left side of the heatmap show how the variables and the rows are clustered independently (they indicate the degree of similarity between the variables or yeast strains). Color coding is used to show the values of each variable in the dataset and also to show clusters of variables or samples that have similar expression patterns. The color scale indicates the range of values for each variable, with low values represented by dark colors and high values represented by light colors. Variables and/or samples (i.e., yeast strains) that are more similar to each other are grouped in the same cluster/block. The height of the dendrogram branches represents the degree of similarity between the variables or samples, with lower heights indicating a higher degree of similarity.

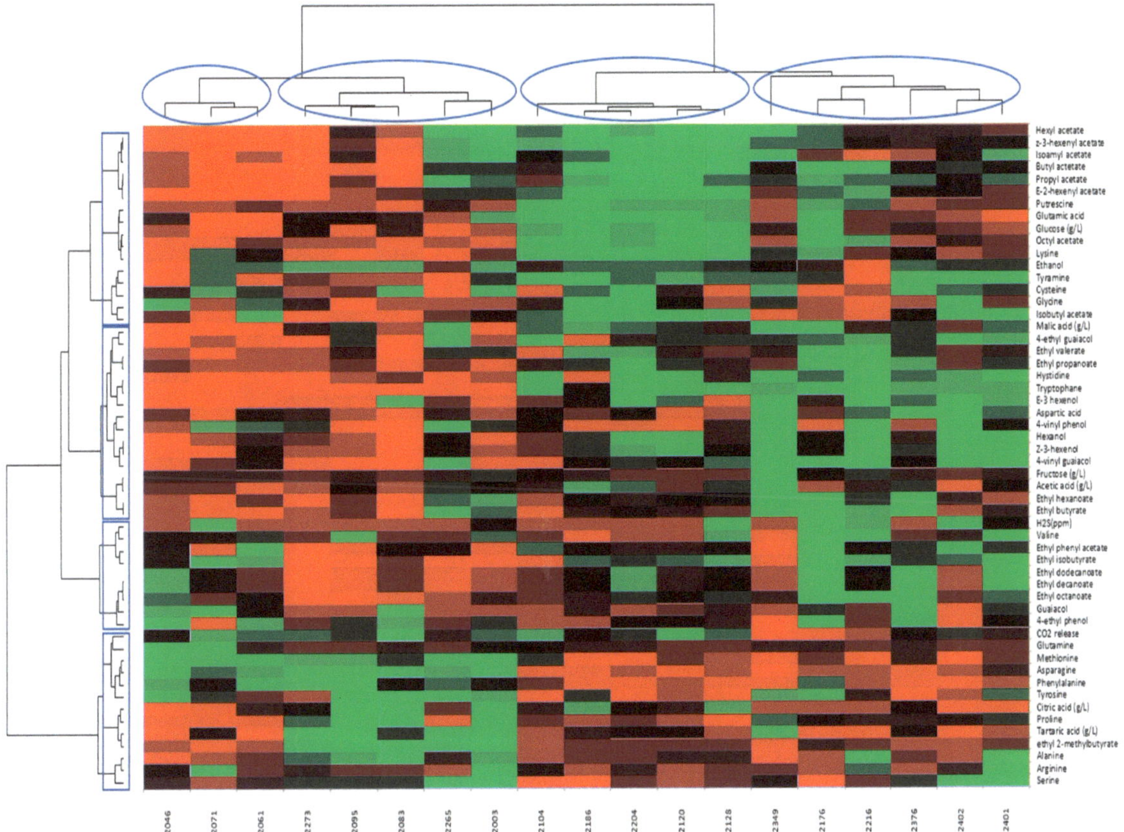

Figure 3. The heatmap and cluster analysis of a total of 54 continuous variables: 26 aroma compounds (i.e., the response variables grouped into 5 chemical classes/groups in Y-block data used in PLS regression) and 28 variables including amino acids, sugars, ethanol, organic acids, etc. (i.e., the predictor variables which made the X-block data used in PLS regression) of 19 *Saccharomyces* strains used in the fermentation trial.

Although all patterns in the heat map may indicate a relationship between rows and columns, we look for rectangular areas that are approximately the same color. This indicates a group of rows correlated with the corresponding group of columns. According to the results of the heatmap analysis, the total of 54 variables associated with the 19 yeasts used for fermentation can be divided into four main classes, shown on the left side of the heatmap. In addition, the yeast strains are also divided into four main groups, shown in the upper part of the heatmap (Figure 3).

The first class of variables (i.e., the first class on the upper left) consists mainly of aromatic compounds (hexyl acetate, Z-3-hexenyl acetate, isoamyl acetate, butyl acetate, octyl acetate, propyl acetate, E-2-hexenyl acetate, octyl acetate); in this class are the amino acids glutamic acid and lysine and the biogenic amines putrescine and glucose. The second class of variables (i.e., the second class on the upper left) includes malic acid, 4-ethyl guaiacol, 4-ethyl valerate and ethyl butyrate. The third class of variables (i.e., the third class at the top left) includes H_2S, the amino acid valine, and the aromatic compounds ethyl phenyl acetate and 4-ethyl phenol. The fourth class of variables (i.e., the fourth class at the top left) consists of CO_2 release, glutamine, methionine, and serine.

It was found that the compounds of the first class of variables on the upper left in cider fermented with *S. uvarum* strains 2046, 2071, and 2061 (i.e., the first group of yeast strains on the upper left) had the lowest content. The highest levels of the same class of variables/compounds were found in cider fermented with *S. uvarum* 2104, 2186, 2204, 2120, and 2128 (i.e., the third group of yeasts above). In an analogous analysis, the compounds of the second class of variables are found to have the second lowest content in fermented strains *S. uvarum* 2046, 2071, and 2061 (i.e., the first group at the top left).

The second yeast group with 4 *S. cerevisiae* strains (2273, 2095, 2265, and 2003) and one *S. uvarum* strain (2083) showed predominantly high levels of the fourth class of variables/compounds: relatively high CO_2 release, high levels of the remaining amino acids glutamine, methionine, asparagine, phenylalanine, tyrosine, proline, alanine, serine, and tartaric and citric acids, and the highest levels of the aromatic compound ethyl 2-methylbutyrate.

Based on the observed color scales, it can also be said that the fourth yeast group (i.e., the fourth group at the top from the left), consisting of one *S. cerevisiae* strain (2349) and five *S. uvarum* strains (2176, 2216, 2376, 2402, and 2401), has the highest content of the second class of variables (relatively high contents of malic acid and the other amino acids histidine and tryptophan, high contents of volatile phenols 4-vinylphenol, 4-vinylguaiacol and 4-ethylguaiacol, relatively high contents of all three C6 alcohols and ethyl esters ethyl valerate, ethyl propanoate, ethyl hexanoate, and ethyl butyrate).

3.6. Relationships between Aroma Compounds and Variables

3.6.1. Principal Component Analysis (PCA)

To perceive an initial configuration of our data/variables and to simplify the dataset by identifying possible patterns and relationships between all variables, a principal component analysis (PCA) was performed and validated using the correlation matrix in IBM SPSS Statistics 27 and GraphPad Prism 9.5.1 (see Appendix A for additional description). In the PCA analysis, ethyl esters of fatty acids, ethyl esters of branched acids, acetate esters, C6 alcohols, and volatile phenols were the five chemical classes/groups of aroma compounds, and the other 28 features, including amino acids, sugars, ethanol, organic acids, etc., were entered as other important variables.

From the eigenanalysis of the correlation matrix related to PCA, nine principal components (represented by PC or F) were extracted, and 54.0% of the variance in the data set was explained by the first three components (F1 = PC1 = 27.9%, F2 = PC2 = 13.8%, and F3 = PC3 = 12.4%). When we refer to explained variance in terms of the PCs, we are referring to the proportion of variance in the entire collection of response and predictor variables that is explained by the PCs. In our results, the first nine principal components have eigenvalues greater than 1 (see the scree plot in Figure 4). These nine components explained approximately 89% of the variation in the data (Figure 4). However, since the cumulative variance of 54.1% in the first three components does not report the adequate amount of variation we expected, we performed another statistical analysis, the partial least squares (PLS) regression analysis.

The Component Plot in Rotated Space (CPRS) shown in Figure 4 displays the scores of the first three PCs, which capture most of the variance in our data set. This plot is a graphical representation of the results of PCA with orthogonal rotation. Each point on the plot represents a variable in our data set, and the position of the point in the new rotated space is determined by the scores of the observation on the PCs. In other words, the CPRS shows how the variables in our dataset are related based on the underlying patterns identified by the PCA method. Variables that are close to each other on the graph (e.g., lysine and putrescine) have similar principal component values, indicating that they share similar underlying patterns. Observations that are far apart on the graph (e.g., lysine and serine) have different values on the PCs, indicating that they have different underlying patterns. The CPRS can be useful in identifying clusters or groups of variables in the data set that have similar patterns. The problem of clustering yeast strains can also be explored in the PCA Bootstrap hulls (implemented in XLSTAT 2023) in Figure 5 and the partial least squares (PLS) regression analysis (implemented in Minitab 21 and XLSTAT 2023) in Figure 6.

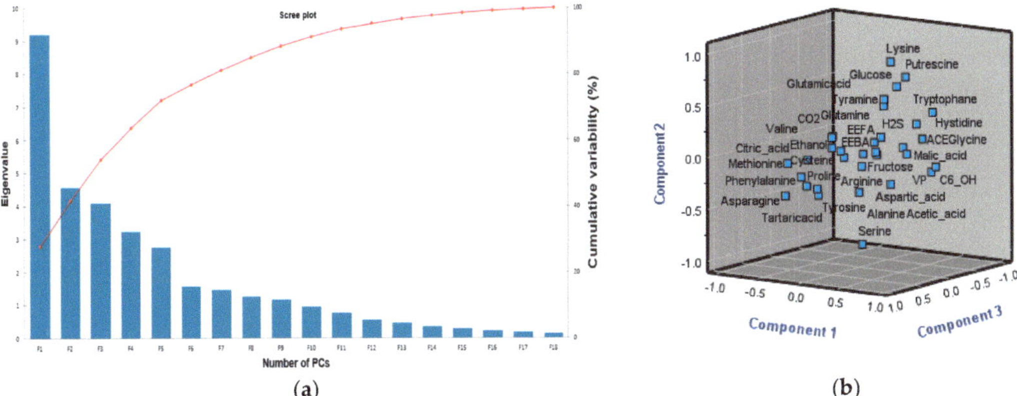

Figure 4. The scree plot (**a**) and the component 3D plot in rotated space (**b**) allied with the PCA. The scree plot orders the eigenvalues from largest to smallest.

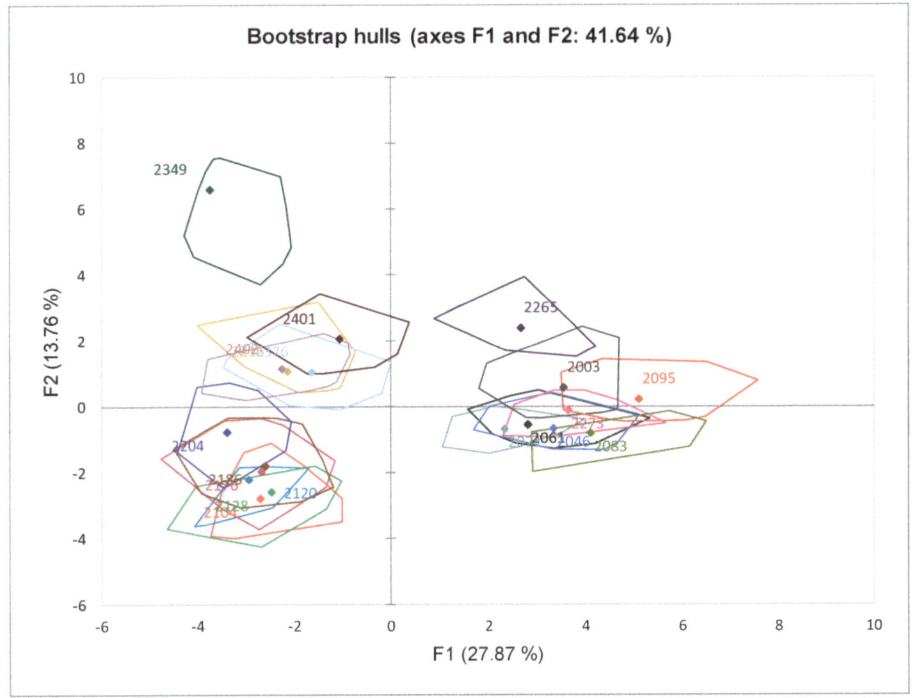

Figure 5. The Bootstrap hulls: the graphical method for assessing the significance of the yeast strains in the PCA method and observing the possible clusters in the fermented yeast strains in our dataset.

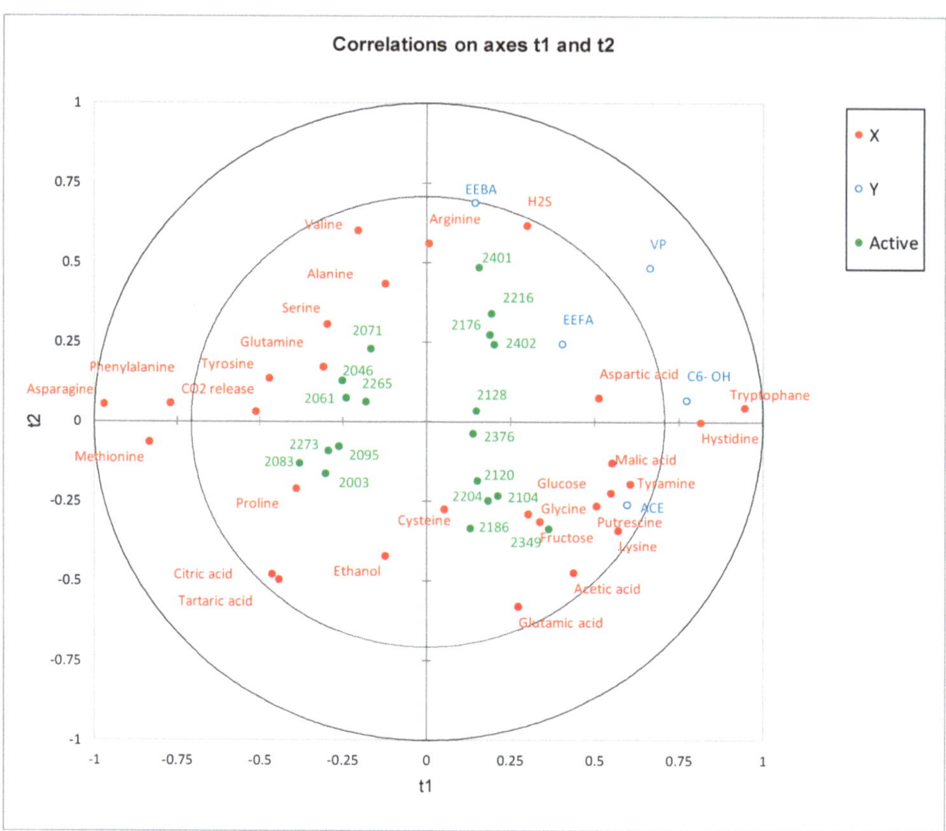

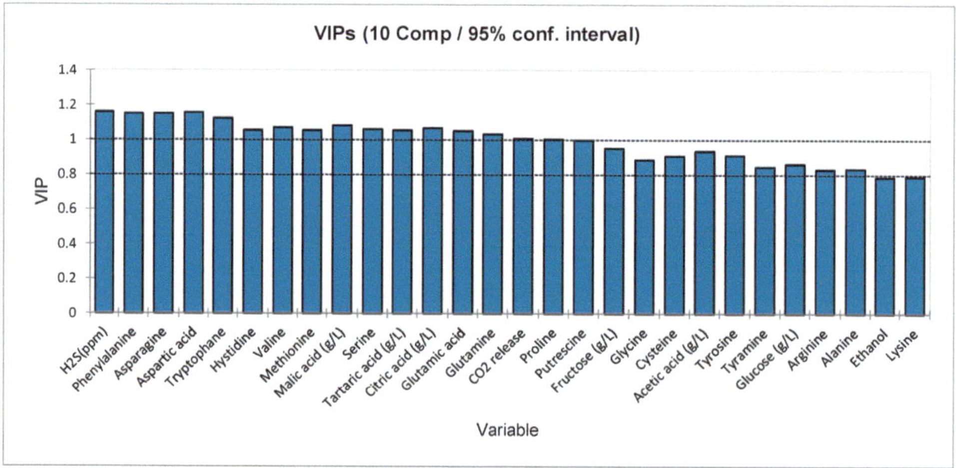

Figure 6. The 2D configuration, correlations, loading, clustering, and projection of observations (yeast strains) and all attributes and aroma groups of compounds ($p < 0.05$) of all 19 different yeasts on the first two components (**top**), and the VIP values associated with the 28 predictor variables from the PLS regression model with 10 LVs (**bottom**). Legend: Ethyl esters from branched acids—EEBA; Ethyl esters from fatty acids—EEFA; Acetate esters—AE; Volatile phenols—VP; C6-alcohols—C6-OH.

Bootstrap hulls involve generating a series of bootstrap samples from the original dataset and then computing the PCs for each bootstrap sample. We use them to assess the significance of individual observations/yeasts in our dataset and to see the possible clusters in our 19 fermented yeast strains (in the first two PCs). Yeast strains that consistently fall within the convex hull of the scores for each bootstrap sample are considered statistically significant, while yeast strains that consistently fall outside the convex hull are considered statistically non-significant. By using bootstrap hulls in PCA, we identified influential observations (i.e., the yeast strains) in the dataset that could affect the results of the analysis. The results of the analysis can be interpreted based on the location of the observations relative to the bootstrap hulls. Observations that consistently fall outside the convex hull may be outliers or noise in the data. By identifying influential observations, it is possible to determine if they are outliers that need to be removed from the analysis or if they represent important patterns in the data that should be investigated further. However, as can be seen in Figure 5, our observations consistently fall within the designated convex envelope. Therefore, they should all be considered statistically important and reliable; this is because we have already identified and removed the outliers from our data set.

3.6.2. Partial Least Squares (PLS) Regression Analysis

In this section, the ferments/ciders produced with 19 different yeast strains were further analyzed. Since this is a high-dimensional data set, to further determine the associations between the aroma groups of attributes (y variables, n = 5) and the other physicochemical parameters (x variables, n = 28) in ciders from 19 tested strains, we used partial least squares (PLS) regression—a multivariate statistical analysis used for both regression and classification tasks. To prepare for the PLS regression, we first had to distinguish between the response variables (y variables) and the predictor variables (x variables). Therefore, the five chemical classes/groups of aroma compounds (ethyl esters of fatty acids, ethyl esters of branched acids, acetate esters, C6 alcohols, and volatile phenols) were considered as the response variables that formed the Y-block in the PLS regression. The amino acids, sugars, ethanol, organic acids, etc., were entered and considered as predictor variables used to develop the X-block.

Thus, the goal was to find out the relationship between two blocks, the Y-block and the X-block (i.e., the set of response variables and the predictor variables of interest). PLS regression was used to identify (1) the underlying factors responsible for the variation in the data; (2) the contributions of 28 traits/predictors, including amino acids, sugars, ethanol, acids, etc. (i.e., the variables forming the X-block) and to correlate and discover the five groups of cider aroma attributes (i.e., the variables forming the five groups in the Y-block) from 19 yeast strains used for fermentation (Table 3).

The method works by finding the linear combinations of the predictor variables X (also called latent variables (LVs)) that are most strongly related to the response variables Y. This is done by maximizing the covariance between the X-block and the Y-block while ensuring that the predictor variables are orthogonal (i.e., uncorrelated) to each other. Subsequently, the PLS regression analysis was used to determine which variables contributed most to the Y-block and the X-block contributed most to the variation in the data. Finally, PLS regression analysis indicated how the variables were correlated with each other and then lumped the variables into a new latent variable (LV). Our results show that our X-block variables mentioned above are strongly correlated with each of the groups of aroma compounds, for which large correlations were always found in different numbers of LVs (Table 3).

The results of our PLS regression model show how the amino acids, sugars, ethanol, acids, etc., contributed to the fermentations of the yeasts as well as to the chemical classes of the aroma compounds, and they also lead to some important findings. Table 3 summarizes the results of this analysis and shows that our PLS regression model (built with 10 LVs) includes 89% of the variation in the X-block data and 93% in the Y-block data. Although the optimal number of latent variables (LVs) was analytically set at ten by the PLS regression method, even with eight LVs, our PLS model can well explain the variation in the X-block and Y-block data, that is, 82% of the variation in the X-block and 88% in the Y-block. PLS regression analysis achieved a correlation coefficient of $R^2 \geq 0.85$ for all aroma classes of compounds and captured $R^2X \geq 85\%$ of the variation in the X-block data (i.e., predictor variables amino acids, sugars, ethanol, acids, etc.) and $R^2Y \geq 85\%$ of the variation in the Y-block data (i.e., aroma classes of compounds), implying a robust linear relationship between the measured aroma classes of compounds and the predictors in all 19 yeast strains we tested in this study.

Table 3. Summary of results from the partial least squares (PLS) regression analysis.

PLS Regression: Model Selection and Validation for Different Families of Aroma Attributes.				
Y-Block (i.e., Aroma Groups of Attributes) (y Variables, n = 5)	**Number of Latent Variables (LVs)**	**R-Sq (R^2)**	**Captured Cumulative X-Block Variance (R^2X cum)**	**Captured Cumulative Y-Block Variance (R^2Y cum)**
Ethyl esters from fatty acids		**0.96**		
Ethyl esters of branched acids		**0.92**		
Acetate esters	10 *	**0.95**	0.89	0.93
C6-alcohols		**0.86**		
Volatile phenols		**0.94**		
Ethyl esters from fatty acids		0.96		
Ethyl esters of branched acids		0.91		
Acetate esters	9	0.87	0.86	0.90
C6-alcohols		0.86		
Volatile phenols		0.92		
Ethyl esters from fatty acids		0.87		
Ethyl esters of branched acids		0.91		
Acetate esters	8	0.86	0.82	0.88
C6-alcohols		0.86		
Volatile phenols		0.92		

* The optimal number of latent variables (LVs) was systematically chosen to 10 by the PLS regression method.

Besides much useful information that Figure 6 can provide, it shows the correlation coefficients between each predictor variable and the response variable, as well as the loading weights that indicate the importance of each predictor variable in the model.

As can be seen in Figure 6, the aroma attributes of ethyl esters of fatty acids, ethyl esters of branched acids, volatile phenols, and C6 alcohols can be used to distinguish the aroma of ciders produced with yeast strains *S. uvarum* 2401, 2216, 2176, and others in the same cluster. Similarly, the aroma properties of acetate esters can be used to discriminate the aroma of ciders produced with yeast strains *S. cerevisiae* 2349, *S. uvarum* 2186, and others in the same cluster. At the same time, yeast strain 2349 was well associated not only with acetate esters but also with the amino acids glutamic acid and glycine, the biogenic amine putrescine, and the sugars fructose and glucose.

The ciders of *S. uvarum* 2046 and *S. cerevisiae* 2265 showed a high correlation with some attributes, such as the amino acids tyrosine, serine, and glutamine.

The analysis showed that the amino acids aspartic acid, tryptophan, and histidine were highly correlated, with tryptophan having greater importance in the model because it had greater (positive) loadings in the first component. It was also observed that for CO_2 release, the amino acids asparagine and methionine were also strongly correlated, with asparagine being of greater importance in the model as it had greater (negative) loadings in the first component. In addition, glucose, tyramine, and malic acid were correlated, but only malic acid was important in the model. On the first component, predictors such as aspartic acid, malic acid, and serine had similar absolute loadings, suggesting that they were equally important. On the second component, H_2S production, glutamic acid, valine, tartaric acid, and citric acid had similar absolute loadings, indicating that they were equally important (see Figure A4).

In addition, the values for the importance of the variables in the projection (VIP) were also obtained in the PLS regression model and shown in Figure 6. These values are a good measure of the importance of each predictor variable in our PLS regression model. They are calculated by considering both the amount of variation in the response variable (five chemical groups of aroma compounds) explained by each latent variable (LV) and the importance (i.e., loading) of each predictor variable in this LV. The VIP values for the 16 predictor variables were ≥ 1, including H_2S production, phenylalanine, asparagine, aspartic acid, tryptophan, histidine, valine, methionine, malic acid, serine, tartaric acid, citric acid, glutamic acid, glutamine, CO_2 release, and proline. These predictor variables with high VIP scores are more important to us than those with low VIP scores. Moreover, these VIP scores are also used for variable selection by selecting only the variables with the highest VIP scores for inclusion in the model (namely, those with VIP ≥ 1).

Finally, we proceeded with error analysis of our PLS model using the x-residual matrix plot. To this aim, we examined general patterns in the residuals and identified areas where problems exist. We then examined the x-residuals displayed in the output to determine which observations and predictors the model may be poorly describing. As can be seen in Figure 7, the PLS residual X-plot shows that the residuals are close to zero, indicating that our model does a good job of describing most of the variance in the predictors in our experimental analysis with cider made with 19 *Saccharomyces* strains. There is no specific line on the graph that deviates dramatically from the other lines; therefore, the model describes all observations/yeast strains (represented by lines) very well. At both points 16 and 28, which correspond to the two predictors lysine and H_2S, respectively, the lines are far apart. At point 16, the lines are slightly apart at the same point on the x-axis. Therefore, the model can still be considered a good statistical tool to describe the predictor at this point (i.e., lysine). However, at point 28, most of the lines diverge at the same point on the x-axis, and this means that the model can poorly describe the corresponding predictor at this point (i.e., H_2S). This could be related to the values obtained for H_2S, a sparse vector where 63% of the elements had a value of zero.

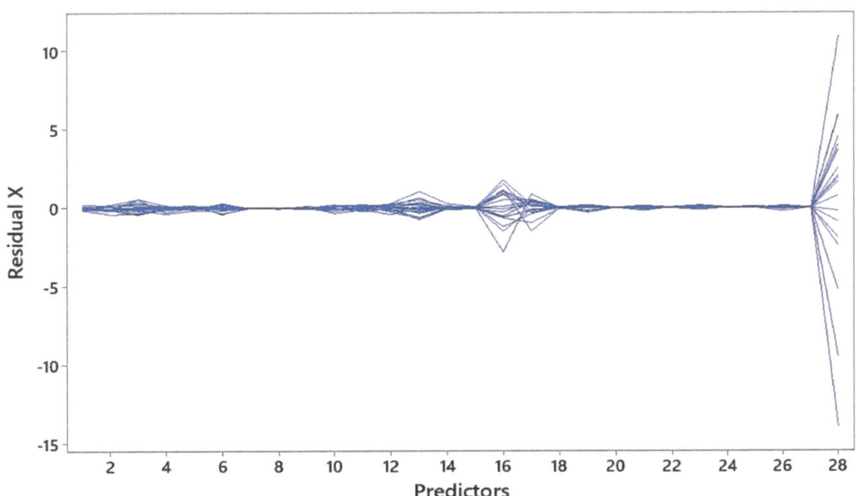

Figure 7. Error analysis using the Residual X-plot associated with our PLS regression model (#LV = 10). Each line in the graph represents an observation/yeast strain and has as many points as it has predictors.

4. Conclusions

A fermentation screening of six *S. cerevisiae* and fourteen *S. uvarum* strains isolated from ciders produced in Hardanger [4] provided information on metabolic capabilities with an emphasis on aroma production. This is a first selection small-scale fermentation experiment with chemical characterization and is a stepping stone for the selection process of indigenous yeasts from Hardanger with respect to their suitability for cider production.

Using the HPLC–UV/RI and HS–SPME/GC–MS methods for chemical characterization of cider produced with different strains tested, we determined seventeen AAs, two BAs, four organic acids, ethanol, glucose, fructose, and twenty-six volatile compounds, including eighteen esters (seven ethyl esters of fatty acids, two ethyl esters of branched acids, and nine acetate esters), three C6 alcohols, and five volatile phenols. In addition, in the current study, we also successfully implemented a new analytical approach for the simultaneous determination of AAs and BAs using the HPLC–UV system.

When statistical analyses were applied to the obtained chemical data, cluster analysis allowed us to divide the *Saccharomyces* strains into four main groups. The yeast groups differed in the production of aromatic components. Two groups produced few aromatic compounds. The other two groups, primarily consisting of *S. uvarum* strains, were good producers of aromatic components; one was characterized by the highest production of acetate esters, while the other exhibited the highest production of ethyl esters, volatile phenols, and C6 alcohols. Additionally, with PLS regression, we established a relationship between aroma compounds and predictor variables (AAs, BAs, organic acids, sugars, H_2S production, and CO_2 release), and the obtained VIP scores showed that the most important predictor variables affecting aroma compounds were 16, most of which belong to the following AAs: phenylalanine, asparagine, aspartic acid, tryptophan, histidine, valine, methionine, serine, glutamic acid, glutamine, and proline.

Further detailed studies on the representatives of the four yeast groups identified in our study during the fermentation process on larger scales, including sensory evaluation, are needed to find an alternative *Saccharomyces* yeast for potential cider production from Hardanger.

Supplementary Materials: The following supporting information can be downloaded at https://www.mdpi.com/article/10.3390/fermentation9090824/s1, Table S1: Yeast strains used in the study. Table S2: Calibration parameters for sugars, acids, and ethanol by HPLC–UV/RI method. Table S3: Gradient profile of HPLC-UV method for determination of amino acids and biogenic amines. Table S4: Calibration parameters for amino acids and biogenic amines detected by HPLC–UV method. Table S5: Physicochemical parameters of apple juice used in the study. Values represent the mean ± SD for three replicates. Table S6: Content of amino acids (mg/L) and biogenic amines (mg/L) in ciders produced from apple juice fermentation carried out by 19 *Saccharomyces* strains. Table S7: Content of organic acids, sugars, ethanol, and other measured parameters (CO_2 release) in ciders produced from apple juice fermentation carried out by 19 *Saccharomyces* strains.

Author Contributions: Conceptualization, A.H., B.M.V. and L.B.; Data curation, U.Č., I.Ø., B.M.V. and L.B.; Formal analysis, U.Č., M.M., T.R.V. and A.H.; Funding acquisition, I.Ø.; Investigation, U.Č. and L.B.; Methodology, L.B.; Project administration, I.Ø., B.M.V. and L.B.; Resources, I.Ø., B.M.V. and L.B.; Supervision, B.M.V. and L.B.; Validation, U.Č., M.M., T.R.V., A.H. and L.B.; Visualization, A.H. and L.B.; Writing—original draft, U.Č., A.H., B.M.V. and L.B.; Writing—review and editing, M.M., I.Ø., A.H., B.M.V. and L.B. All authors have read and agreed to the published version of the manuscript.

Funding: This research was funded by Regionalt Forskningsfond Vestland (RFF Vestland), Norway, through the research project Sider—Smaken av Hardanger, grant number 299284.

Institutional Review Board Statement: Not applicable.

Informed Consent Statement: Not applicable.

Data Availability Statement: Data is contained within the article or supplementary material.

Acknowledgments: The HS-SPME/GC-MS methods were previously established within the project Uncorking rural heritage: indigenous production of fermented beverages for local cultural and environmental sustainability benefits from a 1,501,000.00 € grant from Iceland, Liechtenstein and Norway through the EEA and Norway Grants Fund for Regional Cooperation.

Conflicts of Interest: The authors declare no conflict of interest. The funders had no role in the design of the study, in the collection, analyses, or interpretation of data, in the writing of the manuscript, or in the decision to publish the results.

Appendix A

To study the relationship between aroma attributes, amino acids, and other features of interest and to display the pairwise Pearson correlation coefficients (with a significance level alpha = 0.05) between all our 54 continuous variables (including the response and predictor variables), we employ the correlation matrix which is visualized in Figure A1. The values range from -1 to 1, with -1 (light red) indicating a strong negative correlation, 1 (light green) indicating a strong positive correlation, and 0 (dark red or green) indicating no correlation. In this image, positive correlations (i.e., green color) indicate that variables tend to increase or decrease together, while negative correlations (i.e., red color) indicate that variables tend to move in opposite directions.

Knowing that outliers can have a strong influence on the correlation coefficients and may distort the overall pattern of the data, we appropriately handled the outliers in our data set using the Grubbs test in XLSTAT 2023 (with a significance level of alpha = 0.05) before calculating the correlation (image) matrix. To interpret a correlation image matrix, we look at the values/colors in the (image) matrix and use them to conclude the relationships between variables. Cells in light green or light red indicate a strong relationship between the two corresponding variables. Cells in dark green or dark red indicate a weak or no relationship. As seen in Figure A1, there are not many cells that are colored light green or red, and therefore, there should not exist many groups of variables that have high positive or negative correlations. This indicates that there may not be any underlying patterns in the data.

However, we remark that the Pearson correlation (image) matrix, although it provides valuable insights into the relationships between variables in the data set, can only measure and represent the possible linear relationships between the variables. Therefore, if the relationship between variables is not linear, the Pearson correlation matrix may not accurately capture the true relationship. On the other hand, just because of a light green/red cell, we cannot conclude that the two corresponding variables are highly correlated, and it does not mean that one causes the other (i.e., correlation does not imply causation). Therefore, it is important to consider other factors, interpret the results carefully, and consider the broader context of the data and further statistical (multivariate) analysis, such as PCA and PLS regression analysis.

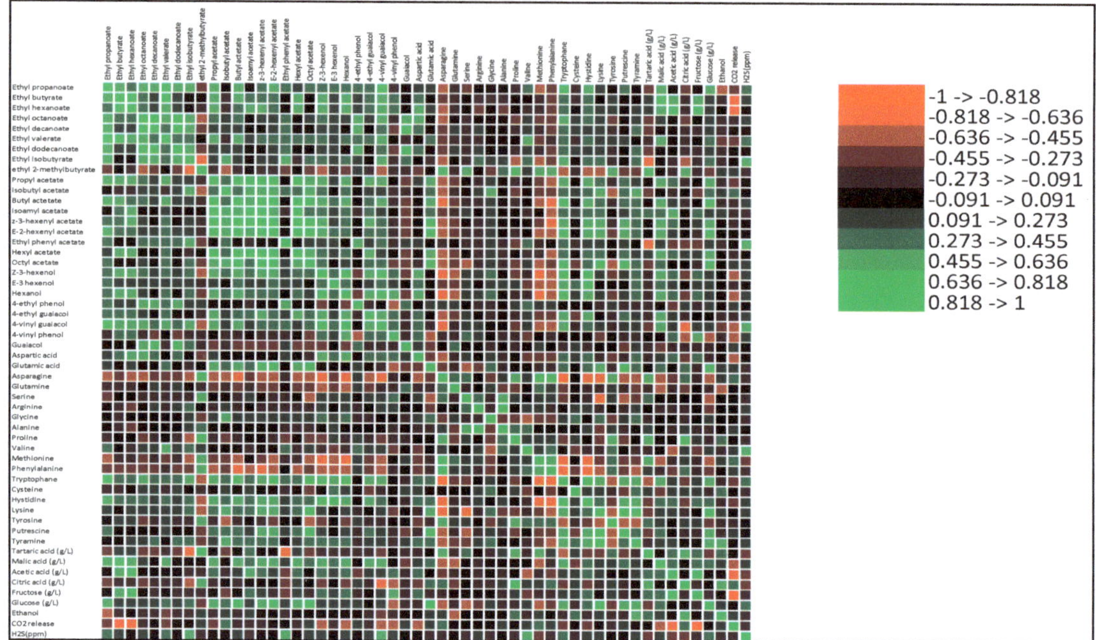

Figure A1. Image of the correlation matrix of a total of 54 continuous variables: 26 aroma compounds (i.e., the variables which will make the 5 groups in Y-block in the PLS regression analysis) and 28 features/predictors including amino acids, sugars, ethanol, acids, etc. (i.e., the variables which will make the X-block in the PLS regression analysis) of 19 yeast strains fermented. The correlations are according to Pearson, and the significance level alpha is considered 0.05.

To display the pairwise Pearson coefficients of determination of our 54 continuous variables (including the response and predictor variables), we use the matrix of R-squared values that are visualized in Figure A2, implemented in XLSTAT 2023 (with a significance level alpha = 0.05). The R-squared (R-Sq or R^2) value is a statistical measure that can be interpreted as the percentage of the variability in one variable that can be explained by the other variable. The matrix of coefficients of determination is similar to the correlation matrix, but instead of displaying the correlation coefficients between variables, it displays the R-squared values. The R-Sq values range from 0 to 1. An R-Sq value of 0 indicates that there is no relationship between the two variables, while an R-squared value of 1 indicates that all of the variability in one variable can be explained by the other variable.

Like the correlation matrix, each row and column of the matrix represents a different variable, and the values/color in the (image) matrix represent the R-Sq value between the two corresponding variables. The diagonal of the matrix represents the R-Sq value of each variable with itself, which is always equal to 1. Higher R-Sq values indicate a stronger relationship between variables, while lower R-squared values indicate a weaker relationship. However, it is important to note that the R-Sq value only measures the proportion of variance in one variable that can be explained by the other variable and does not provide information about the direction or causality of the relationship. As seen in Figure A2, there are not many cells that are colored in black/dark blue or dark brown, and hence, there do not seem to exist many groups of variables whose large proportion of their variance can be explained by the other groups of variables. Most of the cells are in the middle range (0.1, 0.3) and (−0.3, 0). This indicates that there may not be any underlying discoverable patterns in the data. However, we highlight that we will employ R-Sq values for model selection. More precisely, we exploit the R-Sq values to select the best model for our data set in the partial least squares (PLS) regression analysis. Models with higher R-Sq values are generally considered to be better at explaining the variation in the data (see Table 2).

Like the Pearson correlation (image) matrix, R-Sq values can be biased by outliers or other factors that affect the relationship between the variables. Therefore, before calculating the matrix of coefficients of determination, we aptly coped with the outliers in our data set using the Grubbs test in XLSTAT 2023 (with a significance level of alpha = 0.05). We remark that the (image) matrix of coefficients of determination, although it provides valuable insights into the factors that are driving variation in the data set, does not indicate causality. Like correlation coefficients, R-Sq values do not provide information about the direction or causality of the relationship between the variables. Therefore, it is important to consider other factors and conduct further (multivariate) analysis before making any conclusions.

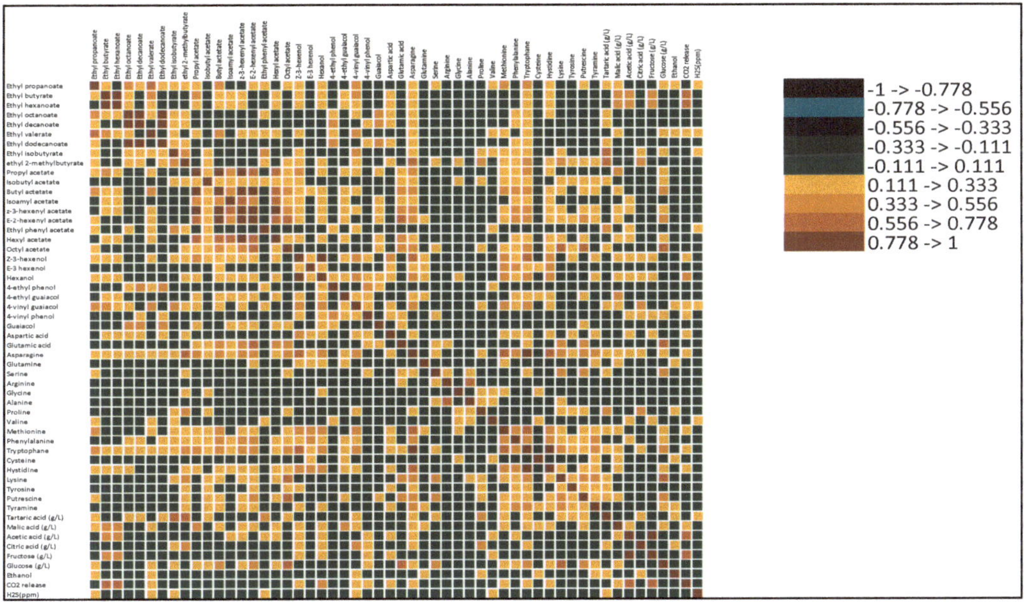

Figure A2. Image of the matrix of coefficients of determination of a total of 54 continuous variables: 26 aroma compounds (i.e., the variables which will make the 5 groups in Y-block in the PLS regression analysis) and 28 features/predictors including amino acids, sugars, ethanol, acids, etc. (i.e., the variables which will make the X-block in the PLS regression analysis) of 19 yeast strains fermented. The correlations are according to Pearson, and the significance level alpha is considered 0.05.

The PCA loading plot depicted in Figure A3 displays the loadings/associations of the variables on the PCs. Loadings represent the correlation between each variable and the PCs and can be used to interpret the underlying patterns in the data set. The correlation monoplot plots vectors pointing away from the origin to represent the original variables. The angle between the vectors is an approximation of the correlation between the variables. A small angle indicates that the variables are positively correlated, an angle of 90 degrees indicates that the variables are not correlated, and an angle close to 180 degrees indicates that the variables are negatively correlated. The length of the line and its closeness to the circle indicate how well the plot represents the variable. It is, therefore, unwise to make inferences about relationships involving variables with poor representation.

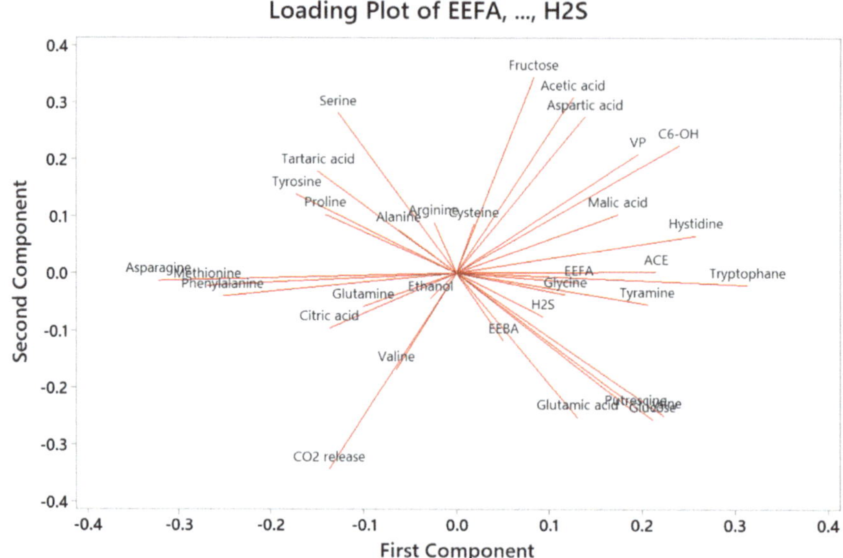

Figure A3. The PCA loading plot: the graphical method for displaying the associations of the variables on the PCs. Legend: Ethyl esters from branched acids—EEBA; Ethyl esters from fatty acids—EEFA; Acetate esters—AE; Volatile phenols—VP; C6-alcohols—C6-OH.

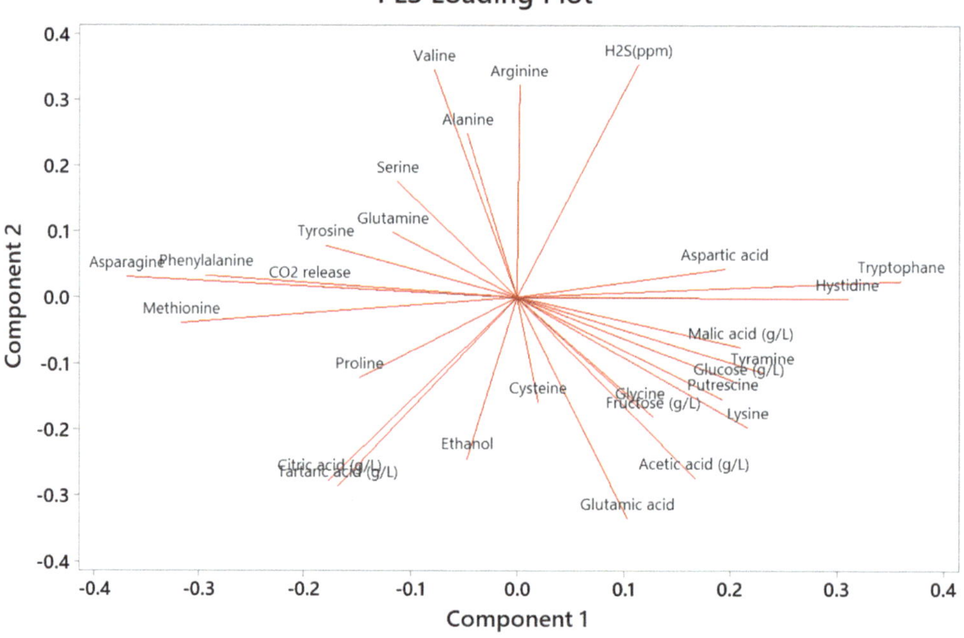

Figure A4. The PLS loading plot: the graphical method for displaying the associations of the variables on the LVs.

The PLS normal probability plot of the residuals (Figure A5) displays the standardized residuals versus their expected values when the distribution is normal. We use the normal probability plot of the residuals to verify the assumption that the residuals are normally distributed. All our obtained points in the graph fall randomly on both sides of the normal line, with no recognizable patterns in the points. This is verified in our analysis and is demonstrated in the picture.

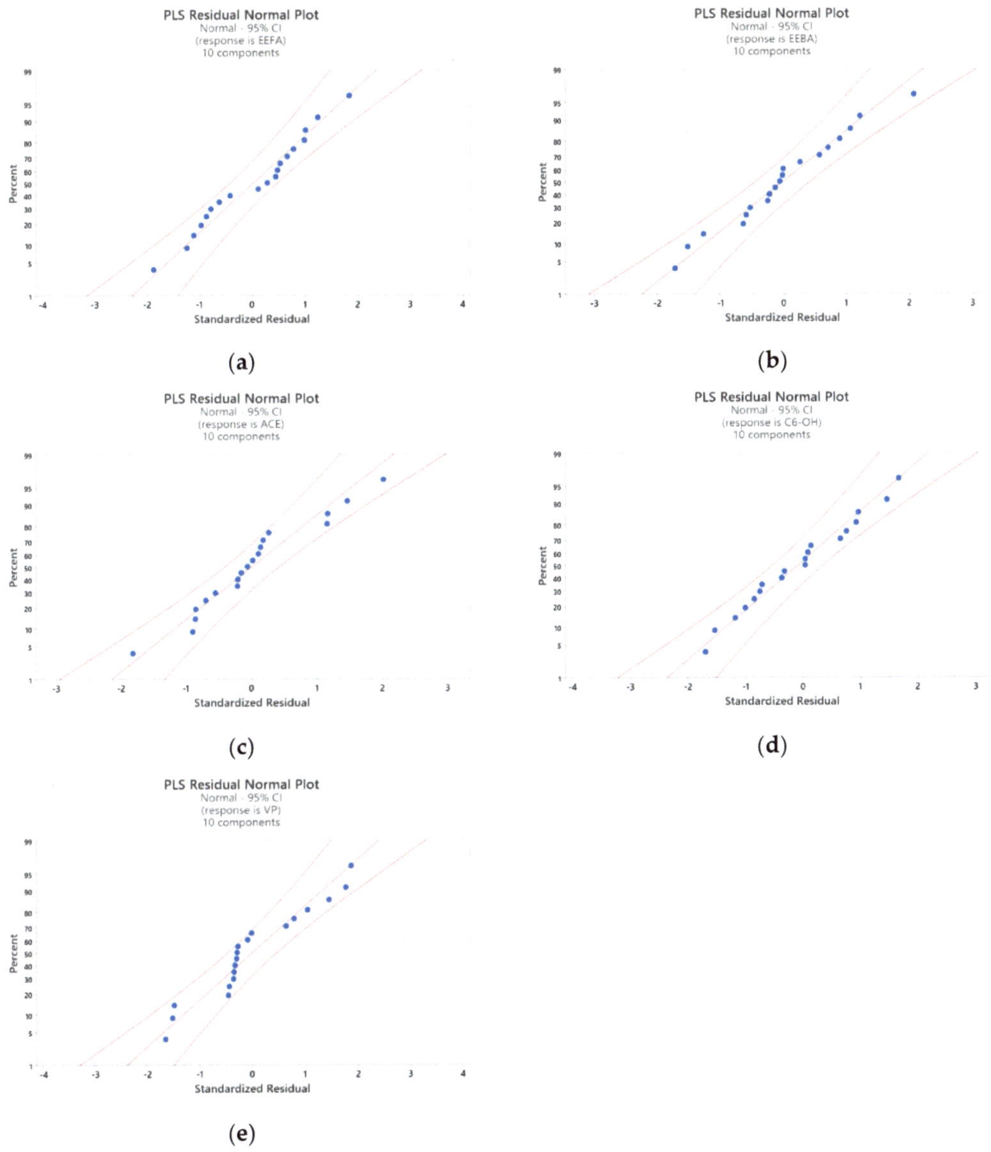

Figure A5. The PLS normal probability plot of the residuals with respect to each aroma group of compounds: (**a**)—ethyl esters from fatty acids; (**b**)—ethyl esters from branched acids; (**c**)—acetate esters; (**d**)—C6-alcohols; (**e**)—volatile phenols. The standardized residuals are on the x-axis, and the expected values are on the y-axis.

References

1. Øvsthus, I.; Martenalc, M.; Radovanović Vukajlović, T.; Lesica, M.; Butinar, L.; Mozetič Vodopivec, B.; Antalick, G. Chemical Composition of Apple Cider: A Comparative Study of Norwegian and French Ciders. *Acta Hortic.* 2023; *In press.*
2. Vangdal, E.; Kvamm-Lichtenfeld, K. Ciders Produced from Norwegian Fresh Consumption Apple Cultivars. *Acta Hortic.* **2018**, *1205*, 527–532. [CrossRef]
3. Wicklund, T.; Skottheim, E.R.; Remberg, S.F. Various Factors Affect Product Properties in Apple Cider Production. *Int. J. Food Stud.* **2020**, *9*, SI84–SI96. [CrossRef]
4. Česnik, U.; Øvsthus, I.; Martelanc, M.; Mozetič Vodopivec, B.; Butinar, L. Yeast Diversity from Ciders Produced in Hardanger. *Foods*, 2023; *Article in preparation.*
5. Nguyen, H.; Gaillardin, C. Evolutionary Relationships between the Former Species Saccharomyces Uvarum and the Hybrids Saccharomyces Bayanus and Saccharomyces Pastorianus; Reinstatement of Saccharomyces Uvarum (Beijerinck) as a Distinct Species. *FEMS Yeast Res.* **2005**, *5*, 471–483. [CrossRef]
6. Libkind, D.; Hittinger, C.T.; Valério, E.; Gonçalves, C.; Dover, J.; Johnston, M.; Gonçalves, P.; Sampaio, J.P. Microbe Domestication and the Identification of the Wild Genetic Stock of Lager-Brewing Yeast. *Proc. Natl. Acad. Sci. USA* **2011**, *108*, 14539–14544. [CrossRef]
7. Almeida, P.; Gonçalves, C.; Teixeira, S.; Libkind, D.; Bontrager, M.; Masneuf-Pomarède, I.; Albertin, W.; Durrens, P.; Sherman, D.J.; Marullo, P.; et al. A Gondwanan Imprint on Global Diversity and Domestication of Wine and Cider Yeast Saccharomyces Uvarum. *Nat. Commun.* **2014**, *5*, 4044. [CrossRef]
8. Marullo, P.; Dubourdieu, D. 12—Yeast Selection for Wine Flavor Modulation. In *Managing Wine Quality*, 2nd ed.; Reynolds, A.G., Ed.; Woodhead Publishing Series in Food Science, Technology and Nutrition; Woodhead Publishing: Sawston, UK, 2022; pp. 371–426, ISBN 978-0-08-102065-4.
9. McCarthy, G.C.; Morgan, S.C.; Martiniuk, J.T.; Newman, B.L.; McCann, S.E.; Measday, V.; Durall, D.M. An Indigenous Saccharomyces Uvarum Population with High Genetic Diversity Dominates Uninoculated Chardonnay Fermentations at a Canadian Winery. *PLoS ONE* **2021**, *16*, e0225615. [CrossRef]
10. Naumov, G.I.; Nguyen, H.-V.; Naumova, E.S.; Michel, A.; Aigle, M.; Gaillardin, C. Genetic Identification of *Saccharomyces Bayanus* Var. Uvarum, a Cider-Fermenting Yeast. *Int. J. Food Microbiol.* **2001**, *65*, 163–171. [CrossRef]
11. Valles, B.; Bedrinana, R.; Tascon, N.; Simon, A.; Madrera, R. Yeast Species Associated with the Spontaneous Fermentation of Cider. *Food Microbiol.* **2007**, *24*, 25–31. [CrossRef]
12. Blein-Nicolas, M.; Albertin, W.; Valot, B.; Marullo, P.; Sicard, D.; Giraud, C.; Huet, S.; Bourgais, A.; Dillmann, C.; De Vienne, D.; et al. Yeast Proteome Variations Reveal Different Adaptive Responses to Grape Must Fermentation. *Mol. Biol. Evol.* **2013**, *30*, 1368–1383. [CrossRef] [PubMed]
13. McKay, M.; Buglass, A.J.; Gook Lee, C. Cider and Perry. In *Handbook of Alcoholic Beverages*; Buglass, A.J., Ed.; John Wiley & Sons, Ltd.: Chichester, UK, 2011; pp. 231–265, ISBN 978-0-470-97652-4.
14. Wei, J.; Wang, S.; Zhang, Y.; Yuan, Y.; Yue, T. Characterization and Screening of Non-Saccharomyces Yeasts Used to Produce Fragrant Cider. *LWT* **2019**, *107*, 191–198. [CrossRef]
15. Wei, J.; Zhang, Y.; Wang, Y.; Ju, H.; Niu, C.; Song, Z.; Yuan, Y.; Yue, T. Assessment of Chemical Composition and Sensorial Properties of Ciders Fermented with Different Non-Saccharomyces Yeasts in Pure and Mixed Fermentations. *Int. J. Food Microbiol.* **2020**, *318*, 108471. [CrossRef] [PubMed]
16. Aung, M.T.; Lee, P.-R.; Yu, B.; Liu, S.-Q. Cider Fermentation with Three Williopsis Saturnus Yeast Strains and Volatile Changes. *Ann. Microbiol.* **2015**, *65*, 921–928. [CrossRef]
17. Gutiérrez, A.; Boekhout, T.; Gojkovic, Z.; Katz, M. Evaluation of Non-Saccharomyces Yeasts in the Fermentation of Wine, Beer and Cider for the Development of New Beverages. *J. Inst. Brew.* **2018**, *124*, 389–402. [CrossRef]
18. McKay, M.; Buglass, A.J.; Gook Lee, C. Fermented Beverages: Beers, Ciders, Wines and Related Drinks. In *Handbook of Alcoholic Beverages: Technical, Analytical and Nutritional Aspects*; John Wiley & Sons, Ltd.: Chichester, UK, 2010; ISBN 978-0-470-51202-9.
19. Jiranek, V.; Langridge, P.; Henschke, P.A. Validation of Bismuth-Containing Indicator Media for Predicting H2S-Producing Potential of Saccharomyces Cerevisiae Wine Yeasts Under Enological Conditions. *Am. J. Enol. Vitic.* **1995**, *46*, 269–273. [CrossRef]
20. Ugliano, M.; Henschke, P.A. Comparison of Three Methods for Accurate Quantification of Hydrogen Sulfide during Fermentation. *Anal. Chim. Acta* **2010**, *660*, 87–91. [CrossRef]
21. Layne, J.; Lomas, S. *Simple Sugars Analysis from Wine by LC-RI Using Luna® Omega SUGAR LC Columns*; Phenomenex, Inc.: Torrance, CA, USA, 2018; p. 2.
22. Martelanc, M. Simultaneous Determination of Chemical Parameters for Wine and Cider Quality Control Using HPLC-UV/RI System Based on Reverse-Phase Separation Capabilities. In Proceedings of the Analytical and Biomedical Applications, Kemomed, Kranj, Slovenia, 20 October 2022.
23. Nguyen, N.; McGinley, M. Real-Time Response to Bacteria Infection of Bioethanol Fermentation Using a Short Rezex™ ROA Column. *Appl. Noteb.* **2008**, *1*, 1–4.
24. Topić Božič, J.; Butinar, L.; Bergant Marušič, M.; Korte, D.; Mozetič Vodopivec, B. Determination of Biogenic Amines Formation by Autochthonous Lactic Acid Bacteria from 'Refošk' Grapes Using Different Analytical Methods. *LWT* **2022**, *156*, 112908. [CrossRef]
25. Antalick, G.; Perello, M.-C.; De Revel, G. Esters in Wines: New Insight through the Establishment of a Database of French Wines. *Am. J. Enol. Vitic.* **2014**, *65*, 293–304. [CrossRef]

26. Gobert, A.; Tourdot-Maréchal, R.; Sparrow, C.; Morge, C.; Alexandre, H. Influence of Nitrogen Status in Wine Alcoholic Fermentation. *Food Microbiol.* **2019**, *83*, 71–85. [CrossRef]
27. Ljungdahl, P.O.; Daignan-Fornier, B. Regulation of Amino Acid, Nucleotide, and Phosphate Metabolism in *Saccharomyces Cerevisiae*. *Genetics* **2012**, *190*, 885–929. [CrossRef] [PubMed]
28. Liu, S.; Jiang, J.; Ma, Z.; Xiao, M.; Yang, L.; Tian, B.; Yu, Y.; Bi, C.; Fang, A.; Yang, Y. The Role of Hydroxycinnamic Acid Amide Pathway in Plant Immunity. *Front. Plant Sci.* **2022**, *13*, 922119. [CrossRef] [PubMed]
29. Tronchoni, J.; Gamero, A.; Arroyo-López, F.N.; Barrio, E.; Querol, A. Differences in the Glucose and Fructose Consumption Profiles in Diverse Saccharomyces Wine Species and Their Hybrids during Grape Juice Fermentation. *Int. J. Food Microbiol.* **2009**, *134*, 237–243. [CrossRef] [PubMed]
30. Schütz, M.; Gafner, J. Lower Fructose Uptake Capacity of Genetically Characterized Strains of *Saccharomyces Bayanus* Compared to Strains of *Saccharomyces Cerevisiae*: A Likely Cause of Reduced Alcoholic Fermentation Activity. *Am. J. Enol. Vitic.* **1995**, *46*, 175–180. [CrossRef]
31. Wang, D.; Xu, Y.; Hu, J.; Zhao, G. Fermentation Kinetics of Different Sugars by Apple Wine Yeast Saccharomyces Cerevisiae. *J. Inst. Brew.* **2004**, *110*, 340–346. [CrossRef]
32. Wicklund, T.; Guyot, S.; Le Quéré, J.-M. Chemical Composition of Apples Cultivated in Norway. *Crops* **2021**, *1*, 8–19. [CrossRef]
33. Waterhouse, A.L.; Sacks, G.L.; Jeffery, D.W. *Understanding Wine Chemistry*; John Wiley & Sons, Ltd.: Chichester, UK, 2016; ISBN 978-1-118-62780-8.
34. Qin, Z.; Petersen, M.A.; Bredie, W.L.P. Flavor Profiling of Apple Ciders from the UK and Scandinavian Region. *Food Res. Int.* **2018**, *105*, 713–723. [CrossRef]
35. He, W.; Liu, S.; Heponiemi, P.; Heinonen, M.; Marsol-Vall, A.; Ma, X.; Yang, B.; Laaksonen, O. Effect of *Saccharomyces Cerevisiae* and *Schizosaccharomyces* Pombe Strains on Chemical Composition and Sensory Quality of Ciders Made from Finnish Apple Cultivars. *Food Chem.* **2021**, *345*, 128833. [CrossRef]
36. Medina, S.; Perestrelo, R.; Pereira, R.; Câmara, J.S. Evaluation of Volatilomic Fingerprint from Apple Fruits to Ciders: A Useful Tool to Find Putative Biomarkers for Each Apple Variety. *Foods* **2020**, *9*, 1830. [CrossRef]
37. Satora, P.; Sroka, P.; Duda-Chodak, A.; Tarko, T.; Tuszyński, T. The Profile of Volatile Compounds and Polyphenols in Wines Produced from Dessert Varieties of Apples. *Food Chem.* **2008**, *111*, 513–519. [CrossRef] [PubMed]
38. Sousa, A.; Vareda, J.; Pereira, R.; Silva, C.; Câmara, J.S.; Perestrelo, R. Geographical Differentiation of Apple Ciders Based on Volatile Fingerprint. *Food Res. Int.* **2020**, *137*, 109550. [CrossRef] [PubMed]
39. Peng, B.; Li, F.; Cui, L.; Guo, Y. Effects of Fermentation Temperature on Key Aroma Compounds and Sensory Properties of Apple Wine. *J. Food Sci.* **2015**, *80*, S2937–S2943. [CrossRef]
40. Pour Nikfardjam, M.; May, B.; Tschiersch, C. 4-Ethylphenol and 4-Ethylguaiacol Contents in Bottled Wines from the German 'Württemberg' Region. *Eur. Food Res. Technol.* **2009**, *230*, 333–341. [CrossRef]

Article

Effects of Main Nutrient Sources on Improving *Monascus* Pigments and Saccharifying Power of *Monascus purpureus* in Submerged Fermentation

Yingying Huang [1,2,3], Jiashi Chen [1], Qing Chen [1] and Chenglong Yang [1,2,3,*]

1   Institute of Agricultural Engineering Technology, Fujian Academy of Agricultural Sciences, Fuzhou 350003, China; hyy_202@163.com (Y.H.)
2   Key Laboratory of Subtropical Characteristic Fruits, Vegetables and Edible Fungi Processing (Co-Construction by Ministry and Province), Ministry of Agriculture and Rural Affairs, Fuzhou 350003, China
3   Fujian Key Laboratory of Agricultural Products (Food) Processing, Fuzhou 350003, China
*   Correspondence: yclmail1227@163.com

Abstract: *Hong Qu* (HQ), obtained through fermentation of various grains using *Monascus* spp., has been widely utilized as the main and characteristic initial saccharification and traditional fermentation starter in the food brewing industry. The quality, color, and flavor of HQ and HQ wine are closely related to the saccharifying power (SP) and *Monascus* pigments (MPs) of *Monascus* spp. In this study, to optimize the culture medium in submerged fermentation by *M. purpureus* G11 for improving SP and MPs, the effects of carbon source, nitrogen source, inorganic salts, and vitamins on SP activity and biosynthesis of MPs were explored through single-factor analysis and response surface Box–Behnken experiments. The results showed that the optimal medium composition was 6.008% rice powder, 1.021% peptone, 0.0049% $CuSO_4$, and 0.052% vitamin B1. Validation experiments performed under the optimized fermentation conditions showed a significant increase in MPs and SP by 14.91% and 36.24%, with maximum MPs and SP reaching 112.61 and 365.12 u/mL, respectively. This study provides a theoretical basis for enhancing MPs and SP in *M. purpureus* for HQ production, to improve the production efficiency and shorten the production cycle of HQ-related fermentation products.

Keywords: *Hong Qu*; saccharification starter; *Monascus* pigments; response surface methodology; culture medium optimization

Citation: Huang, Y.; Chen, J.; Chen, Q.; Yang, C. Effects of Main Nutrient Sources on Improving *Monascus* Pigments and Saccharifying Power of *Monascus purpureus* in Submerged Fermentation. *Fermentation* **2023**, *9*, 696. https://doi.org/10.3390/fermentation9070696

Academic Editors: Roberta Comunian and Luigi Chessa

Received: 22 June 2023
Revised: 17 July 2023
Accepted: 20 July 2023
Published: 24 July 2023

1. Introduction

Hong Qu (HQ) is one of the four traditional fermentation starters used in China, and is the main and characteristic initial saccharification and fermentation agent for fermented foods, such as *Hong Qu* glutinous rice wine (HQ wine), vinegar, and soy sauce [1–3]. HQ wine is the second most distinctive yellow rice wine in China. Unlike other types of yellow rice wine, HQ wine does not require the addition of colorants because the *Monascus* strains incorporated into the HQ wine during the brewing process can produce *Monascus* pigments (MPs), which give the wine a unique natural red–orange color. *Monascus* is widely used in the field of food fermentation, which can produce various important metabolites, including MPs, monacolin-like substances, and γ-aminobutyric acid, as well as various enzymes, such as amylase, glucoamylase, protease, pectinase, and esterification enzymes [4,5]. HQ used in the brewing industry must possess high saccharifying power (SP) and esterifying power [6,7]. The quality and safety of HQ play crucial roles in determining the flavor, color, functional activity, and overall quality of HQ wine [8]. Hence, enhancement of HQ quality is an important strategy for the development of the HQ wine industry. Although HQ has been widely used in the wine-making industry, the availability of only a few varieties of HQ starters and low fermentability significantly limit its applications [9].

Wine fermentation starters are the primary driving force for brewing yellow rice wine. Studies on HQ wine brewing have mainly focused on the microbial communities of the starters, fermentation processes, active substances, and volatile flavor components [10–12]. In recent years, there has been a gradual increase in research on esterification enzymes and saccharifying enzymes of *Monascus* spp., especially on the isolation and identification of α-amylase (liquefying power) and its fermentation optimization, analysis of genes related to esterification enzymes and their biosynthetic pathways, enhancement of production efficiency, and shortening of the production cycle of HQ-related fermentation foods [13–17]. Although most of the studies on saccharifying enzymes of *Monascus* spp. have mainly focused on its applications, only a few studies have investigated their fundamental characteristics [18]. Saccharifying enzymes can fully convert starch into fermentable sugar and provide necessary substrates for the fermentation process in the food brewing industry [19]. Although many researchers have attempted to use saccharifying fermentation agents directly in the saccharification stage of the wine-making process and obtained good outputs, the agent can cause rapid hydrolysis of starch, resulting in microecological imbalance during the fermentation process, which can decrease the wine quality and create a bitter taste [20]. Thus, improving the SP of HQ has gained increasing attention in recent years [21]. Because *Monascus* plays an important role in the HQ brewing industry, it is particularly critical to obtain *Monascus* strains that not only have high SP but also adequate quantities of MPs. The use of such efficient *Monascus* strains in brewing HQ can improve the quality of HQ, enhance the color and flavor of HQ wine, and reduce the amount of red HQ used in brewing. Although a representative strain, *Monascus purpureus* G11, which can produce high yield of MPs and strong SP, was isolated in our previous study, the fermentation process was optimized.

It is well known that optimization of medium composition can increase the production of beneficial metabolites, such as MPs and lovastatin, in *Monascus* spp., and reduce the generation of harmful products, such as citrinin [22,23]. Many researchers have studied the effects of changes in carbon and nitrogen sources as well as the addition of vitamins and inorganic salts on the metabolism of *Monascus* spp. For instance, Yang Dongcheng et al. found that Fe^{2+} ions can increase the production of MPs in *Monascus* spp. [24]. However, the effects of medium optimization on SP and MPs in *M. purpureus* have not been extensively investigated. Response surface methodology (RSM) has been used for optimization of fermentation conditions based on the sophisticated interactions among multiple process variables [25,26]. However, the use of RSM for the optimization of fermentation conditions for improving SP is limited.

In the present study, RSM was employed for the optimization of fermentation conditions for both SP activity and the production of MPs in *M. purpureus*. Submerged liquid fermentation (SmF), with the advantages of short cultivation period, easy control of conditions, and good reproducibility, was employed to investigate systematically the effects of several categories of main nutrient sources on the brewing strain, *M. purpureus* G11. The medium composition was optimized for improving SP and MPs of liquid seed production and solid-state fermentation by identifying the preferred nutrients, including carbon sources, nitrogen sources, inorganic salts, and vitamins (and precursor substances). A mathematical model was established using the response surface Box–Behnken experimental design to optimize the culture medium and further improve SP and MPs. The main indicators of HQ quality, namely, SP and color value, were used as the evaluation standard. The results of this study provide systematic technical support for research on liquid seed production and solid-state fermentation processes for brewing HQ.

2. Materials and Methods

2.1. Strain and Culture Conditions

M. purpureus G11, which can produce both a high yield of MPs and SP activity, was isolated from HQ collected from HQ wine manufacturers and preserved in our laboratory and at the China Center for Type Culture Collection (CCTCC, Wuhan, China; CCTCC No. M 2023550). For the initial growth, *M. purpureus* G11 was inoculated onto MEA slant (6% malt extract and 2% agar; pH 6–7; sterilized at 121 °C for 30 min) and incubated at 35 °C for

10 d. The seed and liquid media consisted of 6% glucose, 3% soluble starch, 1% peptone, 0.15% KH_2PO_4, 0.2% K_2HPO_4, 0.2% $MgSO_4$, and 0.1% $NaNO_3$ without pH adjustment, and were sterilized at 121 °C for 30 min. The seed culture was prepared as described previously, and 6% of the seed was inoculated onto 50 mL of aseptic liquid fermentation medium and incubated at 35 °C and 180 rpm for 96 h [27].

2.2. Metabolite Detection

Color value (yield of MPs, u/mL) was determined according to the Chinese National Standard (GB1886.19-2015), with three biological replicates. The fermentation broth in each flask (0.05 mL) was mixed well and then extracted with 25 mL of 70% ethanol at 60 °C for 1 h. After filtration and cooling to room temperature, the extracted liquid or filtrate was serially diluted (A times), and the optical density (OD) was measured using an ultraviolet–visible spectrophotometer (UV-240, Shimadzu, Japan) against a 70% ethanol blank at 505 nm. The yield of MPs was calculated as follows: MPs = A $\times$ OD $\times$ V $\times$ 25, where A is the number of dilutions, OD is the optical density, and V is the total volume of the fermentation broth after fermentation (mL).

SP (u/mL) was determined according to the Chinese National Standard (QB/T 5188-2017), with three biological replicates. One unit (U) of enzyme activity was defined as the ability of 1.0 g of HQ to convert soluble starch into 1 mg of glucose in 1 h at 40 °C and pH 4.6. First, the fermentation broth in each flask (3 mL) was mixed well and then extracted with 22 mL of acetic acid–sodium acetate buffer solution at 35 °C for 2 h. After filtration and cooling to room temperature, the extracted liquid was used as the sample. Second, both 2% soluble starch solution (25 mL) and acetic acid–sodium acetate solution (5 mL) were added to two 50 mL colorimetric tubes in bottles A and B at 40 °C for 10 min. The extracted liquid sample (2 mL) was added to bottle A at 40 °C for 30 min, then 20% sodium hydroxide (0.2 mL) was added and mixed well, and cooled at 1–3 °C for 2 min. In bottle B (blank bottle), 20% sodium hydroxide (0.2 mL) was added and mixed well, cooled to room temperature, then 2 mL of the extracted liquid sample was added. Third, the solutions in bottles A and B (5 mL) were mixed well with iodine solution (10 mL) and 0.1 mol/L of sodium hydroxide (15 mL) for dark reaction (15 min), then 2 mol/L of sulfuric acid (2 mL) was added and the solution titrated with sodium thiosulfate solution until the blue color disappeared.

SP was calculated as follows: SP = $(V - V_1) \times C_1 \times 90.05 \times \frac{32.2}{5} \times \frac{1}{2} \times N \times 2$, where V (mL) is the volume of thiosulfate sodium solution in the blank in bottle B, V_1 (mL) is the volume of thiosulfate sodium solution in the sample in bottle A, C_1 (mol/L) is the concentration of thiosulfate sodium standard solution, 90.05 (mg) is the mass of glucose equivalent to 1.00 mL of thiosulfate sodium, 32.2 (mL) is the total volume of the reaction solution, 5 (mL) is the volume of the reaction solution used, 1/2 is the sample volume conversion factor from 2 mL to 1 mL, N is the dilution factor (25 mL/3 mL), and 2 is the reaction time conversion factor from 30 min to 1 h.

2.3. Effects of Carbon and Nitrogen Sources on MPs and SP by M. purpureus G11 in SmF

The medium composition was optimized for improving high MPs and SP in *M. purpureus* G11 by identifying the preferred nutrients. Accordingly, the effects of different carbon and nitrogen sources on *M. purpureus* G11 were determined using single-factor analysis by incubating *M. purpureus* G11 in basal medium containing different carbon sources (6%) (glucose, sucrose, corn starch, lactose, fructose, glycerin, maltose, rice powder, and galactose) and nitrogen sources (1%) (soybean powder, sodium glutamate, beef extract, fish meal, KNO_3, peptone, yeast extract, soy powder, NH_4NO_3, $(NH_4)_2SO_4$, and corn steep liquor), respectively, with the other factors remaining constant.

2.4. Effect of Inorganic Salt Sources on MPs and SP by M. purpureus G11 in SmF

Inorganic salts, including $CuSO_4$ (0.001% and 0.005%), $FeSO_4$ (0.005% and 0.01%), $LaCl_3$ (0.04% and 0.08%), $MnCl_2$ (0.005% and 0.01%), $MgSO_4$ (0.1%), KH_2PO_4 (0.2%), $CoCl_2$

(0.005%), $ZnSO_4$ (0.05% and 0.1%), and $CaCl_2$ (1% and 1.5%), were added to the basal medium to ascertain their effects on MPs and SP in *M. purpureus* G11 using single-factor analysis. After determining the optimal inorganic salt source, the effect of inorganic salt concentration (0.001%, 0.0025%, 0.005%, 0.0075%, and 0.01%) on MPs and SP in *M. purpureus* G11 was investigated.

2.5. Effect of Vitamins on MPs and SP by M. purpureus G11 in SmF

Vitamins, including vitamin B1 (VB1; 0.03% and 0.08%), vitamin B3 (VB3; 0.03% and 0.08%), vitamin B5 (VB5; 0.03% and 0.08%), vitamin B6 (VB6; 0.03% and 0.08%), vitamin B9 (VB9; 0.07% and 0.1%), erythorbic acid (EVC; 0.03% and 0.05%), and vitamin H (H; 0.03% and 0.08%), were individually added to the basal medium to investigate their effects on MPs and SP in *M. purpureus* G11 using single-factor analysis. After determining the optimal vitamin, the effect of vitamin concentration (0.01%, 0.03%, 0.05%, 0.07%, and 0.09%) on MPs and SP in *M. purpureus* G11 was analyzed.

2.6. Optimization of Medium Composition Using RSM

Based on the abovementioned results, the four selected medium components (rice powder, peptone, Cu^{2+}, and VB1) were further optimized through RSM to achieve high MPs and SP in *M. purpureus* G11 (Table 1).

Table 1. Factors and their code levels optimized using RSM experimental design.

Independent Factors	Symbol	Range and Code Levels		
		−1	0	1
Rice powder (%)	A	5	6	7
Peptone (%)	B	0.5	1	1.5
Cu^{2+} (%)	C	0.003	0.005	0.007
VB1 (%)	D	0.03	0.05	0.07

2.7. Statistical Analysis

All data were analyzed using SPSS and ANOVA to determine significant differences. All the experiments were performed in triplicate, and the average value was employed. The data were plotted using Origin Pro 8.6.

3. Results and Discussion

3.1. Effects of Carbon Source on MPs and SP by M. purpureus G11

The carbon source is one of the most important components in the fermentation medium, which mainly provides the carbon skeleton for the microbial cells and synthesized products and supplies energy for metabolic activities [28,29]. The properties and utilization limitations of carbon sources can directly affect *M. purpureus* metabolism, which in turn can influence material synthesis and degradation, having a certain impact on the production of MPs [30,31]. The results of the present study indicated that the type of carbon source had a significant impact on MPs and SP by *M. purpureus* G11 (Figure 1). The presence of fructose, maltose, and rice powder as carbon sources in *M. purpureus* G11 culture caused higher production of MPs and faster mycelial growth, with the yield of MPs reaching 95.81, 96.2, and 104.15 u/mL, respectively. Furthermore, cultivation of *M. purpureus* G11 in the presence of rice powder as carbon source caused higher SP (267.8 u/mL), which was significantly higher than that noted in the presence of other carbon sources ($p < 0.05$).

Glycosylation capacity is one of the important indicators reflecting the enzymatic ability of *M. purpureus*. The level of glycosylation capacity directly reflects the ability of *M. purpureus* to produce fermentable sugars, which ultimately affects wine production. *M. purpureus* exhibited the best saccharifying ability as well as higher production of MPs (104.15 u/mL) when cultivated in the presence of rice powder as carbon source. This obvious increase in MPs and SP in the presence of rice powder could possibly be owing to the ability of *M. purpureus* G11 to completely utilize the various nutrients in rice powder during cultivation. Therefore, rice powder was selected as the suitable carbon source for achieving high MPs and SP in *M. purpureus* G11.

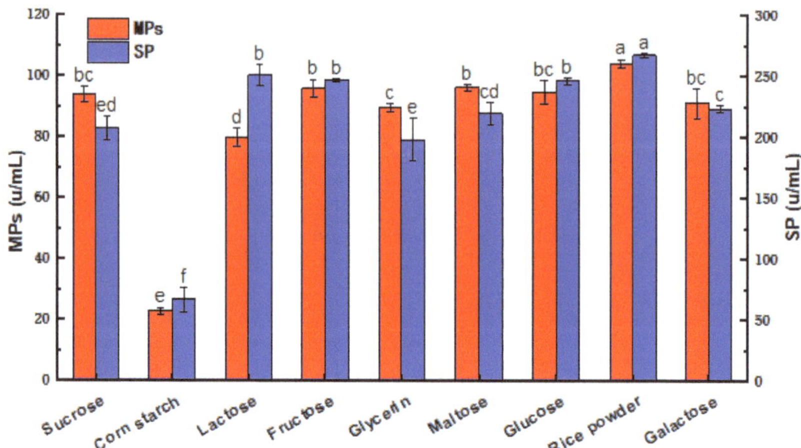

Figure 1. Effects of different carbon sources on MPs and SP by *M. purpureus* G11. Values are expressed as means $\pm$ SD (n = 3). Different lowercase letters (a, b, c, d, e and f) in the figure indicate a statistically significant difference ($p < 0.05$).

It is worth noting that when corn starch was used as the carbon source, both MPs and SP were significantly low. We speculated that except for corn starch and rice flour, all the other carbon sources are small molecules that can be quickly dissolved and absorbed as nutrients. Only corn starch and rice powder are mixed nutrients, and rice powder contains abundant fast nutrients, such as monosaccharides and disaccharides, which can also provide the necessary nutrients for the growth and metabolism of *M. purpureus* G11. However, corn starch is the only pure long-chain starch among the carbon sources without fast nutrient components. The strain cannot grow rapidly, thus affecting the production of MPs and SP activity. In addition, the addition of corn starch results in a semisolid and viscous state of the culture medium, leading to low dissolved oxygen levels, which also restricts the initial growth of the strain. Therefore, neither the composition of corn starch nor the state of the culture medium can improve the growth of the strain and promote production of MPs and SP activity.

3.2. Effects of Nitrogen Source on MPs and SP by M. purpureus G11

A nitrogen source is essential for microbial growth and synthesis of metabolites, and plays a crucial role in the process of MPs and SP by *M. purpureus* during fermentation [32]. The optimal nitrogen source for *M. purpureus* can vary depending on the strain [33–35]. As shown in Figure 2, the type of nitrogen source had a significant impact on the production of MPs and the saccharifying ability of *M. purpureus* G11. Among them, peptone and corn steep liquor had higher saccharification ability than other nitrogen sources, with SP of 263.55 and 258.55 u/mL, respectively, with no significant difference ($p > 0.05$). By contrast, production of MPs by *M. purpureus* G11 in the presence of peptone and corn steep liquor was 103.04 and 15.48 u/mL, respectively, with peptone causing significantly higher yield of MPs than corn steep liquor ($p < 0.05$). Hence, peptone was selected as the suitable nitrogen source for obtaining high MPs and SP in *M. purpureus* G11.

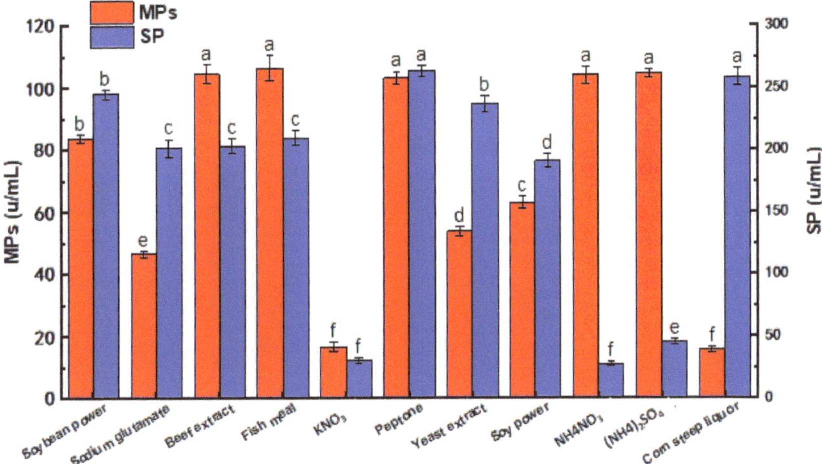

Figure 2. Effects of different nitrogen sources on MPs and SP by *M. purpureus* G11. Values are expressed as means $\pm$ SD (n = 3). Different lowercase letters (a, b, c, d, e and f) in the figure indicate a statistically significant difference ($p < 0.05$).

3.3. Effects of Inorganic Salts on MPs and SP by M. purpureus G11

Inorganic salts are essential for microbial growth and reproduction, and metal ions also play an important regulatory role in the secondary metabolism of microorganisms [36]. Different metal ions have different effects on the production of MPs and the saccharifying activity in *M. purpureus*. Accordingly, in the present study, the optimal inorganic salt and its appropriate concentration for achieving high MPs and SP in *M. purpureus* G11 was screened among Cu^{2+}, Fe^{2+}, La^{3+}, Mn^{2+}, Mg^{2+}, K^+, Co^{2+}, Zn^{2+}, and Ca^{2+} as exogenous metal ion additives [37]. Based on the results of preliminary experiments, $CuSO_4$ (0.001% and 0.005%), $FeSO_4$ (0.005% and 0.01%), $LaCl_3$ (0.04% and 0.08%), $MnCl_2$ (0.005% and 0.01%), $MgSO_4$ (0.1%), KH_2PO_4 (0.2%), $CoCl_2$ (0.005%), $ZnSO_4$ (0.05% and 0.1%), and $CaCl_2$ (1% and 1.5%) were added to the culture medium of *M. purpureus* G11 to explore their effects on MPs and the saccharifying capacity of the strain.

As shown in Figure 3a, the type of inorganic salt had a significant effect on MPs and the saccharifying ability of *M. purpureus* G11. Addition of 0.001% Cu^{2+} significantly promoted the SP activity (309.62 u/mL) of *M. purpureus* G11 in comparison with other inorganic salts ($p < 0.05$). Moreover, 0.001% Cu^{2+} addition achieved higher yield of MPs (104.2 u/mL) in *M. purpureus* G11, which was only slightly lower than those noted with the addition of 0.1% Mg^{2+} and 0.2% K^+, but did not show any significant difference ($p > 0.05$). Based on these results, Cu^{2+} was selected as the optimal exogenous metal ion additive for achieving high MPs and SP in *M. purpureus* G11.

Subsequently, the optimal Cu^{2+} concentration for achieving high MPs and SP in *M. purpureus* G11 was determined by adding different concentrations of Cu^{2+} (0.001%, 0.003%, 0.005%, 0.007%, and 0.009%) to the culture medium. The blank control comprised *M. purpureus* G11 culture medium without Cu^{2+} addition. As shown in Figure 3b, the yield of MPs decreased with the increase in Cu^{2+} concentration, and the highest yield of MPs (105.86 u/mL) was achieved with the addition of 0.003% Cu^{2+}. Similarly, the SP activity decreased with the increase in Cu^{2+} concentration, and the highest SP activity (335.2 u/mL) was observed in the presence of 0.005% Cu^{2+}. Based on these results, 0.005% Cu^{2+} was selected as the optimal Cu^{2+} concentration for subsequent experiments.

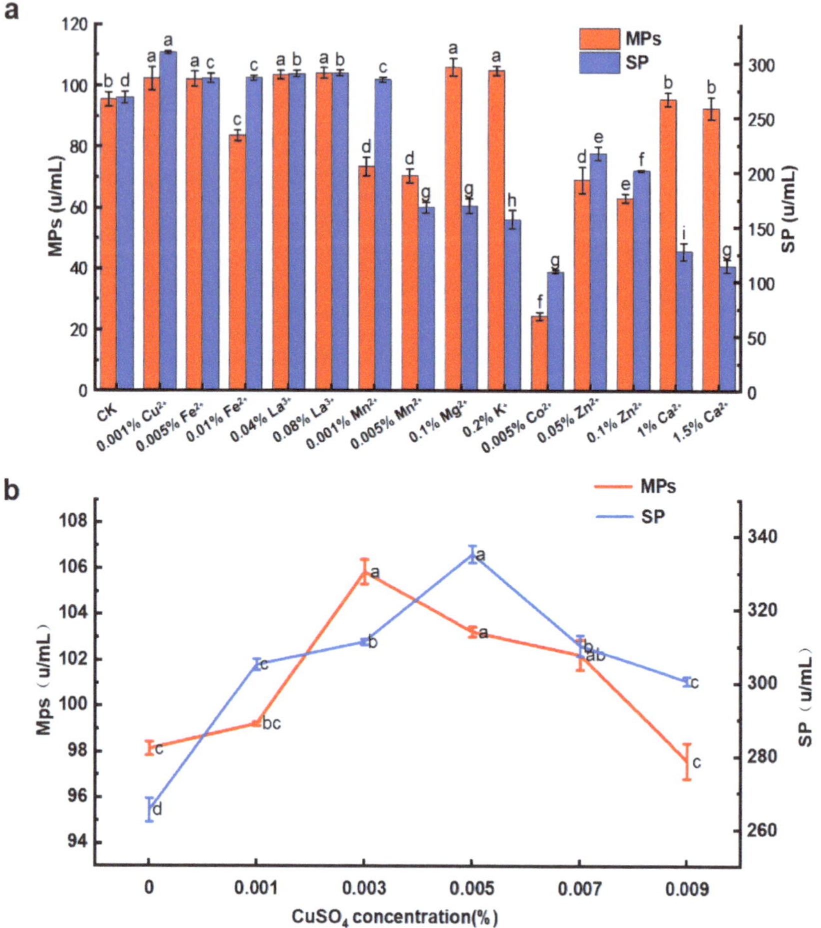

Figure 3. Effects of (**a**) different inorganic salts and (**b**) Cu^{2+} concentrations on MPs and SP in *M. purpureus* G11. Values are expressed as means $\pm$ SD (n = 3). Different lowercase letters (a, b, c, d, e, f, g, and h) in the same figure indicate a statistically significant difference ($p < 0.05$).

3.4. Effects of Vitamins on MPs and SP by M. purpureus G11

Based on the results of preliminary experiments, VB1 (0.03% and 0.08%), VB3 (0.03% and 0.08%), VB5 (0.03% and 0.08%), VB6 (0.03% and 0.08%), VB9 (0.07% and 0.1%), EVC (0.03% and 0.05%), and H (0.03% and 0.08%) were selected as exogenous vitamin additives and added to the culture medium to determine their effects on MPs and SP in *M. purpureus* G11. As shown in Figure 4a, vitamins had a significant effect on the production of MPs and SP activity in *M. purpureus* G11. In particular, 0.03% VB1 and 0.08% VB1 significantly increased SP, reaching 316.5 and 320.21 u/mL, respectively, in comparison with other vitamins ($p < 0.05$). Similarly, addition of 0.03% VB1 and 0.08% VB1 increased the yield of MPs to 99.16 and 100.09 u/mL, respectively. Therefore, VB1 was selected as the optimal vitamin for achieving high MPs and SP in *M. purpureus* G11.

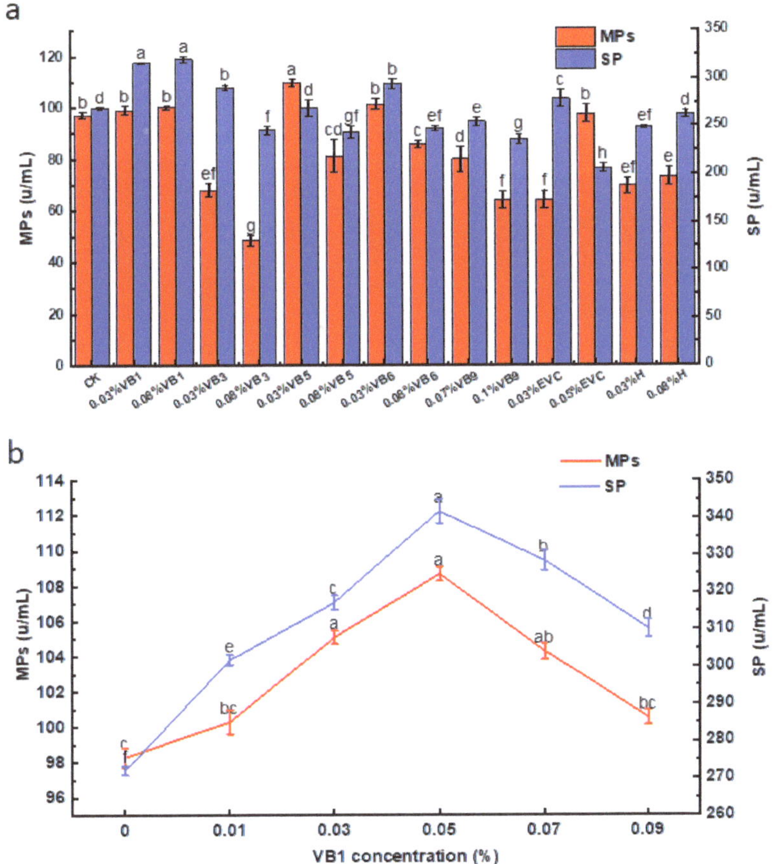

Figure 4. Effect of (**a**) vitamins and (**b**) different concentrations of VB1 on MPs and SP by *M. purpureus* G11. Values are expressed as means ± SD (n = 3). Different lowercase letters (a, b, c, d, e, f, g, and h) in the same figure indicate a statistically significant difference ($p < 0.05$).

Subsequently, the optimal concentration of VB1 for obtaining high MPs and SP in *M. purpureus* G11 was ascertained by adding different concentrations of VB1 (0.01%, 0.03%, 0.05%, 0.07%, and 0.09%) to the culture medium. The blank control comprised *M. purpureus* G11 culture medium without VB1 addition. As shown in Figure 4b, the yield of MPs first increased and then decreased with the increasing VB1 concentration. The yield of MPs was the highest (108.7 u/mL) in the presence of 0.05% VB1. Similarly, the SP activity first increased and then decreased with increasing VB1 concentration. The highest SP activity of 341.75 u/mL was achieved with the addition of 0.05% VB1. Therefore, 0.05% VB1 was chosen as the optimal concentration for subsequent experiments.

3.5. Analysis of RSM Results for MPs and SP by M. purpureus G11

Based on the abovementioned results, the four selected medium components (rice powder, peptone, Cu^{2+}, and VB1) were further optimized using response surface Box–Behnken experiments by establishing regression equations between nutrient sources and MPs and SP, with yield of MPs and SP activity employed as response variables. The design and results of the response surface Box–Behnken experiments are presented in Table 2. The quadratic polynomial regression equations for MPs (Y1) and SP (Y2) were fitted using Design-Expert 13 software as follows:

Table 2. Design and results of response surface Box–Behnken experiments.

No.	Rice Powder/%	Peptone/%	Cu^{2+}/%	VB1/%	Observed MP Yield/(u/mL)	Observed SP Activity/(u/mL)
1	5	0.5	0.005	0.05	106.4	339.56
2	7	0.5	0.005	0.05	105.6	344.63
3	5	1.5	0.005	0.05	106.2	347.17
4	7	1.5	0.005	0.05	104.4	349.78
5	6	1	0.003	0.03	106.5	323.66
6	6	1	0.007	0.03	102.5	327.63
7	6	1	0.003	0.07	106.2	334.25
8	6	1	0.007	0.07	103.2	340.14
9	5	1	0.005	0.03	103.1	329.45
10	7	1	0.005	0.03	103.2	332.64
11	5	1	0.005	0.07	105.7	338.16
12	7	1	0.005	0.07	105.2	341.57
13	6	0.5	0.003	0.05	107.5	332.72
14	6	1.5	0.003	0.05	106.2	337.80
15	6	0.5	0.007	0.05	103.4	337.84
16	6	1.5	0.007	0.05	104.4	342.25
17	5	1	0.003	0.05	107.5	334.42
18	7	1	0.003	0.05	106.5	337.80
19	5	1	0.007	0.05	103.8	339.63
20	7	1	0.007	0.05	102.9	342.76
21	6	0.5	0.005	0.03	102.1	327.04
22	6	1.5	0.005	0.03	101.5	332.16
23	6	0.5	0.005	0.07	103.1	336.59
24	6	1.5	0.005	0.07	103.2	341.77
25	6	1	0.005	0.05	113.42	364.15
26	6	1	0.005	0.05	114.2	365.86
27	6	1	0.005	0.05	113.5	360.85
28	6	1	0.005	0.05	113.2	364.26
29	6	1	0.005	0.05	114.2	358.13

$$Y1 = -0.4083A - 0.1833B - 1.68C + 0.6417D - 0.25AB + 0.025AC - 0.15AD + 0.575BC + 0.175BD + 0.25CD - 3.89A^2 - 4.7B^2 - 3.87C^2 - 5.76D^2,$$

$$Y2 = 1.73A + 2.71B + 2.47C + 4.99D - 0.6140AB - 0.0633AC + 0.056AD - 0.1668BC + 0.0167BD + 0.481CD - 8.77A^2 - 9.8B^2 - 14.6C^2 - 17.83D^2.$$

Table 3 shows the variance analysis of the MPs' regression model using Design-Expert 13 software. Factors C (Cu^{2+}), A^2, B^2, C^2, and D^2 had highly significant effects on the yield of MPs ($p < 0.01$), whereas factor D (VB1) had a significant effect on MPs ($p < 0.05$). However, factors A (rice powder), B (peptone), AB, AC, AD, BC, BD, and CD had no significant effects on the yield of MPs ($p > 0.05$). Based on the magnitude of the first-order coefficients in the quadratic regression equation, the order of the factors in terms of their effects on the yield of MPs was Cu^{2+} > VB1 > rice powder > peptone, with Cu^{2+} having the most significant effect on the yield of MPs. With the yield of MPs as the response variable, the model presented $p < 0.01$, indicating that the quadratic equation model was highly significant. Meanwhile, the lack of fit of the model was not significant ($p = 0.07 > 0.05$), suggesting that the experimental results and mathematical model fitted well, and that the model selection was appropriate. Therefore, this model can be used to predict the test results. The coefficient of determination, R^2, of the regression equation was 0.9722, implying that 97.22% of the variation in color value could be attributed to the selected variables.

Table 3. ANOVA results of the MPs' regression model.

Source	Sum Sq	Df	Mean Sq	F Value	p Value	Significant
Model	405.76	14	28.98	34.93	<0.0001	**
A	2	1	2	2.41	0.1428	
B	0.4033	1	0.4033	0.4861	0.4971	
C	34	1	34	40.98	<0.0001	**
D	4.94	1	4.94	5.95	0.0286	*
AB	0.25	1	0.25	0.3013	0.5917	
AC	0.0025	1	0.0025	0.003	0.957	
AD	0.09	1	0.09	0.1085	0.7468	
BC	1.32	1	1.32	1.59	0.2274	
BD	0.1225	1	0.1225	0.1476	0.7066	
CD	0.25	1	0.25	0.3013	0.5917	
A^2	97.92	1	97.92	118.02	<0.0001	**
B^2	143.15	1	143.15	172.54	<0.0001	**
C^2	97.29	1	97.29	117.26	<0.0001	**
D^2	215.23	1	215.23	259.41	<0.0001	**
Residual	11.62	14	0.8297			
Lack-of-fit	10.75	10	1.07	4.95	0.0685	
Pure error	0.8683	4	0.2171			
Cor. total	417.38	28				

Note: **, highly significantly different ($p < 0.01$), * significantly different ($p < 0.05$). A: Rice powder; B: Peptone; C: Cu^{2+}; D: VB1.

The ANOVA results of the SP regression model obtained using Design-Expert 13 software are shown in Table 4. Factors B (peptone), C (Cu^{2+}), D (VB1), A^2, B^2, C^2, and D^2 had a highly significant effect on SP activity ($p < 0.01$), whereas factor A (rice powder) had a significant effect on SP activity ($p < 0.05$). By contrast, factors AB, AC, AD, BC, BD, and CD had no significant effect on SP activity ($p > 0.05$). Based on the magnitude of the first-order coefficients of the quadratic regression equation, the order of the factors in terms of their effects on SP activity was VB1 > peptone > Cu^{2+} > rice powder, with VB1 having the greatest effect on SP activity. When the SP activity was used as the response variable, the model presented $p < 0.01$, indicating that the quadratic equation model was highly significant. At the same time, the lack of fit of the model was not significant ($p = 0.97 > 0.05$), suggesting that the experimental results and mathematical model fitted well, and that the model was suitable for predicting the test results. The coefficient of determination, R^2, of this regression equation was 0.9832, implying that 98.32% of the variation in enzyme activity could be attributed to the selected variables.

3.6. Model Validation and Confirmation

The response surface graph clearly showed the interaction between the different factors, with steeper slope indicating higher impact of the factors on the response value, and shallower slope denoting lower impact of the factors on the response value [38]. Figure 5 illustrates the interactive effects of various factors on the yield of MPs, with steep response surfaces indicating a significant influence of the interaction between factors on the yield of MPs. Figure 6 reveals the interactive effects of various factors on SP activity. The steep response surfaces (Figure 6b–e) implied that the SP activity increased and then decreased with the increasing levels of rice powder and Cu^{2+}, rice powder and VB1, peptone and Cu^{2+}, peptone and VB1, and VB1 and Cu^{2+}, denoting a significant influence of the interaction between factors on SP activity. By contrast, the relatively flat response surface (Figure 6a) suggested that the interaction between peptone and rice powder had little effect on SP activity.

Table 4. ANOVA results of the SP regression model.

Source	Sum Sq	DF	Mean Sq	F Value	p Value	Significant
Model	3621.32	14	258.67	58.41	<0.0001	**
A	36.01	1	36.01	8.13	0.0128	*
B	88.23	1	88.23	19.92	0.0005	**
C	73.08	1	73.08	16.5	0.0012	**
D	298.98	1	298.98	67.51	<0.0001	**
AB	1.51	1	1.51	0.3405	0.5688	
AC	0.016	1	0.016	0.0036	0.9529	
AD	0.0125	1	0.0125	0.0028	0.9583	
BC	0.1112	1	0.1112	0.0251	0.8763	
BD	0.0011	1	0.0011	0.0002	0.9877	
CD	0.9254	1	0.9254	0.209	0.6546	
A^2	498.87	1	498.87	112.65	<0.0001	**
B^2	623.57	1	623.57	140.81	<0.0001	**
C^2	1383.4	1	1383.4	312.39	<0.0001	**
D^2	2063.08	1	2063.08	465.88	<0.0001	**
Residual	62	14	4.43			
Lack-of-fit	23.23	10	2.32	0.2396	0.9696	
Pure error	38.77	4	9.69			
Cor. total	3683.32	28				

Note: **, highly significantly different ($p < 0.01$), * significantly different ($p < 0.05$). A: Rice powder; B: Peptone; C: Cu^{2+}; D: VB1.

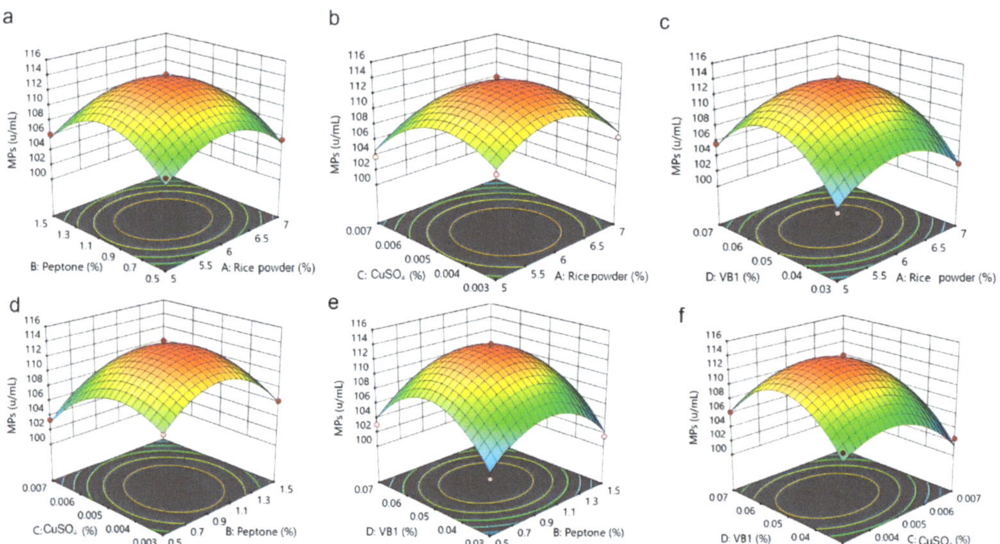

Figure 5. Response surface diagram of the influence of factors interaction (A: Rice powder; B: Peptone; C: $CuSO_4$; D: VB1) on the response of the yield of MPs.

Through optimization using Design-Expert 13 data analysis software, the optimal process parameters for obtaining a high yield of MPs and SP activity in *M. purpureus* G11 were determined to be 6.008% rice powder, 1.021% peptone, 0.0049% Cu^{2+}, and 0.052% VB1. Under these optimal conditions, the theoretical yields of MPs and SP were 113.77 and 362.91 u/mL, respectively. However, considering the operability, the optimal conditions were adjusted to 6% rice powder, 1% peptone, 0.005% Cu^{2+}, and 0.05% VB1, and the adjusted parameters were validated through experiments. The MPs and SP under these adjusted conditions were 112.61 and 365.12 u/mL, respectively, which were close to the theoretical predicted values, indicating that the equation obtained through RSM optimization had practical significance and that the model can correctly predict the production of MPs and SP activity in *M. purpureus* G11.

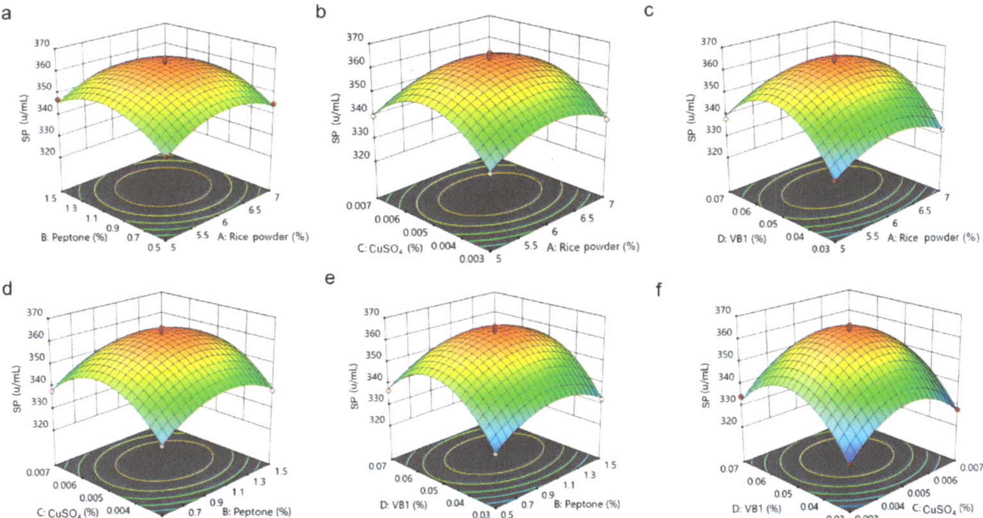

Figure 6. Response surface diagram of the influence of factors interaction (A: Rice powder; B: Peptone; C: CuSO$_4$; D: VB1) on the response of on SP activity.

4. Conclusions

SP activity is one of the most important indicators for assessing the quality and brewing process of HQ. The SP directly determines the utilization rate of raw materials and yield of HQ wine. *Monascus* is one of the key fungal genera that affects the saccharification ability and color of HQ. The difference in the SP activity and production of MPs can lead to variations in the characteristics of HQ, which in turn can affect its fermentation quality. In this study, different medium compositions were examined using single-factor experiments to select the optimal concentrations of medium components for liquid-state fermentation. RSM revealed that the optimum medium components for achieving a high yield of MPs and SP activity were 6.008% rice powder, 1.021% peptone, 0.0049% Cu^{2+}, and 0.052% VB1, which were verified by a liquid-state fermentation experiment. Under these optimal conditions, the yield of MPs and SP activity of 112.61 and 365.12 u/mL, respectively, were achieved. The results of this study provide systematic technical support for HQ production and further research on solid-state fermentation technology. Moreover, exploration and development of production technology for achieving high-quality standardized HQ can offer a theoretical basis and technical guarantee for upgrading the products in HQ-related industries.

Author Contributions: Conceptualization, Y.H.; methodology, J.C.; validation, Q.C.; investigation, Q.C.; resources, Y.H.; data curation, J.C.; writing—original draft preparation, Y.H.; writing—review and editing, Y.H.; visualization, C.Y.; project administration, C.Y. All authors have read and agreed to the published version of the manuscript.

Funding: This research was funded by the National Natural Science Foundation of China (No. 31901803 to Y.H.), the Public Scientific Research Program of Fujian Province, China (No. 2022R1032002 to Y.H.; No. 2021R10320012 to C.Y.), the Natural Science Foundation of Fujian Province, China (No. 2023J01200 to Y.H.), and the Project of Fujian Academy of Agricultural Sciences (No. CYPT202207 to C.Y.).

Data Availability Statement: The data used to support the findings of this study are included within the article.

Fermentation **2023**, 9, 696

Acknowledgments: We would like to express our thanks to all the participants in the present research.

Conflicts of Interest: The authors declare no conflict of interest.

References

1. Huang, Y.Y.; Liang, Z.C.; Lin, X.Z.; He, Z.G.; Ren, X.Y.; Li, W.X.; Molnar, I. Fungal community diversity and fermentation characteristics in regional varieties of traditional fermentation starters for Hong Qu glutinous rice wine. *Food Res. Int.* **2021**, *141*, 110146. [CrossRef]

2. Shao, Y.; Lei, M.; Mao, Z.; Zhou, Y.; Chen, F. Insights into *Monascus* biology at the genetic level. *Appl. Microbiol. Biotechnol.* **2014**, *98*, 3911–3922. [CrossRef] [PubMed]

3. Shin, H.M.; Lim, J.W.; Shin, C.G.; Shin, C.S. Comparative characteristics of rice wine fermentations using *Monascus* koji and rice nuruk. *Food Sci. Biotechnol.* **2017**, *26*, 1349–1355. [CrossRef]

4. Chen, J.Z.; Lin, J.; Ye, X.Y.; Xu, M.A.; Li, W.J.; Deng, X.Z. Study on the liquid fermentation conditions of glucoamylase by *Monascus purpureus*. *Sci. Technol. Food Ind.* **2017**, *38*, 146–150.

5. Yang, C.L.; Wu, X.P.; Chen, B.; Deng, S.S.; Chen, Z.E.; Huang, Y.Y.; Jin, S.S. Comparative analysis of genetic polymorphisms among *Monascus* strains by ISSR and RAPD markers. *J. Sci. Food Agric.* **2017**, *97*, 636–640. [CrossRef] [PubMed]

6. Agboyibor, C.; Kong, W.; Chen, D.; Zhang, A.; Niu, S. *Monascus* pigments production, composition, bioactivity and its application: A review. *Biocatal. Agric. Biotechnol.* **2018**, *16*, 433–447. [CrossRef]

7. Zhao, W.; Qian, M.; Dong, H.; Liu, X.; Bai, W.; Liu, G.; Lv, X. Effect of Hong Qu on the flavor and quality of Hakka yellow rice wine (Huangjiu) produced in Southern China. *LWT* **2022**, *160*, 113264. [CrossRef]

8. Lin, X.; Ren, X.; Huang, Y.; Liang, Z.; Li, W.; Su, H.; He, Z. Regional characteristics and discrimination of the fermentation starter Hong Qu in traditional rice wine brewing. *Int. J. Food Sci. Technol.* **2021**, *56*, 5664–5673. [CrossRef]

9. Chen, S.; Huang, Y.Y.; Yang, C.L. Research status and prospect of brewing Hongqu. *China Brew.* **2022**, *41*, 8–12.

10. Huang, Z.R.; Guo, W.L.; Zhou, W.B.; Li, L.; Xu, J.X.; Hong, J.L.; Liu, H.P.; Zeng, F.; Bai, W.D.; Liu, B.; et al. Microbial communities and volatile metabolites in different traditional fermentation starters used for Hong Qu glutinous rice wine. *Food Res. Int.* **2019**, *121*, 593–603. [CrossRef]

11. Liu, Z.; Wang, Z.; Lv, X.; Zhu, X.; Chen, L.; Ni, L. Comparison study of the volatile profiles and microbial communities of Wuyi Qu and Gutian Qu, two major types of traditional fermentation starters of Hong Qu glutinous rice wine. *Food Microbiol.* **2018**, *69*, 105–115. [CrossRef]

12. Chen, G.M.; Huang, Z.R.; Wu, L.; Wu, Q.; Guo, W.L.; Zhao, W.H.; Liu, B.; Zhang, W.; Rao, P.F.; Lv, X.C.; et al. Microbial diversity and flavor of Chinese rice wine (Huangjiu): An overview of current research and future prospects. *Curr. Opin. Food Sci.* **2021**, *42*, 37–50. [CrossRef]

13. Padmavathi, T.; Rashmi, D.; Anusha, J.; Umme, S. Partial purification and characterization of amylase enzyme under solid state fermentation from *Monascus* sanguineus. *J. Genet. Eng. Biotechnol.* **2017**, *15*, 95–101.

14. Li, M.; Yang, L.R.; Xu, G.; Wu, J.P. Screening, purification and characterization of a novel cold-active and organic solvent-tolerant lipase from Stenotrophomonas maltophilia CGMCC 4254. *Bioresour. Technol.* **2013**, *148*, 114–120. [CrossRef] [PubMed]

15. Yumiko, Y.; Tomoka, S.; Kazunori, T.; Hisanori, T.; Kiyoshi, I.; Yoshihiro, S. Characterization of glucoamylase and α-amylase from *Monascus* anka: Enhanced production of α-amylase in red koji. *J. Biosci. Bioeng.* **2010**, *110*, 670–674.

16. Xu, Y.; Wang, X.; Liu, X.; Li, X.; Zhang, C.; Li, W.; Sun, X.; Wang, W.; Sun, B. Discovery and development of a novel short-chain fatty acid ester synthetic biocatalyst under aqueous phase from *Monascus purpureus* isolated from Baijiu. *Food Chem.* **2021**, *338*, 128025. [CrossRef]

17. Guo, H.; Zhang, Y.; Shao, Y.; Chen, W.; Chen, F.; Li, M. Cloning, expression and characterization of a novel cold-active and organic solvent-tolerant esterase from *Monascus ruber* M7. *Extrem. Life Under Extrem. Cond.* **2016**, *20*, 451–459. [CrossRef]

18. Yasuda, M.; Tachibana, S.; Kuba-Miyara, M. Biochemical aspects of red koji and tofuyo prepared using *Monascus* fungi. *Appl. Microbiol. Biot.* **2012**, *96*, 49–60. [CrossRef]

19. Tachibana, S.; Yasuda, M. Purification and characterization of heterogeneous glucoamylases from *Monascus purpureus*. *Biosci. Biotechnol. Biochem.* **2007**, *71*, 2573–2576. [CrossRef] [PubMed]

20. Zeng, C.; Yoshizaki, Y.; Yin, X.; Wang, Z.; Okutsu, K.; Futagami, T.; Tamaki, H.; Takamine, K. Additional moisture during koji preparation contributes to the pigment production of red koji (*Monascus*-fermented rice) by influencing gene expression. *J. Food Sci.* **2021**, *86*, 969–976. [CrossRef] [PubMed]

21. Raveendran, S.; Parameswaran, B.; Ummalyma, S.B.; Abraham, A.; Mathew, A.K.; Madhavan, A.; Rebello, S.; Pandey, A. Applications of microbial enzymes in food industry. *Food Technol. Biotech.* **2018**, *56*, 16. [CrossRef] [PubMed]

22. Tong, A.; Lu, J.; Huang, Z.; Huang, Q.; Zhang, Y.; Farag, M.A.; Liu, B.; Zhao, C. Comparative transcriptomics discloses the regulatory impact of carbon/nitrogen fermentation on the biosynthesis of *Monascus* kaoliang pigments. *Food Chem. X* **2022**, *13*, 100250. [CrossRef] [PubMed]

23. Huang, Y.Y.; Chen, S.; Yang, C.L.; Lu, D.H. A review on metabolism regulation methods of pigment and citrinin in *Monascus*. *China Condiment* **2018**, *43*, 188–194.

24. Yang, D.C.; Qiang, X.H.; Huang, Y.Q.; Wang, W.P. Effect of Cu^{2+}, Fe^{2+} and Fe^{3+} on the yield of *Monascus* pigments. *China Food Addit.* **2017**, *4*, 100–102.

25. Desai, K.M.; Survase, S.A.; Saudagar, P.S.; Lele, S.S.; Singhal, R.S. Comparison of artificial neural network (ANN) and response surface methodology (RSM) in fermentation media optimization: Case study of fermentative production of scleroglucan. *Biochem. Eng. J.* **2008**, *41*, 266–273. [CrossRef]

26. Daud, N.S.; Azam, Z.M.; Othman, N.Z. Optimization of medium compositions and functional characteristics of exopolysaccharide from Paenibacillus polymyxa ATCC 824. *Biocatal. Agric. Biotechnol.* **2023**, *49*, 102656. [CrossRef]

27. Huang, Y.; Yang, C.; Molnar, I.; Chen, S. Comparative transcriptomic analysis of key genes involved in citrinin biosynthesis in *Monascus purpureus*. *J. Fungi* **2023**, *9*, 200. [CrossRef]

28. Shang, X.P.; Xu, S.J.; Chen, L.H.Z.; Chen, L.J. Research progress and prospect of submerged fermentation of functional *Monascus*. *Food Ferment. Ind.* **2022**, *48*, 320–327.

29. Jiang, X.; Qiu, X.; Yang, J.; Zhang, S.; Liu, J.; Ren, J.; Lu, D.; Zhou, X.; Zhou, B. A mutant of *Monascus purpureus* obtained by carbon ion beam irradiation yielded yellow pigments using various nitrogen sources. *Enzym. Microb. Technol.* **2023**, *162*, 110121. [CrossRef]

30. Embaby, A.M.; Hussein, M.N.; Hussein, A. *Monascus* orange and red pigments production by *Monascus purpureus* ATCC16436 through co-solid state fermentation of corn cob and glycerol: An eco-friendly environmental low cost approach. *PLoS ONE* **2018**, *13*, e207755. [CrossRef]

31. Said, F.M.; Brooks, J.; Chisti, Y. Optimal C: N ratio for the production of red pigments by *Monascus ruber*. *World J. Microbiol. Biotechnol.* **2014**, *30*, 2471–2479. [CrossRef] [PubMed]

32. Chen, W.; Chen, R.; Liu, Q.; He, Y.; He, K.; Ding, X.; Kang, L.; Guo, X.; Xie, N.; Zhou, Y.; et al. Orange, red, yellow: Biosynthesis of azaphilone pigments in *Monascus* fungi. *Chem. Sci.* **2017**, *8*, 4917–4925. [CrossRef] [PubMed]

33. Patrovsky, M.; Sinovska, K.; Branska, B.; Patakova, P. Effect of initial pH, different nitrogen sources, and cultivation time on the production of yellow or orange *Monascus purpureus* pigments and the mycotoxin citrinin. *Food Sci. Nutr.* **2019**, *7*, 3494–3500. [CrossRef] [PubMed]

34. Guang, H.W.; Liu, T.T.; Chen, L.; Xu, G.R.; Zhang, B.B. Optimization of conditions for producing natural yellow pigment by submerged fermentation of *Monascus* sp. *Food Ferment. Ind.* **2020**, *46*, 150–156.

35. Shi, K.; Song, D.; Chen, G.; Pistolozzi, M.; Wu, Z.; Quan, L. Controlling composition and color characteristics of *Monascus* pigments by pH and nitrogen sources in submerged fermentation. *J. Biosci. Bioeng.* **2015**, *120*, 145–154. [CrossRef]

36. Huang, S.N.; Gao, M.X.; Liu, Y.B. Regulatory effect of 5 metal lons on the spectrum of major secondary metabolites of *Monascus purpureus*. *Sci. Technol. Food Ind.* **2021**, *42*, 89–99.

37. Liu, H.H.; Yang, F.; Li, J.Z.; Wang, X.F.; Xue, Y.B.; Chen, M.H.; Wang, C.L. Screening and identification of high-yield esterase *Monascus* and the enzymatic property. *China Brew.* **2019**, *38*, 49–53.

38. Kalil, S.J.; Maugeri, F.; Rodrigues, M.I. Response surface analysis and simulation as a tool for bioprocess design and optimization. *Process Biochem.* **2000**, *35*, 539. [CrossRef]

Article

Autochthonous Microbes to Produce Ligurian Taggiasca Olives (Imperia, Liguria, NW Italy) in Brine

Grazia Cecchi [1,*], Simone Di Piazza [1], Ester Rosa [1], Furio De Vecchis [2], Milena Sara Silvagno [2], Junio Valerio Rombi [2], Micaela Tiso [2] and Mirca Zotti [1]

[1] Department of Environmental Life and Earth Science, University of Genoa, Corso Europa 26, 16132 Genoa, Italy; simone.dipiazza@unige.it (S.D.P.); ester.rosa.bio@gmail.com (E.R.); mirca.zotti@unige.it (M.Z.)

[2] MICAMO LAB srl, Spinoff of University of Genoa, Via XX Settembre 33/10, 16121 Genova, Italy; f.devecchis@micamo.com (F.D.V.); m.silvagno@micamo.com (M.S.S.); j.rombi@micamo.com (J.V.R.); m.tiso@micamo.com (M.T.)

* Correspondence: grazia.cecchi@edu.unige.it

Abstract: Table olives are considered high-quality food, and Italy has a wealth of varieties and typical features that are truly unique in the world (about eighty cultivars of table olives or dual-purpose olives, four of which are protected by the protected designation of origin—PDO), and it is the second largest European consumer, behind Spain. The Taggiasca olive does not have a PDO, but it is very appreciated not only in the region of production (Liguria), but also in all the Italian regions and abroad. Autochthonous microbes (bacteria, yeasts, and filamentous fungi) are essential in the fermentative processes for brine olive production. However, these microbial communities that colonised the olive drupes are affected by the environmental conditions and the fermentation treatments. Hence the importance of studying and comparing olive microbes from different farms and investigating the relationships between bacteria, yeasts, and filamentous fungi to speed up the deamarisation process. Our results showed that yeasts are dominant relative to lactobacteria in all three brines studied, and *Wickerhamomyces anomalus* was the most performant fungus for the oleuropein degradation. The latter represents the best candidate for the realisation of a microbial starter.

Keywords: Taggiasca olive; microbial characterisation; *Wickerhamomyces anomalus*; oleuropein degradation

Citation: Cecchi, G.; Di Piazza, S.; Rosa, E.; De Vecchis, F.; Silvagno, M.S.; Rombi, J.V.; Tiso, M.; Zotti, M. Autochthonous Microbes to Produce Ligurian Taggiasca Olives (Imperia, Liguria, NW Italy) in Brine. *Fermentation* **2023**, *9*, 680. https://doi.org/10.3390/fermentation9070680

Academic Editors: Roberta Comunian and Luigi Chessa

Received: 30 June 2023
Revised: 15 July 2023
Accepted: 18 July 2023
Published: 19 July 2023

1. Introduction

The Taggiasca black olive variety is typical of the Liguria region (Northern Italy), where it is cultivated to produce both oil and table olives [1–3]. This variety is recognised as high-quality and takes its name from Taggia, a small town in the province of Imperia; its cultivation is limited to the provinces of Imperia and Savona [2,4]. To date, the production of table olives includes three main methods: i. Spanish process for green olives, with a lye chemical treatment of the drupes before fermentation in brine; ii. Californian process that incorporates lye treatment and air oxidation; iii. Greek process that consists of the natural fermentation of black olives in brine [1,5]. In terms of Taggiasca olives, they are harvested and sorted when they become black. Every producer follows the same general protocol to produce Taggiasca olives in brine, but some steps may be affected by the farm's climatic exposure and structure and by the autochthonous microbial colonisation of the drupes. In general, olives are rinsed with water on site and then placed in barrels. These are later filled with freshly prepared brine with a salt concentration of 8–12% (w/v) [6,7]. For safety reasons, the practice requires a reduction in pH by the addition of lactic or citric acid [6,7]. Olives are marinated in the brine until they lose their bitter taste, and after 6 or more months, they are placed in jars filled with fresh brine and pasteurised [6,7]. During the processing period, the barrels are stored at room temperature and can also be stocked outdoors. In this method, the removal of bitter compounds, mainly represented by oleuropein and its aglycons, occurs due to the enzymatic activity (β-glucosidase and esterase) of the fruits and the microorganisms (bacteria and yeasts), as well as the diffusion of the phenolic compounds into the brine [8,9].

This procedure, however, is often affected by problems. For example, the processing period is strongly affected by climatic conditions, and during cold winters, the fermentation activity can slow down, increasing the time of storage of the olives in the barrels (sometimes up to 8 months) [6,7]. Another problem is represented by the possible growth of undesirable microbes (mainly yeasts and moulds), which can produce "gas pockets" or biofilm on the top layer of the brine [9]. Hence the importance of studying the microbial (bacteria, yeasts, and moulds) community of the drupes as well as the microbial composition of the brine, which can change from farm to farm. In fact, the knowledge of autochthonous microorganisms is essential to avoid and prevent "crazed brines" or taste defects, and to allow the creation of microbial starters employable in fermentation processes for the decreasing of the olives' storage time in barrels [7]. Moreover, the investigation of the Taggiasca drupes funga led to an understanding of the fungal role in the protection of the olive tree against biotic adversities through the production of bioactive compounds and the stimulation of the defence reaction, as well as the application of these microorganisms as potential biopesticides and biofertilisers [10].

This work aims to: i. characterise the fungal community of Taggiasca olives harvested in three different farms in Liguria in winter 2020 and 2021, ii. characterise the microbial community of the brine of these farms, and iii. investigate the potentiality of autochthonous strains as starters.

2. Materials and Methods

2.1. Olives Sampling

Sampling sites were selected by altitude, exposure, and location. The reference sites for the investigation coincide with the olive-processing locations and their respective agricultural farms. Samples were, therefore, collected in the olive groves in the towns of Pompeiana (Imperia, Liguria, Italy—43.85331° N, 7.88898° E), Lucinasco (Imperia, Liguria, Italy—43.96766° N, 7.96472° E), and Diano Arentino (Imperia, Liguria, Italy—43.94692° N, 8.04026° E) in three farms identified by the codes DB, CS, and RF, respectively. The collected olives were characterised based on the indications provided by the producers themselves.

The sampling activities were carried out during November 2020 and 2021 and did not only concern the drupes, but also the processing conditions of the olives in brine. In fact, brine samples were taken to contextualise the presence and development of the microorganisms responsible for the product's fermentation process.

During the sampling operations, environmental parameters were recorded, and information was collected on the operational methods for the production of the brine used by each company, in order to identify technical and technological differences between different producers.

For each farm, two samples were prepared: the first of drupes washed in situ before stocking, and the second of drupes harvested without washing.

After sampling, stocks were prepared for medium-term storage. The olives, as received, were distributed into containers placed at a temperature of 4 °C in the laboratory of Active Cells S.r.l. at the Center for Advanced Biotechnologies. Another part was stored, under similar conditions, in the laboratories of MICAMO and Mycology at DISTAV—UNIGE. A further fraction was placed at −20 °C for long-term storage.

2.2. Fungal Characterisation of Drupes and Brines

All the harvested olives were briefly washed in sterile water and then inoculated in plates of 150 mm in diameter. The culture media employed were Agar Water (AW, SigmaAldrich®, St. Louis, MO, USA) and Potato Dextrose Agar (PDA, SigmaAldrich®). Olives were plated in duplicate for each farm and incubated in the dark at 24 ± 1 °C. The plates were monitored weekly for 21 days.

As for olives in brine, they were inoculated in AW and PDA plates (150 mm diameter) enriched with 5% salt concentration. Moreover, 0.5 mL of brine from each farm was inoculated on PDA plates (90 mm diameter) enriched with 10% salt concentration. All the samples were plated in duplicate, incubated in the dark at $24 \pm 1\ °C$, and monitored weekly for 21 days.

After the fungal growth, colonies were isolated in axenic culture on PDA and Malt Extract Agar (MEA, SigmaAldrich®) plates (60 mm diameter) and finally cryopreserved at $-20\ °C$ in the culture collection of the Mycological Laboratory of the Department of Earth, Environment, and Life Sciences of the University of Genoa (ColD-DISTAV-UNIGE JRU MIRRI-IT).

All fungal morphotypes were identified by a polyphasic approach (morphological and molecular). Macro-micromorphological characteristics were studied by stereomicroscopy ($\times 10$–50) and optical microscopy ($\times 40$–100).

Genomic DNA was extracted from 100 mg of fresh fungal culture using the cetyltrimethylammonium bromide (CTAB) method modified by [11]. The PCR amplification of the ITS region was performed using universal primers ITS1F and ITS4 [12,13]. The PCR protocol was as follows: 1 cycle of 5 min at 95 °C; 40 s at 94 °C; 45 s at 55 °C; 35 1 min cycles at 72 °C; 1 10 min cycle at 72 °C. Later, PCR products were purified and sequenced using Macrogen Inc. (Seoul, Republic of Korea). The sequence assembly and editing were performed using Sequencer® (Gene Codes Corporation, Centerville, MA, USA, version 5.2). The taxonomic assignment of the sequenced samples was carried out using the BLASTN algorithm to compare the sequences obtained against the GenBank database. We took a conservative approach to a species-level assignment (identity $\geq 97\%$) and verified the accuracy of the results by also studying the macro- and micro-morphological features of the colonies. The nomenclature of the species was checked by Index Fungorum (http://www.indexfungorum.org, accessed on 27 February 2023) and Mycobank (https://www.mycobank.org, accessed on 27 February 2023). The sequences obtained were deposited in GenBank with accession numbers ranging from SUB12938672 006_D9 OQ589871 to SUB12938672 008_F1 OQ589911.

2.3. Bacteria Characterisation of the Brines

The microbiological analysis of the brines was carried out in non-selective conditions both aero- and anaerobically.

A pre-enrichment test was used to isolate the bacteria present in the early stages. Since the olive production procedure involves treatment in 10–12% brine, pre-enrichment is used by means of brines at different concentrations to isolate the bacterial flora capable of resisting and proliferating at high brine concentrations. For this purpose, all samples were pre-enriched with 10% brine.

Furthermore, to simulate the natural evolution of the bacterial flora, two series of containers were prepared with a greater quantity of olives from each farm, 120 g of olives and 110 g of brine.

The first experimental series was placed in aerobic conditions. Subsequently, a second experimental series was prepared in airtight jars with limited headspace. Samples were taken from these jars during the first days of the debittering phase and cultivated on special nutrient media. In particular, the deMan Rogosa Sharpe (MRS) culture medium (specific for *Lactobacillus* spp.) and the Thioglycollate Fluid Medium were used.

Once grown on the different media for the isolation of the lactobacilli, we operated according to the following scheme:

A total of 1 mL of the liquid sample, or 1 mL of the stock suspension if solid, and 1 mL of the successive decimal dilutions were placed in Petri dishes, and 10–15 mL of medium was added.

Depending on the type of lactobacilli, the following were incubated:

- Thermophilic lactobacilli: 42 °C for 48 h;
- Mesophilic lactobacilli: 35 °C for 48 h;
- Psychrophilic lactobacilli: 25 °C for 5 days;

- Mesophiles and psychrophiles: 30 °C for 48 h + 22 °C for 24 h.

Microbial strains developed earlier in a sodium thioglycolate broth are potentially anaerobic, as the medium limits oxygen concentration. Then, the positive cultures were transferred to MRS medium and kept in a confined incubation in a CO_2-enriched GasPack. In parallel, the positive cultures were grown on a MRS medium in liquid form.

The GasPack anaerobic system was used to create a low-oxygen, low-CO_2 environment for the growth of anaerobic microorganisms.

The colonies grown on MRS agar or grown in MRS broth were subjected to biochemical tests for the identification of lactobacilli according to the scheme suggested by Sharpe, Fryer, and Smith [14].

2.4. Evaluation of the Autochthonous Microbial Strains' Properties of Oleuropein Degradation

To select the microorganisms capable of debittering the olives (both bacteria and yeasts), a method was developed to evaluate the effectiveness of the selected strains in eliminating oleuropein. This method allowed us to highlight the enzymatic activity of β-D-glucosidase, responsible for the hydrolysis of oleuropein. The detection principle is based on the specific visualisation of β-D-glucosidase through a chromogenic reaction of 5-bromo-4-chloro-3-indolyl-β-D-glucopyranoside, which is diffused to the surface in the isolation medium itself, MRS agar. The medium used in the method is under patent, and it was developed thanks to the project. The nutritional aspect is guaranteed by the enzymatic digestion of casein, glucose, meat, and yeast extract, while the growth stimulus consists of polyoxyethylene sorbitan monooleate, magnesium, and manganese phosphate.

The method developed is under patent and is inspired by the test by Kneifel and Pacher [15], who developed an agar medium, designated X-Glu agar, for the selective counting of *Lb. acidophilus* in yogurt-related dairy products containing a mixed microflora of lactobacilli, streptococci, and bifidobacteria.

The enzyme splits 5-bromo-4-chloro-3-indolyl-β-D-glucopyranoside, a chromogenic substrate included in the formulation of the medium; therefore, the colonies showing the active enzyme β-D-glucosidase are identified because they take on the colour blue.

To accurately evaluate the degradation activity of oleuropein implemented by the isolated strains, high-performance liquid chromatography (HPLC) analyses were also carried out in addition to the colourimetric test, as reported by Servili et al. [16].

The phenolic extract was obtained from the olive brines using liquid–liquid extraction by ethyl acetate: 100 mL of fresh brines filtered through a 0.45-μm CA syringe filter (Whatman, Clifton, NJ, USA) were mixed with 100 mL of ethyl acetate, and after the two separation phases, the organic solvent was recovered, dehydrated through the passage of a column filled with anhydrous sodium sulphate, then evaporated by rotavapor, and finally, the phenolic extract was recovered with 5 mL of methanol and separated by HPLC.

The brines were first filtered through a 0.45-μm CA syringe filter (Whatman). The SPE procedure was followed for both olive brine loading with 2 mL of sample and a 5 g/25 mL Extraclean highload C18 cartridge (Alltech Italia S.r.l., Sedriano, Italy) using 200 mL of methanol as eluting solvent. An Inertsil ODS-3 column (150 mm, 4.6 mm i.d.) (Alltech) was employed for HPLC analysis.

The HPLC system was composed of a Varian 9010 solvent delivery system (Varian Associates, Inc., Walnut Creek, CA, USA) with a 150×4.6 mm i.d. Inertsil ODS-3 column (Alltech Italia Srl) coupled with a Varian Polychrom 9065 ultraviolet (UV) diode array detector operating in the UV region. The samples were dissolved in methanol, and a sample loop of 20 μL capacity was used. The mobile phase was a mixture of solution A (0.2% acetic acid, pH 3.1) and methanol (B), and the flow rate was 1.5 mL/m. The total run time was 55 min, and the gradient changed as follows: 95% A/5% B for 2 min; 75% A/25% B for 8 min; 60% A/40% B for 10 min; 50% A/50% B for 10 min; 0% A/100% B for 10 min; the mixture was maintained for 5 min, and then returned to 95% A/5% B for 10 min.

3. Results

3.1. Fungal Characterisation

The list of fungal species isolated both from drupes and brines is reported in Table 1.

Table 1. List of fungal species isolated from the olives' surface (washed in situ and not washed) and from the brine of the three farms studied during November 2020 and 2021.

Species	Year 2020	Year 2021	Farm 1 (DB)			Farm 2 (CS)			Farm 3 (RF)		
			Wash	Not Wash	Brine	Wash	Not Wash	Brine	Wash	Not Wash	Brine
Acrodontium crateriforme (J.F.H. Beyma) de Hoog	X						X				
Alternaria alternata (Fr.) Keissl.	X	X	X	X		X				X	
Alternaria infectoria E.G. Simmons		X								X	
Alternaria longipes (Ellis and Everh.) E.W. Mason	X					X	X				
Apiospora sacchari (Speg.) Pintos and P. Alvarado	X			X			X			X	
Ascochyta rabiei (Pass.) Labr.	X			X							
Aspergillus heyangensis Z.T. Qi, Z.M. Sun and Yu X. Wang		X								X	
Aspergillus niger Tiegh.	X	X		X						X	
Aspergillus pseudoustus Frisvad, Varga and Samson		X								X	
Aureobasidium microstictum (Bubák) W.B. Cooke	X			X			X		X		
Aureobasidium pullulans (de Bary) G. Arnaud		X	X			X	X		X		
Chaetomium sp.		X					X				
Cladosporium cladosporioides (Fresen.) G.A. de Vries		X	X				X				
Cladosporium perangustum Bensch, Crous and U. Braun		X					X			X	
Cladosporium sp.	X	X		X			X			X	
Didymella pinodella (L.K. Jones) Qian Chen and L. Cai		X					X				
Epicoccum nigrum Link	X	X		X					X		
Fusarium acuminatum Ellis and Everh.	X						X				
Fusarium brachygibbosum Padwick		X	X								
Fusarium oxysporum Schltdl.	X	X	X								
Fusarium sp.	X	X	X	X			X			X	
Mucor racemosus Bull.	X	X		X						X	
Neocucurbitaria juglandicola Jaklitsch and Voglmayr	X									X	
Nigrospora sp.	X					X					
Penicillium brevicompactum Dierckx	X									X	
Penicillium carneum (Frisvad) Frisvad		X							X		X
Penicillium sp.	X	X		X						X	
Pyrenophora avenicola Y. Marín and Crous		X							X		
Rhodotorula sp.	X					X					
Trichoderma gamsii Samuels and Druzhin.		X		X							
Trichoderma sp.	X	X		X		X			X		
Wickerhamomyces anomalus (E.C. Hansen) Kurtzman, Robnett and Basehoar-Powers		X			X		X	X			X

A total of 19 species were isolated from the 2020 samples, while 21 species were isolated from the 2021 samples.

Nine species were isolated in both years: *Alternaria alternata, Aspergillus niger, Cladosporium* sp., *Epicoccum nigrum, Fusarium oxysporum, Fusarium* sp., *Mucor racemosus, Penicillium* sp., and *Trichoderma* sp.

Concerning the farms, DB showed a total of 12 species on the drupes not washed, 6 species were isolated from the washed drupes, and only 1 yeast species (*Wickerhamomyces anomalus*) was isolated from the brine samples. Regarding the CS farm, 13 species were isolated from the not washed samples, while 6 were from the washed drupes, and from the brine samples, 2 species were isolated: a yeast (*Wickerhamomyces anomalus*) and a filamentous fungus (*Penicillium carneum*).

As far as the RF farm, 13 species were isolated from the not-washed drupes, 5 from the washed ones, and from the brine samples, the same 2 species were found isolated from the brine samples of the CS farm.

In general, seven species were isolated from both washed and not washed samples, such as *Alternaria alternata, Aureobasidium pullulans*, and *A. microstictum*, and species belonging to the genera *Cladosporium* and *Trichoderma*. A total of 17 species were only isolated from not-washed olives (*Acrodontium crateriforme, Alternaria infectoria, Apiospora sacchari, Ascochyta rabiei, Aspergillus* species, *Chaetomium* sp., *Didymella pinodella, Fusarium acuminatum, Mucor racemosus, Neocucurbitaria juglandicola, Penicillium brevicompactum*, and *Trichoderma gamsii*), while 5 species were isolated only from washed drupes (*Fusarium brachygibbosum, F. oxysporum, Nigrospora* sp., *Pyrenophora avenicola*, and *Rhodotorula* sp.).

3.2. Bacteria

Aero- and anaerobic microbiological analyses highlighted a heterogeneous load with a high concentration of fungi, moulds, and yeasts, which mask the bacterial component, while the pre-enrichment treatment showed a remarkable superficial growth of fungi, which, in our case, limited the possibility of isolating bacterial strains. Tables 2 and 3 summarise the results of the cultures.

Table 2. Samples cultured on MRS medium jars.

Jar-DB22-Wash	Jar-DB2-Not Wash	Jar-CS05-Wash	Jar-CS02-Not Wash	Jar-RF05-Wash	Jar-RF02-Not Wash
Presence of fermentation and microbial film	Presence of few bubbles and little microbial film	Few bubbles and microbial film	bubbles and microbial film on the surface	Presence of filaments on the surface: moulds	Presence of filaments on the surface: moulds
Growth in abundant culture after 7 days in MRS broth	Abundant growth in culture after 7 days in MRS broth, presence of strong fermentation	Growth in abundant culture after 7 days in MRS broth	NOT Growth in culture after 7 days in MRS broth	Growth in abundant culture after 24 h in MRS broth	Growth in abundant culture after 4 days in MRS broth

Table 3. Samples cultured on tubes of Thioglycollate Fluid Medium and then plated on MRS.

DB2-Not Wash	CS05-Wash	RF05-Wash	RF05-Wash
High growth in tubes after 4 days in MRS broth.	High growth in tubes after 4 days in MRS broth. Presence of filaments on the surface.	High growth in tubes after 48 h in MRS broth. Presence of filaments on the surface.	Weak growth in tubes after 7 days in MRS broth. Presence of filaments on the surface.
MRS plate: diffuse growth after 7 days.	MRS plate: diffuse growth after 7 days.	MRS plate: diffuse growth after 7 days.	MRS plate: no growth after 7 days.

As for the anaerobic positive cultures analysed by the GasPack system, after the incubation period, many colonies have grown on the specific MRS medium for *Lactobacillus* sp. At the same time, growth with the probable presence of lactic flora was observed in liquid culture.

As far as the batch of olives kept in aerobic conditions is concerned, a non-bacterial component is mainly evident: transparent soil with a superficial growth of fungi.

The enzymatic tests showed negative results, underlining the presence of only three lactobacilli strains with different colonies' morphologies: i. smooth, ii. little smooth, iii. wrinkled.

3.3. Evaluation of the Autochthonous Microbial Strains' Properties of Oleuropein Degradation

As far as yeasts, the strains isolated from the drupes and brines samples were identified by the following codes: A5R, A5L, B2, C6, and D5, were tested.

The percentage of oleuropein degradation was measured by evaluating the decrease in the concentration of the molecule of interest in solutions containing the isolated strains, starting from a known concentration identified by HPLC. As it is possible to see first in the summary table (Table 4) and then in the graph, even if A5R is able to split oleuropein, the yeast strains B2, C6, and D5 are certainly more effective in this activity, while the A5L strain does not show a good ability to hydrolyse oleuropein in time.

Table 4. Percentage of oleuropein degradation by the selected yeast strains during time.

	% of Oleuropein Degradation				
Hours	**A5 R**	**A5 L**	**B2**	**C6**	**D5**
0 h	0%	0%	0%	0%	0%
22 h	0%	12%	3%	1%	26%
46 h	0%	12%	15%	30%	34%
70 h	0%	9%	41%	41%	42%
136 h	0%	25%	65%	68%	52%
186 h	7%	63%	85%	75%	77%
280 h	11%	88%	94%	87%	90%

The graph (Figure 1) shows the degradation of oleuropein by yeasts starting from a known concentration and decreasing over time (as the graph visually suggests).

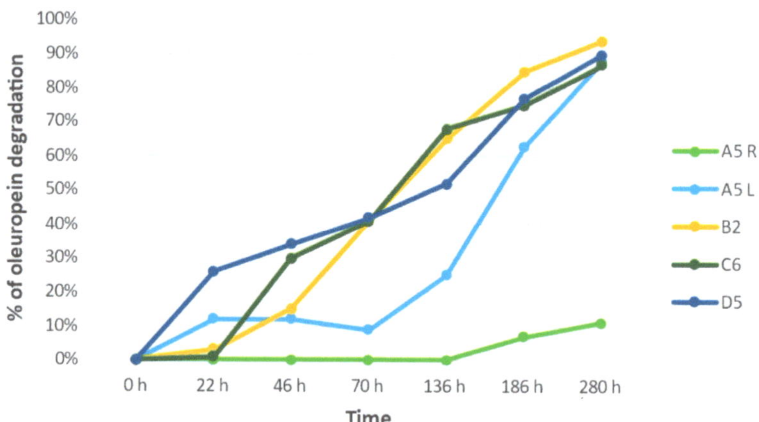

Figure 1. Yeast oleuropein degradation trend.

With regards to lactic bacteria, the first following graph (Figure 2) shows the degradation of oleuropein by lactic acid bacteria, which starts at a known concentration and decreases over time (as the graph visually suggests).

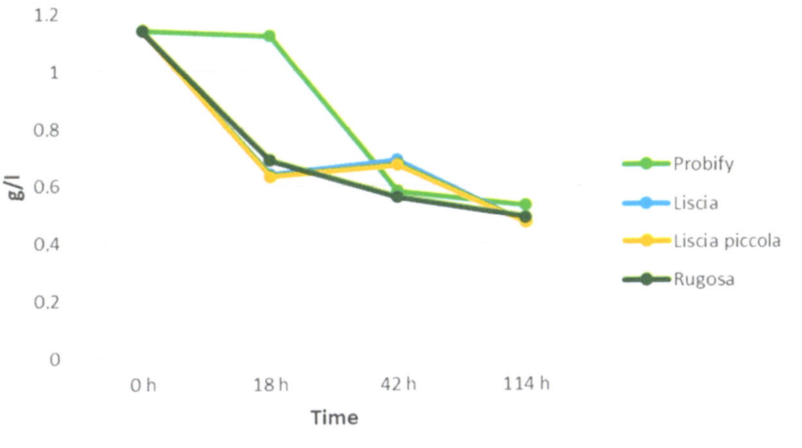

Figure 2. Bacteria degradation trend of oleuropein. Colour lines refer to: light green—Probify strain; light blue—smooth colony; orange—little smooth colony; green—wrinkled colony.

The other two graphs (Figures 3 and 4) refer to two peaks due to the presence of molecules that may probably be the degradation products of oleuropein (hydroxytyrosol and elenolic acid). Their increase over time coincides with the decrease in the oleuropein peak.

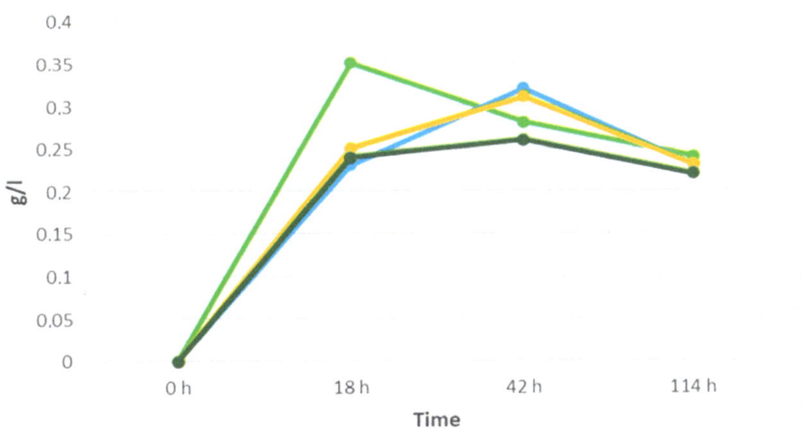

Figure 3. Oleuropein degradation and presence of degradation product: Hydroxytyrosol rates. Colour lines refer to: light green—Probify strain; light blue—smooth colony; orange—little smooth colony; green—wrinkled colony.

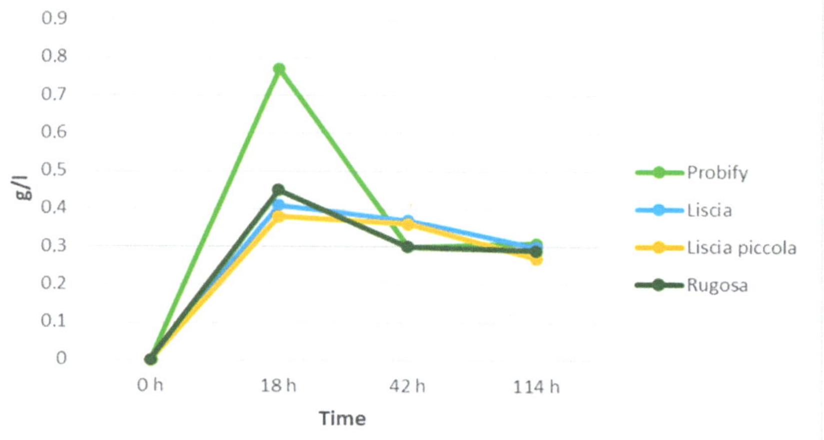

Figure 4. Oleuropein degradation and presence of degradation product: Elenolic acid rates. Colour lines refer to: light green—Probify strain; light blue—smooth colony; orange—little smooth colony; green—wrinkled colony.

Figure 5 shows the outputs of the HPLC analyses.

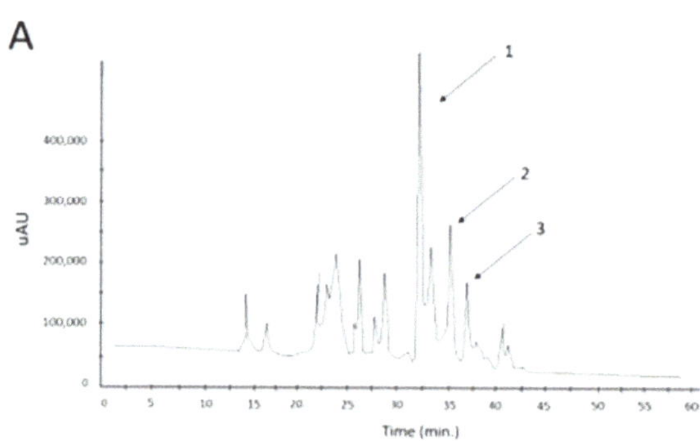

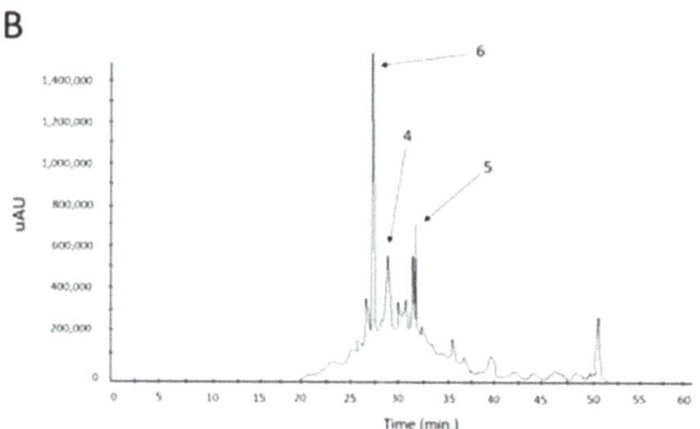

Figure 5. HPLC chromatograms of ethyl acetate extract 1, hydroxytyrosol; 2, tyrosol; 3, caffeic acid; 4, elenolic acid; 5, oleuropein; 6, luteolin. (**A**) is the enlargement portion of (**B**) spectrum.

A commercial strain, Probify—*L. plantarum*, was used as a control.

The results are also reported in the summary tables (Tables 5–7).

Table 5. Oleuropein degradation by bacteria strains.

	Total			
Hours	**Probify g/L**	**Smooth g/L**	**Little Smooth g/L**	**Wrinkled g/L**
0 h	1.14	1.14	1.14	1.14
18 h	1.12	0.64	0.63	0.69
42 h	0.58	0.69	0.67	0.56
114 h	0.53	0.47	0.47	0.49

Table 6. Oleuropein degradation expressed by the presence of Hydroxytyrosol.

	First Peak			
Hours	**Probify g/L**	**Smooth g/L**	**Little Smooth g/L**	**Wrinkled g/L**
0 h	0	0	0	0
18 h	0.35	0.23	0.25	0.24
42 h	0.28	0.32	0.31	0.26
114 h	0.24	0.23	0.23	0.22

Table 7. Oleuropein degradation expressed by the presence of Elenolic acid.

	Second Peak			
Hours	**Probify g/L**	**Smooth g/L**	**Little Smooth g/L**	**Wrinkled g/L**
0 h	0	0	0	0
18 h	0.77	0.41	0.38	0.45
42 h	0.3	0.37	0.36	0.3
114 h	0.31	0.3	0.27	0.29

4. Discussion

Results showed that some fungal species found were common to both sampling years, evidencing that there is a typical funga of the Taggiasca olive. Many of these species, in fact, were also common to the three farms studied (*Alternaria* species, *Cladosporium* species, *Fusarium* species, and *Trichoderma* species). Nicoletti et al. [10] in their work listed the main species of endophytic fungi of *Olea europea*, and many of those isolated in our work were present. Moreover, results showed that the in situ washing of olives significantly reduced the number of isolates, highlighting that only a few species were typical epiphytic fungi. Among these, some species were well-known potential pathogens or parasitic fungi (e.g., *Aspergillus niger*, *Didymella pinodella*, *Fusarium* sp., *Pyrenophora avenicola*), while others were noted biocontrol fungi (e.g., *Aureobasidium pullulans*, *Epicoccum nigrum*, *Trichoderma* sp.) [10,17].

Many studies show that the microbiota of processed olives and/or brine are composed of a complex association of bacteria, such as lactic acid bacteria, Enterobacteriaceae, *Clostridium*, *Staphylococcus*, yeasts, and, occasionally, moulds [18,19].

During the ripening phases of the olives in brine, the presence of autochthonous microorganisms conferred an assimilative or degrading capacity which was naturally used, without control, with the possible development of anomalous ripening processes.

The maturation process derives from the interaction between the microbial population and the concentration of brine used [20]. It had to be emphasised that both the microbial population and the substances present in the drupes were heterogeneous groups made up of many families of microorganisms and different substances [21]. Among the microorganisms, we found species of bacteria, fungi, and yeasts, all endowed with complex metabolic activities that often acted in syntrophic conditions. The products of such interactions led to debittering.

Oleuropein was the main polyphenol present in the leaves and fruits of the olive tree; its activity in humans was anti-inflammatory, antioxidant, and immunomodulatory, but it also had antimicrobial properties, which have proven effective in the course of infections with Gram-positive and Gram-negative bacteria [22,23].

Oleuropein is the main constituent responsible for the bitter taste of olives and olive leaves [22,24]. It, like all phytoalexins, possessed antimicrobial, fungicidal, and insecticidal activities, acting as a defence against infections and infestations [22].

In our study, several nutritional substrates were tested to allow the isolation of most of the species present. In fact, it is known that by isolating microorganisms directly from the environment on a synthetic substrate, the fastest microorganisms were often chosen, leaving out those that, due to their slowness, had the best degradation activities of oleuropein. The samples we tested fall into this category; in fact, there were numerous culture media, but the composition of the flora found indicates that the olives have a very low presence of lactic flora.

The main characteristics sought in the bacteria isolated that could potentially be used as starter cultures to produce table olives, included the ability to degrade oleuropein (Figure 2), the ability to grow in the presence of high concentrations of chloride of sodium (NaCl) and phenols, and the ability to withstand low temperatures [25,26]. These bacteria had to exhibit homolactic fermentation for carbohydrates and the ability for rapid acidification of brines. Other important characteristics for strain selection were represented by the expression of specific enzymes such as β-glucosidase, by the expression of antimicrobial substances and flavouring metabolites [26]. Recently, mixed starter cultures of lactobacilli and enterococci have been used for this type of fermentation, but the use of enterococci has been severely limited due to the possible presence of transferable factors of resistance to antibiotics [27].

However, the vitality of the lactic acid bacteria was found to be low, confirming the information found in the scientific literature, which sees yeasts as protagonists of the natural fermentation of Taggiasca olives [26,27].

Regarding yeast isolation in brine, during olive fermentation, they can be associated with the production of volatile compounds (e.g., alcohols, ethyl acetate, and acetaldehyde), metabolites that improve the taste and aroma, and olives preservation characteristics [7]. It was very interesting how only one species was found: *Wickerhamomyces anomalus* (anamorph *Candida pelliculosa*). However, this was later isolated from all the farms' brines. Many studies have highlighted how this yeast is essential in the fermentation process of many products, in particular, the olives in brine [5,9]. Moreover, this species is not only characterised by β-glucosidase enzyme, but also by the production of antioxidant compounds and lethal toxins against human pathogens and biodeteriogen microorganisms [9]. It was noted for its ability to grow under stressful environmental conditions, such as extremes of pH, low water activity, and anaerobic conditions [28]. This yeast had many roles in the agricultural and food industries; it was often among the "film-forming" yeasts associated with beer spoilage and it had been extensively tested for biocontrol of mould growth that developed during post-harvest storage of apples and airtight-storage of grain [28]. Despite these important characteristics, the results evidenced how these properties, and in particular, the debittering capability, were strain specific. In fact, among the isolated strains, only two yeast strains (D5 and B2) showed a considerable degradation of oleuropein, among them the fastest was D5 which was the most efficient. Hence the importance of testing and conserving each isolated strain. The choice and selection, in fact, of the most performant strain were essential for the preparation of microbial starters employable in the olive brine processes. The use of starter cultures for table olive fermentation was highly recommended [9,29]. The inoculum reduced the effects of spoilage microorganisms, inhibited the growth of pathogenic microorganisms, and helped to achieve a controlled process, reducing debittering time and improving the sensory and hygienic quality of the final product [8,30,31]. The employment of local and autochthonous microbial strains was important to produce not reproducible starters. They enriched the final product with unique and specific sensory characteristics [32]. However, only a few studies reported the application of autochthonous starter cultures [6,32,33]. Three main stages of table olive fermentation can be identified: i. high pH level (6–11) with Enterobacteriaceae as the predominant microbial group together with few Gram-positive bacteria; ii. the reduction in pH level up to 5 and the beginning of the fermentation phase due to *Lactobacillus* species, which are dominant; iii. pH levels are reduced below 5 and some strains of yeast species, especially *Candida*, *Pichia*, and *Saccharomyces* are dominant [9,19]. So, the development of mixed starters (bacteria and yeasts) for the acceleration of the process and the reduction in storage time for olives, could have positive effects compared to the use of a single strain. This method, in fact, mimicked the real succession of microorganisms involved in the fermentation process [30,34]. However, recent studies focused mainly on the development mainly of yeast starter cultures, probably due to their better adaptability to the pH level's strong variation during the fermentation processes [9] and the lower vitality of lactic bacteria.

5. Conclusions

This work investigated the microbial flora of Taggiasca olives and its brine to discover and select high-performing bacteria and yeasts employable in the deamarisation and fermentation processes of brine. Moreover, the study of the funga that inhabit and colonise the olives' drupes allows us to understand that there are some similarities in the fungal communities of Taggiasca olive trees located in different towns, and how many fungi and which ones survive during the fermentation processes. The results showed that yeasts are dominant in all phases of Taggiasca brine production, while Lactobacteria are weaker and cannot tolerate the low pH values.

Author Contributions: Conceptualisation, M.Z. and M.T.; methodology, S.D.P., G.C., E.R., F.D.V. and M.S.S.; validation, M.Z. and M.T.; formal analysis, S.D.P., G.C. and E.R.; investigation, S.D.P., G.C. and E.R.; resources, S.D.P., G.C. and E.R.; data curation, G.C.; writing—original draft preparation, G.C.; writing—review and editing, M.Z. and M.T.; visualisation, S.D.P., E.R., F.D.V., J.V.R. and M.S.S.; supervision, M.Z.; project administration, M.T.; funding acquisition, M.T. All authors have read and agreed to the published version of the manuscript.

Funding: This research was funded by European Agricultural Fund for Rural Development: Regione Liguria, Programma di Sviluppo Rurale 2014–2020, Fondo Europeo Agricolo per lo Sviluppo Rurale: l'Europa investe nelle zone rurali Misura 16.2—cooperazione. Supporto per progetti pilota e per lo sviluppo di nuovi prodotti, pratiche, processi e tecnologie Progetto STAMOIL: STARTER DAL MICROBIOTA DELLE OLIVE domanda n. 12786.

Institutional Review Board Statement: Not applicable.

Informed Consent Statement: Not applicable.

Data Availability Statement: Data are private but can be available asking to the corresponding author.

Acknowledgments: Coldiretti Imperia (Capofila), Micamo Lab srl, Active Cells srl, Az. Agr. Siffredi, Az. Agr. Fazio, Az. Agr. Bassan.

Conflicts of Interest: The authors declare no conflict of interest.

References

1. Pistarino, E.; Aliakbarian, B.; Casazza, A.A.; Paini, M.; Cosulich, M.E.; Perego, P. Combined effect of starter culture and temperature on phenolic compounds during fermentation of Taggiasca black olives. *Food Chem.* **2013**, *138*, 2043–2049. [CrossRef]
2. Aceto, M.; Calà, E.; Musso, D.; Regalli, N.; Oddone, M. A preliminary study on the authentication and traceability of extra virgin olive oil made from Taggiasca olives by means of trace and ultra-trace elements distribution. *Food Chem.* **2019**, *298*, 125047. [CrossRef]
3. Senizza, B.; Ganugi, P.; Trevisan, M.; Lucini, L. Combining untargeted profiling of phenolics and sterols, supervised multivariate class modelling and artificial neural networks for the origin and authenticity of extra-virgin olive oil: A case study on Taggiasca Ligure. *Food Chem.* **2023**, *404*, 134543. [CrossRef]

4. Rellini, I.; Demasi, M.; Scopesi, C.; Ghislandi, S.; Salvidio, S.; Pini, S.; Stagno, A. Evaluation of the environmental components of the taggiasca "terroir" olive (imperia, Italy). *BELS-Bull. Environ. Life Sci.* **2022**, *4*, 1.

5. Penland, M.; Pawtowski, A.; Pioli, A.; Maillard, M.B.; Debaets, S.; Deutsch, S.M.; Coton, M. Brine salt concentration reduction and inoculation with autochthonous consortia: Impact on Protected Designation of Origin Nyons black table olive fermentations. *Food Res. Int.* **2022**, *155*, 111069. [CrossRef]

6. Ciafardini, G.; Zullo, B.A. Use of air-protected headspace to prevent yeast film formation on the brine of Leccino and Taggiasca black table olives processed in industrial-scale plastic barrels. *Foods* **2020**, *9*, 941. [CrossRef]

7. Ciafardini, G.; Venditti, G.; Zullo, B.A. Yeast dynamics in the black table olives processing using fermented brine as starter. *Food Res.* **2021**, *5*, 92–106. [CrossRef]

8. Corsetti, A.; Perpetuini, G.; Schirone, M.; Tofalo, R.; Suzzi, G. Application of starter cultures to table olive fermentation: An overview on the experimental studies. *Front. Microbiol.* **2012**, *3*, 248. [CrossRef]

9. Perpetuini, G.; Prete, R.; Garcia-Gonzalez, N.; Khairul Alam, M.; Corsetti, A. Table olives more than a fermented food. *Foods* **2020**, *9*, 178. [CrossRef]

10. Nicoletti, R.; Di Vaio, C.; Cirillo, C. Endophytic fungi of olive tree. *Microorganisms* **2020**, *8*, 1321. [CrossRef] [PubMed]

11. Doyle, J.J.; Doyle, J.L. A rapid DNA isolation procedure for small quantities of fresh leaf tissue (No. RESEARCH). *Phytochem. Bull.* **1987**, *19*, 11–15.

12. White, T.J.; Bruns, T.; Lee, S.; Taylor, J. Amplification and direct sequencing of fungal ribosomal RNA genes for phylogenies. In *PCR Protocols: A Guide to Methods and Applications*; Innis, M.A., Gelfand, D.H., Sninsky, J.J., White, T.J., Eds.; Academic Press: San Diego, CA, USA, 1990; pp. 315–322.

13. Gardes, M.; Bruns, T.D. ITS primers with enhanced specificity for basidiomycetes-application to the identification of mycorrhizae and rusts. *Mol. Ecol.* **1993**, *2*, 113–118. [CrossRef]

14. Sharpe, M.E.; Fryer, T.F.; Smith, D.G. Identification of the lactic acid bacteria. In *Identification Methods for Microbiologists Part.A. The Society for Applied Bacteriology*; Gibbs, B.M., Skinner, F.A., Eds.; Technical Series, 1; Academic Press: London, UK, 1966; pp. 65–79.

15. Kneifel, W.; Pacher, B. An X-glu based agar medium for the selective enumeration of Lactobacillus acidophilus in yogurt-related milk products. *Int. Dairy J.* **1993**, *3*, 277–291. [CrossRef]

16. Servili, M.; Baldioli, M.; Selvaggini, R.; Miniati, E.; Macchioni, A.; Montedoro, G.F. HPLC evaluation of phenols in olive fruit, virgin olive oil, vegetation waters and pomace and 1D and 2DNMR characterization. *J. Am. Oil Chem. Soc.* **1999**, *76*, 873882. [CrossRef]

17. Lorenzini, M.; Zapparoli, G. Occurrence and infection of Cladosporium, Fusarium, Epicoccum and Aureobasidium in withered rotten grapes during post-harvest dehydration. *Antonie van Leeuwenhoek* **2015**, *108*, 1171–1180. [CrossRef]

18. Almeida, M.; Hébert, A.; Abraham, A.L.; Rasmussen, S.; Monnet, C.; Pons, N.; Delbès, C.; Loux, V.; Batto, J.M.; Leonard, P.; et al. Construction of a dairy microbial genome catalog opens new perspectives for the metagenomic analysis of dairy fermented products. *BMC Genom.* **2014**, *15*, 1101. [CrossRef]

19. Demirci, H.; Kurt-Gur, G.; Ordu, E. Microbiota profiling and screening of the lipase active halotolerant yeasts of the olive brine. *World J. Microbiol. Biotechnol.* **2021**, *37*, 23. [CrossRef]

20. Hurtado, A.; Reguant, C.; Esteve-Zarzoso, B.; Bordons, A.; Rozès, N. Microbial population dynamics during the processing of Arbequina table olives. *Food Res. Int.* **2008**, *41*, 738–744. [CrossRef]

21. Botta, C.; Cocolin, L. Microbial dynamics and biodiversity in table olive fermentation: Culture-dependent and -independent approaches. *Front. Microbiol.* **2012**, *3*, 245. [CrossRef]

22. Omar, S.H. Oleuropein in olive and its pharmacological effects. *Sci. Pharm.* **2010**, *78*, 133–154. [CrossRef]

23. Sun, W.; Frost, B.; Liu, J. Oleuropein, unexpected benefits! *Oncotarget* **2017**, *8*, 17409. [CrossRef]

24. Ozdemir, Y.; Guven, E.; Ozturk, A. Understanding the characteristics of oleuropein for table olive processing. *J. Food Process. Technol.* **2014**, *5*, 1.

25. Bonatsou, S.; Tassou, C.C.; Panagou, E.Z.; Nychas, G.E. Table Olive Fermentation Using Starter Cultures with Multifunctional Potential. *Microorganisms* **2017**, *5*, 30. [CrossRef] [PubMed]

26. Portilha-Cunha, M.F.; Macedo, A.C.; Malcata, F.X. A Review on Adventitious Lactic Acid Bacteria from Table Olives. *Foods* **2020**, *9*, 948. [CrossRef] [PubMed]

27. Anagnostopoulos, D.A.; Bozoudi, D.; Tsaltas, D. Enterococci Isolated from Cypriot Green Table Olives as a New Source of Technological and Probiotic Properties. *Fermentation* **2018**, *4*, 48. [CrossRef]

28. Kurtzman, C.P. Phylogeny of the ascomycetous yeasts and the renaming of Pichia anomala to Wickerhamomyces anomalus. *Antonie van Leeuwenhoek* **2011**, *99*, 13–23. [CrossRef]

29. Cosmai, L.; Campanella, D.; De Angelis, M.; Summo, C.; Paradiso, V.M.; Pasqualone, A.; Caponio, F. Use of starter cultures for table olives fermentation as possibility to improve the quality of thermally stabilized olive-based paste. *LWT* **2018**, *90*, 381–388. [CrossRef]

30. De Angelis, M.; Campanella, D.; Cosmai, L.; Summo, C.; Rizzello, C.G.; Caponio, F. Microbiota and metabolome of un-started and started Greek-type fermentation of Bella di Cerignola table olives. *Food Microbiol.* **2015**, *52*, 18–30. [CrossRef]

31. Randazzo, C.L.; Todaro, A.; Pino, A.; Corona, O.; Caggia, C. Microbiota and metabolome during controlled and spontaneous fermentation of Nocellara Etnea table olives. *Food Microbiol.* **2017**, *65*, 136–148. [CrossRef]

32. Medina, E.; Ruiz-Bellido, M.A.; Romero-Gil, V.; Rodríguez-Gómez, F.; Montes-Borrego, M.; Landa, B.B.; Arroyo-López, F.N. Assessment of the bacterial community in directly brined Aloreña de Málaga table olive fermentations by metagenetic analysis. *Int. J. Food Microbiol.* **2016**, *236*, 47–55. [CrossRef]
33. Campus, M.; Degirmencioglu, N.; Comunian, R. Technologies and Trends to Improve Table Olive Quality and Safety. *Front. Microbiol.* **2018**, *9*, 617. [CrossRef] [PubMed]
34. Tıraş, Z.E.; Yıldırım, H.K. Application of mixed starter culture for table olive production. *Grasas Y Aceites* **2021**, *72*, e405. [CrossRef]

 *fermentation*

Article

Dynamics of Microbiota in Three Backslopped Liquid Sourdoughs That Were Triggered with the Same Starter Strains

Valentina Tolu [1,†], Cristina Fraumene [1,†], Angela Carboni [1], Antonio Loddo [2], Manuela Sanna [1], Simonetta Fois [1], Tonina Roggio [1] and Pasquale Catzeddu [1,*]

[1] Porto Conte Ricerche S.r.l., Località Tramariglio n.15, 07041 Alghero, Italy
[2] MFM Sunalle, via Ogliastra 10, 08023 Fonni, Italy
* Correspondence: catzeddu@portocontericerche.it
† These authors contributed equally to this work.

Abstract: The preparation of sourdough may include the use of starter microorganisms to address the fermentation process toward specific conditions. The aim of this work was to study the dynamics of the microbial ecosystem in three liquid sourdoughs (SD1, SD2 and SD3) triggered with the same microbial strains. *Lactiplantibacillus plantarum* (formerly known as *Lactobacillus plantarum*), *Saccharomyces cerevisiae* and *Candida lambica* strains were inoculated as starters, and sourdoughs were differentiated for the fermentation conditions and for the method of starter inoculation. The analyses were performed on the three sourdoughs propagated in the laboratory for 22 days and on the sample SD1, which was transferred to a bakery and refreshed over many months. The dynamics of microbial communities were studied by plate-count analysis and metataxonomic approach. The acidity of sourdough was evaluated over time. Metataxonomic analysis highlighted a large heterogeneity of fungi microbiota in all sourdough preparations, many of them probably originated from the flour, being pathogens of plants. Few yeast species were found, and *S. cerevisiae* was plentiful but did not predominate over the other species, whereas the *C. lambica* species decreased over time and then disappeared in all preparations. The bacterial microbiota was less heterogeneous than the fungi microbiota; the species *L. plantarum*, *Leuconostoc citreum* and *Levilactobacillus brevis* (formerly known as *Lactobacillus brevis*) were always present in all sourdoughs, whereas *Fructilactobacillus sanfranciscensis* (formerly known as *Lactobacillus sanfranciscensis*) became the dominant species in bakery-propagated SD1 and in SD2 at the end of the propagation period.

Keywords: metataxonomy; yeast; lactic acid bacteria; DNA

Citation: Tolu, V.; Fraumene, C.; Carboni, A.; Loddo, A.; Sanna, M.; Fois, S.; Roggio, T.; Catzeddu, P. Dynamics of Microbiota in Three Backslopped Liquid Sourdoughs That Were Triggered with the Same Starter Strains. *Fermentation* 2022, 8, 571. https://doi.org/10.3390/fermentation8100571

Academic Editors: Roberta Comunian and Luigi Chessa

Received: 8 September 2022
Accepted: 19 October 2022
Published: 21 October 2022

Publisher's Note: MDPI stays neutral with regard to jurisdictional claims in published maps and institutional affiliations.

1. Introduction

Spontaneous sourdough is a complex biological system obtained after the fermentation of cereal flour and water by means of bacteria and yeast, mostly deriving from the raw ingredients, the bread-making environment, and the bakers [1]. The microorganisms in sourdough mainly belong to lactic acid bacteria (LAB) and yeast, which ferment the carbohydrates in flour, producing the carbon dioxide responsible for the bread dough rise and other metabolites such as organic acids and alcohols, which are responsible for the organoleptic properties of bread (flavor, texture and shelf life). Since ancient times, sourdough has been used as a natural leavening agent in bread production and shared among artisanal bakers and home-baking communities. Commonly, sourdoughs are classified into four types [2]:

Type I sourdough is derived from a spontaneous fermentation process, which is followed by a daily backslopping, consisting in a cyclic reinoculation of a so-called "mother dough" using a newly prepared batch of flour and water. Fermentation temperature is set between 20–30 °C and the fermentation time ranges from 5 to 24 h. This sourdough can be refrigerated at regular intervals or at occurrence, and it is most commonly used in artisanal bakeries. The dough yield (ratio between the dough obtained and the flour used) is not exceeding 200.

Type II sourdough is obtained when a flour–water mixture is inoculated with LAB and yeast. Fermentation is conducted for one or more days and temperature is set above 30 °C. Sometimes, yeast is added at the final stage of the fermentation process. Sourdough type II is a semiliquid product with a dough yield between 200–300, and for this reason it is usually employed at an industrial scale.

When sourdough type II is commercialized in the form of a semidried product, it is called sourdough type III.

Type IV sourdough is a combination of type II and type I; in fact, the sourdough is obtained by inoculating selected microorganisms in a mixture of flour and water, and thereafter it is maintained by a backslopping procedure according to the type I sourdough method.

The fermentation of sourdough by spontaneous microflora can lead to the unpredictable growth of various microorganisms. The mutation of the sourdough microbial community over time is conditioned by the non-sterile and open-batch conditions of sourdough and by the use of ingredients potential microbial sources, as the flour, the water or the devices which are used in the backslopping process. Moreover, the development and the behavior of microbial species and the competitiveness between species depend on a multitude of factors, such as temperature, dough yield, and microbial metabolites [3]. The use of spontaneously fermented sourdough can result in unstable product quality.

The use of selected LAB and yeasts as a starter in sourdough technology has become a common practice, mainly in industrial production. Starter microorganisms can be used as flavor carriers and texture improvers, or for their antifungal or health-promoting properties, in order to improve the performances and the properties of sourdough [4,5].

When starter microorganisms are inoculated, as in sourdough type IV, there is a competition between the starter microorganisms and the spontaneously growing microorganisms, and if the starter cultures cannot adapt to the sourdough substrate, and to the acidic conditions, then the spontaneously growing microorganisms can dominate [6]. Moreover, De Vuyst and Neysen [7] observed that the persistence of a microbial association over time is dependent on several factors, such as the process parameters (temperature, dough yield, time of fermentation) or microbial competition (bacteriocin production).

Until now, about 70 different species of LAB [8] and 40 species of yeast have been identified in the sourdough environment. Among LAB, *L. plantarum*, *F. sanfranciscensis* and *L. brevis* were the most isolated species in worldwide sourdoughs [4]. Young sourdoughs can harbor a consistent number of spontaneous microorganisms, but they were largely dominated by *L. plantarum* and *L. brevis*, while *F. sanfranciscensis* is considered predominant in traditionally prepared and older sourdoughs [7,9]; in fact, when the sourdough becomes mature, the diversity of microflora decreases, mostly because the organic acids produced by LAB select for acid-tolerant microorganisms [2].

This study describes the dynamics of the microbiota in three liquid sourdoughs that were propagated for 22 days in a laboratory. One of them was also studied after transferring to and propagation in a bakery. The sourdoughs were triggered using the same microbial starter, composed of *L. plantarum*, *S. cerevisiae* and *C. lambica*, with the purpose to guide the fermentation and dominate the mature sourdoughs. The starter strains were inoculated using different methods and sourdoughs were propagated under two different fermentation temperatures, 25 °C and 20 °C, in order to study the behavior of starter strains. To investigate the dynamics of the microbial population, culture-dependent and metataxonomic methods were applied.

2. Materials and Methods

2.1. Microorganisms and Growth Conditions

The strains of *L. plantarum* PCC1034, *S. cerevisiae* PCC1662 and *C. lambica* PCC1649, isolated from Italian sourdoughs and belonging to the culture collection of Porto Conte Ricerche, were used as starters to inoculate a mixture of flour and water. The strains were identified with the instrument MALDI Biotyper (MicroFlex™, Bruker Daltonik GmbH, Bremen, Germany), using the software MBT Compass® 4.1 and the attached libraries (Bruker Daltonik GmbH, Bremen, Germany). The *L. plantarum* PCC1034 was selected among other strains, belonging to the same species, for its acidification capacity at different growth temperatures. Microbial strains were stored at $-80\,^\circ$C. The LAB strain was routinely propagated in MRS liquid medium (Oxoid, Basingstoke, Hampshire, UK), modified [10] with the addition of fresh yeast extract (5%, v/v) and 1% maltose at a final pH of 5.5 (mMRS), and incubated at 28 °C in anaerobic conditions. The yeast strains were propagated in YEPD liquid medium [10] and incubated at 28 °C under stirring conditions.

2.2. Preparation of Liquid Sourdough and Laboratory Propagation

In order to prepare the liquid sourdough, the selected strains of bacteria and yeast were inoculated in mMRS and YEPD liquid media, respectively, and incubated at 28 °C for 24 h. The next day, 400 µL of the bacterial culture and 200 µL of each yeast culture were inoculated in 40 mL of fresh mMRS and in 20 mL of fresh YEPD, respectively, and incubated at 28 °C. After 24 h, the cells were harvested ($6076\times g$ for 10 min at 4 °C), resuspended in 10 mL of physiological solution and used to inoculate a mixture of 1.5 kg re-milled semolina (Mulino Brundu, Torralba, Italy) and 1.5 kg of sterile water. Liquid sourdough was managed in the laboratory using the Automatic Fermenter AFT5 (SITEP S.r.l., Voghiera, Italy). With regard to the cell number of inoculated strains, it was detected in each dough after inoculum. For all samples, the cell concentration was of the order of 10^7 UFC g^{-1} for bacteria and 10^5 UFC g^{-1} for yeast strains. Values were reported in Figure 1 and indicated as "S". The sourdoughs were refreshed over 22 days from Monday to Friday by a daily backslopping procedure, mixing an equal amount (ratio 1:1:1) of mother sourdough, fresh re-milled semolina and water, which was previously autoclaved 15 min at 120 °C. After the fermentation process, the sourdough was stored at a low temperature (5 ± 2 °C). The value of dough yield (DY) was 200.

Three different sourdoughs were prepared, called SD1, SD2 and SD3, using different methods of inoculum for the starter strains and different fermentation conditions, as follows:

- For SD1, the starter strains were inoculated together. The pH of the dough was about 6.5 before starter addition. Fermentation was carried out at 25 °C for 5 h.
- For SD2, the starter strains were inoculated together. The pH of the dough was lowered before starter addition at value 5.5, using lactic acid 90% (Sigma-Aldrich, Milan, Italy). Fermentation was carried out at 20 °C for 8 h.
- For SD3, the pH of the dough was lowered before starter addition, as in SD2. The starter strains were inoculated separately. At first, the *L. plantarum* was inoculated, and it was left to ferment at 20 °C for 17 h. The following day, the yeast strains were inoculated throughout the refreshment step, and fermentation was carried out at 20 °C for 8 h.

The pH was lowered in order to support the growth of *L. plantarum*. The fermentation times were selected based on experience, as such values allowed sourdough pH values in between 4.0 and 4.5.

Samples of sourdoughs used for microbiological, chemical and metataxonomic analyses were collected about 24 h after the refreshment step, during the low-temperature phase, whereas the sample collected on Monday refers to the sourdough refreshed on Friday.

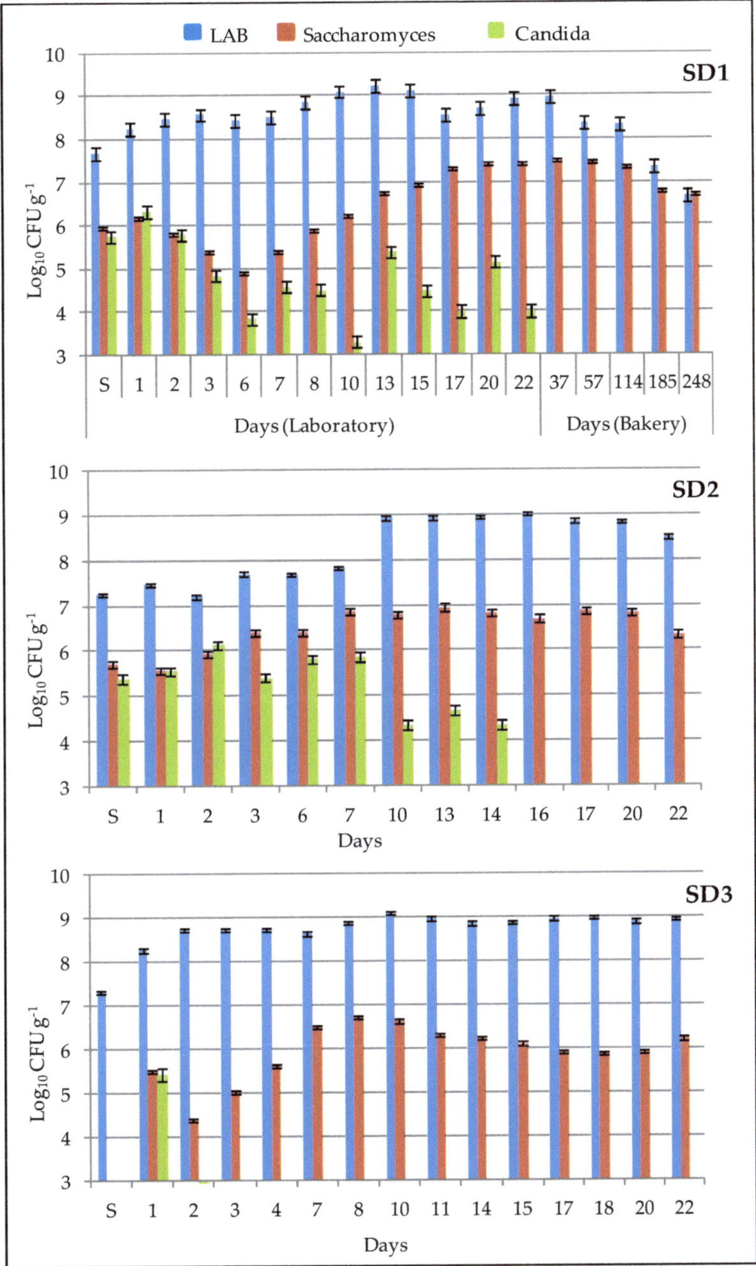

Figure 1. Histogram representation of cell counts ($\log_{10}$ CFU g^{-1}) for the samples SD1, SD2 and SD3. The blue columns indicate the presumptive lactic acid bacteria (LAB) on mMRS agar, the red columns indicate the presumptive *Saccharomyces* and the green columns indicate the presumptive *Candida*, both enumerated on RB agar. "S" refers to the cell number after starter inoculum. Data from duplicate analyses are expressed as mean value. Bars indicate LSD intervals at 95% confidence level.

2.3. Refreshment of Liquid Sourdough in the Bakery

After 22 days, the liquid sourdough SD1 was transferred to an artisanal bakery (MFM Sunalle, Fonni, Italy), referred to as SD1-bak, and used for baking purposes. The sourdough was refreshed twice a week with the aid of the automatic bioreactor Fermentolevain FL80 (Esmach, Italy), mixing sourdough, semolina and water in a ratio of 1:1:1. Fermentation was carried out at 25 °C for 5 h, then the sourdough was cooled to 5 °C. Approximately every 2 months, and up to 8 months, one sample of sourdough was transferred to the Porto Conte Ricerche laboratory and analyzed.

2.4. Determination of pH and Total Titratable Acidity

Ten grams of liquid sourdough were mixed with 90 mL of distillated water and stirred for 30 min, and then an automatic titrator (pH-Matic 23, Crison Instruments, Alella, Spain) was used to measure pH values and total titratable acidity (TTA), the latter was reported as the amount (mL) of NaOH N/10 to achieve pH 8.5 in 10 g of sample. Analyses were done in triplicate.

2.5. Analysis of Sourdough Microorganisms—Culture Dependent Approach

The viable cell number of bacteria and yeast growing in sourdough was estimated by plate-count technique. Ten grams of sourdough were mixed for 2 min with 90 mL of sterile peptone solution (1 g/L of peptone in distilled water) in a sterile stomacher bag, using a Stomacher Lab blender 80 (VWR International PBI, Milano, Italy). Serial dilutions were performed and plated onto mMRS agar for bacterial enumeration and Rose-Bengal Chloramphenicol agar (Oxoid, Basingstoke, UK) for yeast enumeration. Plates of mMRS were incubated under anaerobiosis (AnaeroGen and AnaeroJar, Oxoid, Basingstoke, UK) at 28 °C for 48 h. Rose-Bengal Chloramphenicol agar plates (RB) were incubated at 28 °C for 72 h. The use of RB plates allowed for differentiating *Saccharomyces* from *Candida* based on colony morphology. *Candida* had white colonies with a rugged surface from which "feet" extended from the margins into the surrounding agar, and *Saccharomyces* colonies were circular in shape, violet and had a smooth surface.

2.6. Analysis of Sourdough Microorganisms—Metataxonomic Approach

2.6.1. Nucleic Acid Extraction and High-Throughput Sequencing Analysis

DNA was extracted from the samples collected in the three experiments, SD1, SD2 and SD3. Extraction was performed following the procedure reported in "Manual DNA Extraction from Food Samples" [11] and using the ReliaPrep™ Blood gDNA Miniprep System (Promega, Milano, Italy). DNA quality and yield were evaluated via agarose gel and Qubit fluorometer (Life Technologies, Carlsbad, CA, USA). Libraries were constructed using Illumina's recommendations as implemented in 16S Metagenomic Sequencing Library Preparation guide and Fungal Metagenomic Demonstrated Protocol. Two primers, Lac1 (5′-AGCAGTAGGGAATCTTCCA-3′) and Lac2 (5′-ATTYCACCGCTACACATG-3′), were used to amplify the variable regions 3 and 4 of the bacterial 16S rRNA gene [12]. Among fungi, the gene-specific sequences used in this paper target the fungal ITS1 region between the 18S and 5.8S rRNA genes. They include the ITS1-F and ITS2 primers [13], which are widely used for fungal barcoding studies. All primers were modified to contain adaptors for MiSeq sequencing.

Three separate gene-amplification reactions were performed for each sample, pooled together and cleaned up using AMPure XP (Beckman Coulter, Brea, CA, USA) magnetic beads. The next PCR attached dual index barcodes and sequencing adapters using the Illumina Nextera XT kit so that the PCR products may be pooled and sequenced directly. A final library size and quantification were conducted using a Bioanalyzer 2100 (Agilent Technologies, Santa Clara, CA, USA) and Qubit fluorometer, respectively. DNA sequencing was performed on the Illumina MiSeq platform using v3 chemistry according to the manufacturer's specifications to generate paired-end reads of 251 bases of length in each direction.

2.6.2. Sequence Analyses of the 16S rRNA and ITS Amplicons

For 16S rDNA gene sequencing, data quality control and analyses were performed using the QIIME pipeline package v.1.9.1 [14]. The overlapping paired-end reads were merged using the script join_paired_ends.py inside the QIIME package. Only Illumina reads with a length >200 bp were retained for further analysis. Operational taxonomic units (OTUs) were generated using a pipeline based on USEARCH's OTU clustering recommendations (http://www.drive5.com/usearch/manual/otu_clustering.html (accessed on 26 October 2021)) using the closed-reference OTU picking to allow clustering of 16S sequences, as previously described [14]. Reads were clustered at 97% identity using UCLUST to produce OTUs. The taxonomy classification was determined in accordance with the Greengenes 13_8 database. In addition, taxonomic attribution was completed by searching in the NCBI 16S ribosomal RNA sequences database through the Nucleotide BLASTdatabase [15].

For ITS gene sequencing, data quality control and analyses were performed using the BaseSpace ITS Metagenomics App by Illumina. The ITS Metagenomics workflow performs a taxonomic classification using the UNITE database. Taxonomic identification of strains was completed by comparing the sequences of each sample with those reported in the Nucleotide BLAST database [15].

2.6.3. Data Availability

The sequence data have been deposited in the Sequence Read Archive of the NCBI database (BioProject ID: PRJNA886648).

2.7. Statistical Analysis and Graph Generation

For each sourdough sample, a standard ANOVA procedure was applied to the dataset of acidity and to the cell count of each microbial group. The means were separated by LSD test at a $p = 0.05$ significance level using the Statgraphics Centurion 18 software package (version 18, Statpont Technologies Inc., Warrenton, VA, USA). Some data were subjected to non-parametric statistical analysis by determining the median value. Metataxonomic count data were uploaded to the web application MicrobiomeAnalyst (http://www.microbiomeanalyst.ca (accessed on 26 October 2021)) to assess different statistics through comparative analysis. The relative proportion of read counts was used as a quantitative estimation of the abundance of each taxon of the three sourdoughs. The diversity within each sample (alpha diversity) was estimated with the Shannon diversity index, which takes into account the number of species and their frequency in each sourdough sample. Statistical significance testing between samples was considered and differences were assigned as statistically significant at $p < 0.05$. The differences between microbial samples (beta-diversity) were calculated and visualized as Principal Coordinate Analysis (PCoA). The statistical significance of group clustering was calculated through a permutational multivariate analysis (PERMANOVA) on taxonomic data.

3. Results

Three different liquid sourdoughs, called SD1, SD2 and SD3, were refreshed in a laboratory over a period of 22 days and studied through microbial plate-count, metataxonomic analysis, and pH and TTA determination. The sourdoughs were triggered with the same microbial starter, that was prepared with selected strains of *L. plantarum*, *S. cerevisiae* and *C. lambica*. The fermentation temperature and the method of starter inoculation were modified among the trials, in order to study the performances of the starter strains. Moreover, the first sourdough that was processed in the laboratory, i.e., SD1, was transferred to an artisanal bakery for baking purposes, and this provided the opportunity to analyze its microbiota during the 8 months of refreshments performed in the bakery.

3.1. Lactic Acid Bacteria and Yeast Enumeration

The presumptive LAB and yeast cells were enumerated by plate-count analyses and results were reported in Figure 1. The morphology of the colonies of *C. lambica* and *S. cere-*

visiae was different on RB agar plates (Figure S1), and this allowed us to differentiate the species and to estimate the respective cell number.

Overall, a different evolution of the presumptive LAB and yeasts among sourdoughs was observed. Likely due to the higher number of cells added with the starter, bacteria dominated over yeast cells across samples, and the cell density ($\log_{10}$ CFU g^{-1}) was always higher for LAB than for yeasts, ranging from ca. 7 to 9 $\log_{10}$ CFU g^{-1} for bacteria and from ca. 5 to 7 $\log_{10}$ CFU g^{-1} for yeasts. In spontaneous sourdoughs, yeasts can dominate over bacteria, or cannot appear at all, in the first period of propagation [2]. In this study, the LAB:yeast ratio ranged from 1000:1 to 100:1 during the whole fermentation period, as previously found in sourdoughs [2,7,16]. The number of LAB cells detected 30 min after starter inoculum was ca. 7 $\log_{10}$ CFU g^{-1} for the three sourdoughs, and it increased over time at different rates in the different sourdough samples. Both in SD1 and SD3 the cell density of presumptive LAB exceeded the value of 8 $\log_{10}$ CFU g^{-1} the day after inoculums, whereas in SD2 the presumptive LAB grew slowly in the first week and overcame the 8 $\log_{10}$ CFU g^{-1} after seven days. All samples reached the maximum cell density of ca. 9 $\log_{10}$ CFU g^{-1}, considered the highest value for LAB. The median values of presumptive LAB were 8.56, 8.50 and 8.86 ($\log_{10}$ CFU g^{-1}) for SD1, SD2 and SD3, respectively (Figure S2) and the values corresponding to 25th and 75th percentiles of the data were 8.45 and 8.91 for SD1, 7.68 and 8.92 for SD2, 8.72 and 8.90 for SD3.

The behavior of the presumptive *Candida* strain was different in the sourdough samples, as showed in Figure 1. The cell density, detected 30 min after starter inoculum, was ca. 5 $\log_{10}$ CFU g^{-1} for all sourdough samples. In SD1, the cell density decreased over time but was still detectable up to 22 days, and cells disappeared below the detection limit (considered to be 3 $\log_{10}$ CFU g^{-1}) when the sourdough was refreshed in the artisanal bakery. In SD2, the cell density of presumptive *Candida* decreased after inoculum until it went below the detection limit after 14 days, whereas in SD3 the presumptive *Candida* decreased below the detection limit after the second day.

Among the presumptive number of *Saccharomyces* cells, the values started from 5 $\log_{10}$ CFU g^{-1}, obtained after inoculums of the starter, and increased over time by about 2 logs in the three sourdoughs. The highest number of presumptive *Saccharomyces* cells was found in SD1, where the values ranged from 6.9 to 7.4 from day 15 to 22; the cell number remained high after the sourdough was transferred to the bakery. In SD2 and SD3, the cell density of presumptive *Saccharomyces* always remained below the value of 7 $\log_{10}$ CFU g^{-1}. The median values of presumptive yeast cells, calculated by adding *Candida* and *Saccharomyces* cells, were 6.21, 6.70 and 6.12 ($\log_{10}$ CFU g^{-1}) for SD1, SD2 and SD3, respectively (Figure S3) and the values corresponding to 25th and 75th percentiles of the data were 5.88 and 6.93 for SD1, 6.84 and 6.82 for SD2, 5.78 and 6.32 for SD3.

3.2. pH Values and Total Titratable Acidity (TTA)

The acidifying activity of sourdoughs was reported in Figure 2 and indicated as pH and TTA values. Before starter addition, the value of pH in SD1 was 6.5, corresponding to the value of raw semolina; therefore, the value declined to pH 5.8 the day after the starter fermentation and reached the value of 4.5 after six days. From day 6 to day 22, the pH of SD1 ranged from 4.0 to 4.7 and the TTA values ranged from 5.5 to 9.0 mL NaOH N/10. The acidity increased in SD1 when the sourdough was refreshed in the bakery: pH values ranged from 4.0 to 4.3 and TTA from 7.7 to 11.0 mL NaOH N/10.

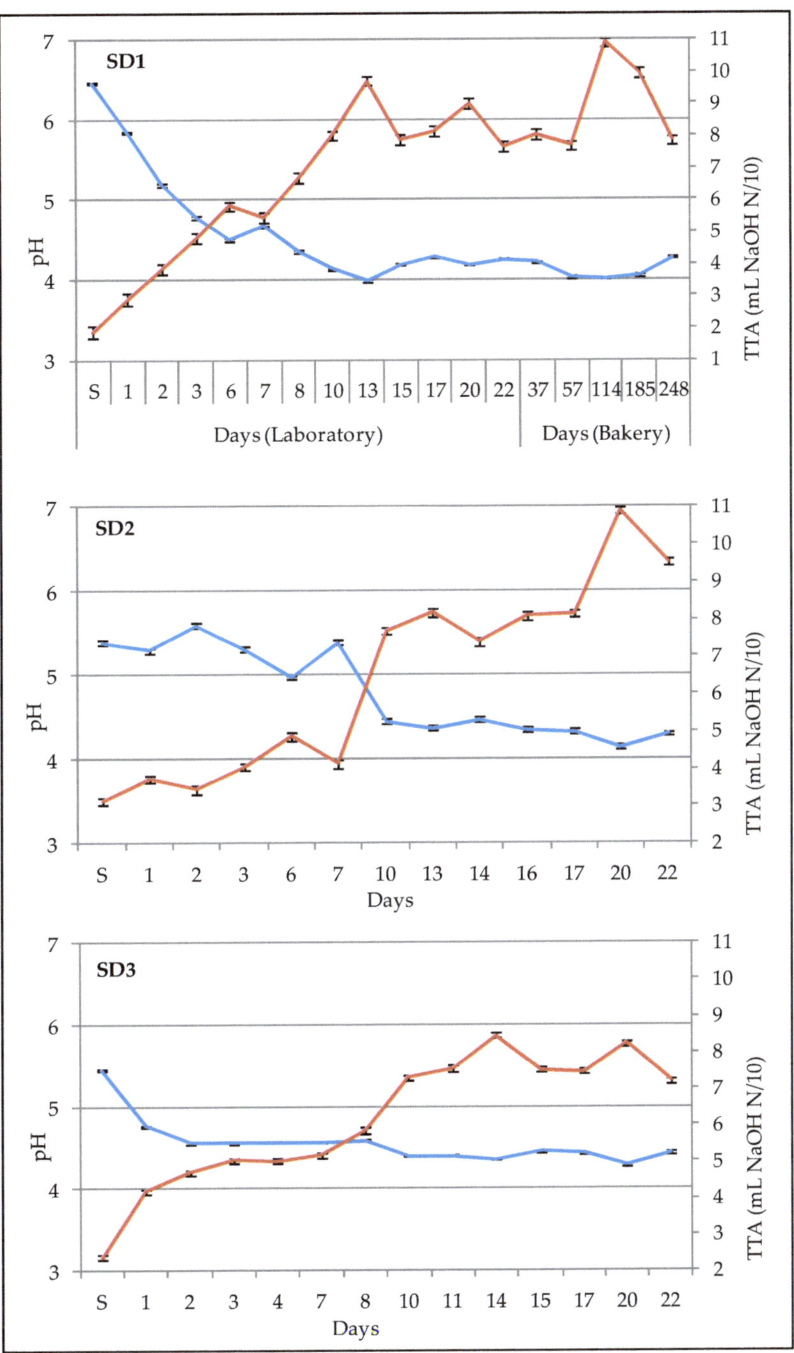

Figure 2. Mean values of pH (blue line) and total titratable acidity (TTA) (red line) for the samples SD1, SD2 and SD3. "S" refers to the pH and TTA values detected after inoculum. Bars indicate LSD intervals at 95% confidence level.

In SD2, the pH of the dough was set down to 5.5 before starter addition using lactic acid, and the value became 5.3 after the starter fermentation; thereafter, the pH value declined quite slowly, reaching the value of 4.5 just after day 10, and ranging from 4.5 to 4.2 until the end. The total acidity, measured as TTA, increased slowly, but at the end reached quite a high value, ca. 11 mL NaOH N/10.

In SD3, the pH was set down at 5.5, as in SD2, but the *L. plantarum* was inoculated alone and left to ferment for 17 h at 20 °C; thereafter, the yeast strains were inoculated. The pH was measured after *L. plantarum* fermentation and revealed a value of 4.8. Later, the pH values ranged from 4.6 to 4.3 until the end. The TTA values increased slowly up to day 14 and did not exceed the value of 8 mL NaOH N/10.

The analyses of ΔpH were reported in Figure 3, referred to as the laboratory-propagated sourdoughs. Actually, the initial lowering of pH by the addition of lactic acid reduced the ΔpH values in SD2 and SD3 with respect to SD1, and the corresponding median values were 0.93, 0.95 and 1.93. The values corresponding to 25th and 75th percentiles of the data were 1.74 and 2.27 for SD1, 0.09 and 1.05 for SD2, 0.88 and 1.04 for SD3, and as a consequence the range of ΔpH displayed between the lower and the upper quartile was wider in SD1 and SD2 compared to SD3. The same phenomenon was observed for the ΔTTA (Figure S4).

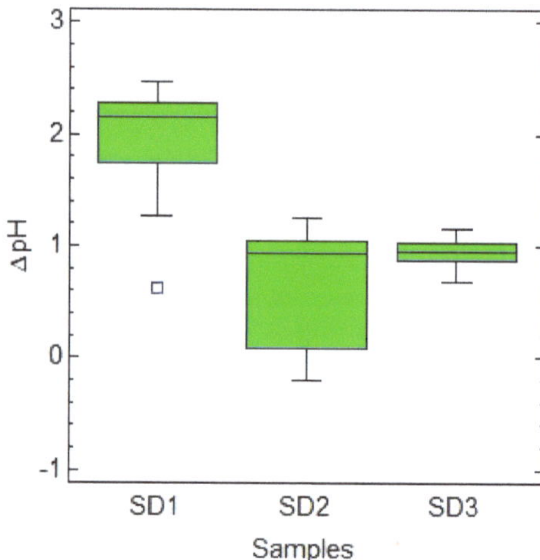

Figure 3. Values of ΔpH (difference in pH units between the initial pH and pH values after sourdough refreshment). Data are the means from three independent experiments (*n* = 3). The inner line of each box represents the median; the top and bottom of the box represent the 75th and 25th percentiles of the data, respectively. The top and bottom of the bars represent the 100% of the data. The square outside the box plot represents the outlier of the data.

3.3. Culture Independent Analysis

The metataxonomic analysis of laboratory-propagated sourdoughs found a total number of 19 species of lactic acid bacteria (Figure 4), belonging to four genera, i.e., *Lactobacillus*, *Leuconostoc*, *Weisella* and *Pediococcus*, one species of *Staphilococcus* and a total of 46 species of fungi, belonging to 27 different genera (Figure 5); most of them are included in the phylum of *Ascomycota* (relative abundance 99.94%) and only four species belong to the *Basidiomycota* phylum. An overall analysis, based on the number of microbial species identified, suggested a greater diversity and heterogeneity of the fungi community compared to the bacterial community.

The most representative microbial species and their relative abundance (%) in sourdough samples were reported in Table 1, and the column of total values refers solely to the laboratory-propagated sourdoughs. Apart from the *L. plantarum* species, inoculated as a starter and with a 55.34% of total relative abundance, the other dominant LAB species were *Leuc. citreum*, *L. brevis* and *F. sanfranciscensis*, and their relative abundances (%) were 14.99, 14.87 and 9.30, respectively (Table 1). Very low values of relative abundance were found for the other LAB species; actually, only *Leuconostoc paramesenteroides* and *Weissella korensis* showed values above 1 (1.77 and 1.15, respectively), and therefore all the other species were considered negligible. Among fungi, the *S. cerevisiae* and *C. lambica* yeast species, which were inoculated as starters, were the predominant fungi species, with a relative abundance of 45.45% and 20.02%, respectively (Table 1). The other yeast species reported in Table 1, i.e., *Wickerhamomyces anomalus*, *Candida santamariae*, *Saccharomyces eubayanus* and *Saccharomyces cariocanus*, showed quite low values of total relative abundance; however, *C. santamariae* and *W. anomalus* were concentrated particularly in SD3, with quite high values of relative abundance: 6.02% and 16.71% respectively. Additionally, *Dipodascus australiensis*, a yeast previously identified in naturally fermented dairy products [17], was detected mostly in SD3, with 20.07% of relative abundance. Other fungi species, quite uncommon for sourdough, were identified in relatively high abundance; the *Microidium phyllanthi*, a pathogenic fungi isolated from plant leaves, was found in all the sourdough samples, with a total value of 7.95% of relative abundance. The *Alternaria infectoria*, a plant pathogenic species, was found in all sourdough samples, but its relative abundance was higher in bakery-propagated sourdough.

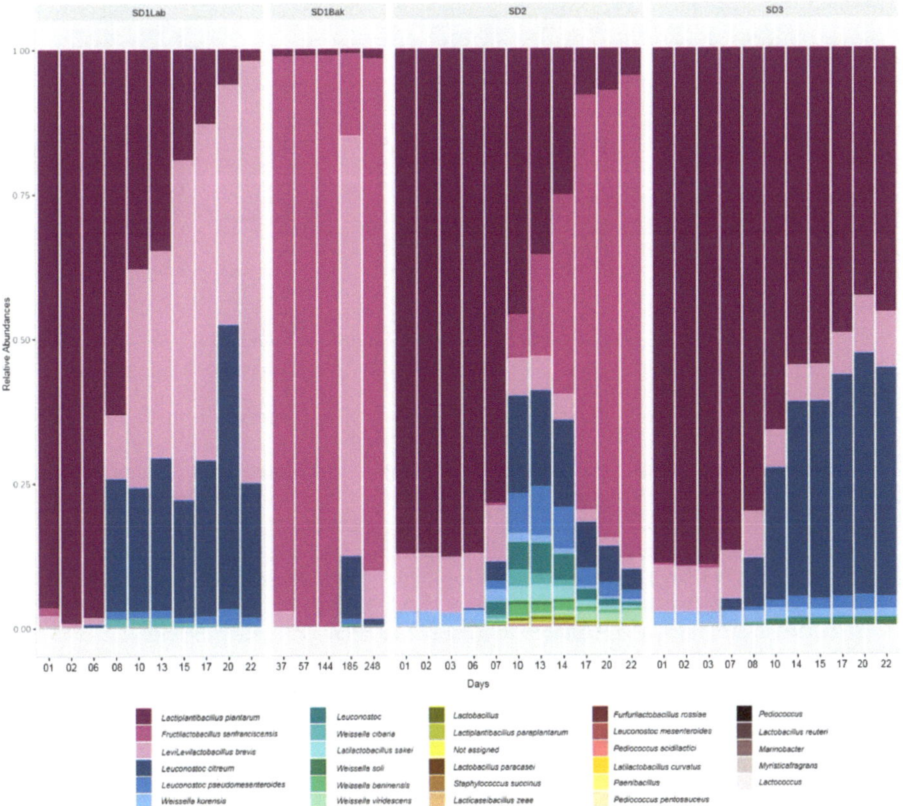

Figure 4. Relative abundance (%) of bacterial OTUs classified at the highest possible taxonomic level (species/genus) found in the three sourdoughs (SD). SD1Lab: laboratory-propagated sourdough. SD1Bak: bakery-propagated sourdough.

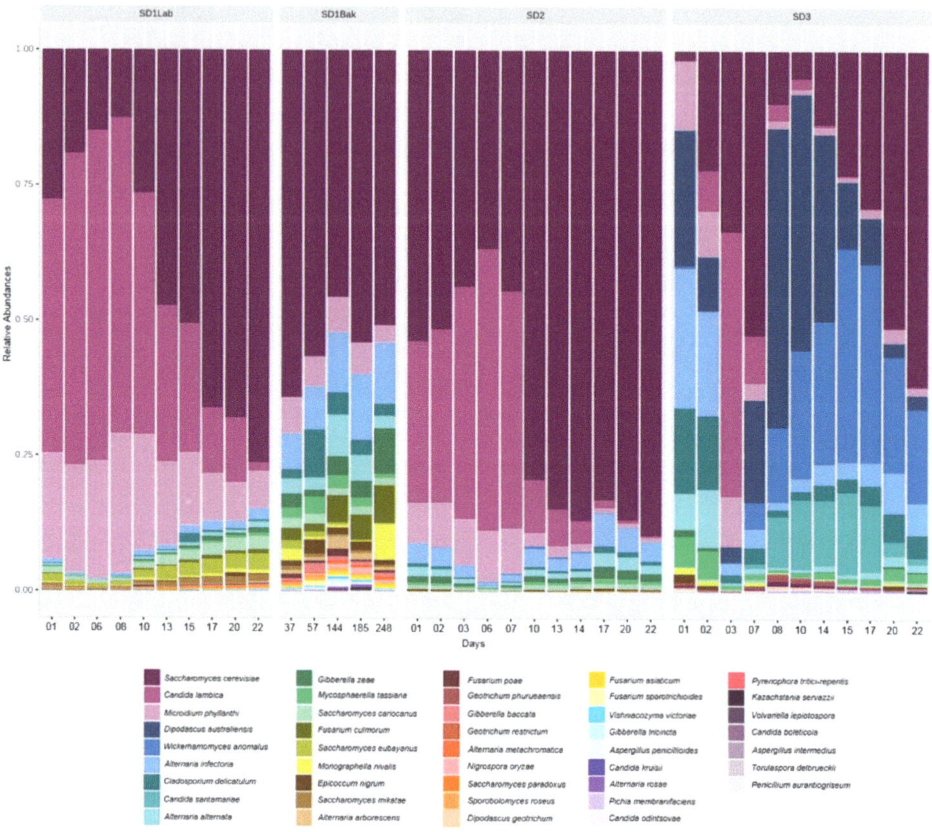

Figure 5. Relative abundance (%) of fungi taxa found in the three sourdoughs (SD). SD1Lab: laboratory-propagated sourdough. SD1Bak: bakery-propagated sourdough.

The dynamics of LAB microbiota can be observed from Figure 4. It is noticeable that *L. plantarum* dominated at day one in the three sourdoughs, and this was certainly due to the high cell number (10^7 CFU g^{-1}) inoculated with the starter strain; nevertheless, other contaminant species, probably coming from semolina, were present at day one. In particular, *L. brevis* (1.91%) and *F. sanfranciscensis* (1.46%) were present in SD1, and *L. brevis* (9.83% and 8.05%) and *W. korensis* (2.45% and 2.29%) were found in SD2 and SD3. The presumptive starter strain *L. plantarum* predominated in all sourdoughs, as evidenced from the relative abundance values (Table 1). In SD1, the *L. plantarum* predominated the first six days and therefore its abundance decreased, reaching very low values at day 22; this could be due to the growth of other two species, *L. brevis* and *Leuc. citreum*, which were present in relatively high numbers. In SD2, *L. plantarum* predominated for the first week, and after that the *F. sanfranciscensis* species appeared, dominating the other species with a high relative abundance, and consequently the abundance of *L. plantarum* decreased until the end of the experiment (4.72% at day 22). In SD3, the decrease of *L. plantarum* after inoculum was lower than in SD1 and SD2, and it was found at high relative abundance throughout the experiment; after 22 days, the relative abundance of the *L. plantarum* species was still 45.8%. The *Leuc. citreum* and the *L. brevis* species were also present in SD3 at quite high relative abundance (20.05% and 7.82%, respectively), but *L. plantarum* dominated the microflora throughout the experiment.

Table 1. Relative abundance (%) of most representative fungi and lactic acid bacteria species, reported for the single sourdough samples and as total value for the three sourdough samples.

	Sourdough Samples				
LAB	**SD1Lab**	**SD1Bak**	**SD2**	**SD3**	**Total**
Lactiplantibacillus plantarum	49.95	1.11	50.47	68.12	55.34
Leuconostoc citreum	19.05	2.39	6.32	20.05	14.99
Fructilactobacillus sanfranciscensis	0.23	79.34	26.73	0.15	9.30
Levilactobacillus brevis	31.99	16.75	6.45	7.82	14.87
Leuconostoc pseudomesenteroides	1.12	0.25	2.93	1.21	1.77
Weissella korensis	0.00	0.01	1.51	1.85	1.15
Fungi					
Saccharomyces cerevisiae	40.96	53.77	67.36	27.61	45.45
Candida lambica	34.63	0.03	20.30	6.45	20.02
Microidium phyllanthi	16.06	5.49	4.57	3.97	7.95
Wickerhamomyces anomalus	0.00	0.00	0.19	16.71	5.81
Dipodascus australiensis	0.00	0.00	0.02	20.07	6.91
Alternaria infectoria	1.01	9.74	2.99	7.15	3.80
Candida santamariae	0.00	0.00	0.00	6.02	2.07
Cladosporium delicatulum	0.37	3.91	0.57	5.06	2.05
Alternaria alternata	0.45	3.90	1.30	2.33	1.39
Mycosphaerella tassiana	0.24	1.99	0.43	1.86	0.86
Gibberella zeae	0.41	3.79	1.13	0.34	0.63
Saccharomyces eubayanus	1.72	0.64	0.00	0.00	0.54
Saccharomyces cariocanus	1.41	1.49	0.72	0.26	0.78
Fusarium culmorum	0.45	4.11	0.00	0.00	0.14

Regarding the yeast species, the evolution of *S. cerevisiae* and *C. lambica* was quite different among the three sourdoughs, as showed in Figure 5. Unlike the *L. plantarum*, the relative abundance of *S. cerevisiae* increased over the experiments, while the *C. lambica* decreased and almost disappeared at the end of the experiments. Despite a similar number of yeast cells (10^5 CFU g^{-1}) being inoculated as starter in the sourdoughs, the relative abundance after the first fermentation step was quite different in the three sourdoughs, at 57.56%, 29.88% and 7.45% for *C. lambica* and 19.21%, 53.73% and 22.10% for *S. cerevisiae* in SD1, SD2 and SD3, respectively. Furthermore, in SD1, *C. lambica* predominated over *S. cerevisae* for the first 10 days (Figure 5), and later *S. cerevisiae* became dominant; in SD2, *S. cerevisiae* was dominant up to 22 days, with 67.4% of relative abundance (Table 1), while *C. lambica* decreased plentifully after 10 days; in SD3, both strains were found at low relative abundance over time (27.6% for *S. cerevisiae* and 6.4% for *C. lambica*) and a multitude of species were present, with *Diplodascus australiensis* and *W. anomalus* found at quite high relative abundance—20.1% and 16.7%, respectively.

The metataxonomic analysis of bakery-propagated SD1 (Figures 4 and 5) highlighted a strong modification of the microbial composition, compared to the laboratory-propagated SD1 sample. Among bacteria, the *L. brevis* and *Leuc. citreum* disappeared after 2 months, and the *F. sanfranciscensis* became the dominant species with a high relative abundance (78.5%). The *L. plantarum* was detected at a very low relative abundance (ca. 1%). Among yeasts, *S. cerevisiae* was present with a relative abundance of 51.4%, but the species diversity increased and numerous species appeared; in particular, fungi of the genera *Alternaria*, *Fusarium*, *Giberella*.

The differences in bacterial composition between sourdoughs were evaluated through beta-diversity analysis. The PCoA plots based on statistically significant (*p*-value < 0.001 PERMANOVA between sourdoughs 16S taxonomic data (Figure 6A) showed that SD1Lab and SD3 samples clustered together and were separated from most of the SD2 and SD1Bak samples. No clear separation among SD1Lab and SD3 could be observed. The two axes explain the 80.6% of variation between samples (PC1 53.1% and PC2 27.5%). The results reported in Figure 6B showed low values of alpha-diversity for the three sourdoughs

(SD1Lab, SD2 and SD3) at the first three points of analysis, according to the microbial composition reported in Figure 4. The alpha-diversity increased throughout the sampling time for SD1Lab and SD3, whereas in SD1Bak and SD2 the diversity decreased in the last three samples due to the dominance of the *F. sanfranciscensis* species.

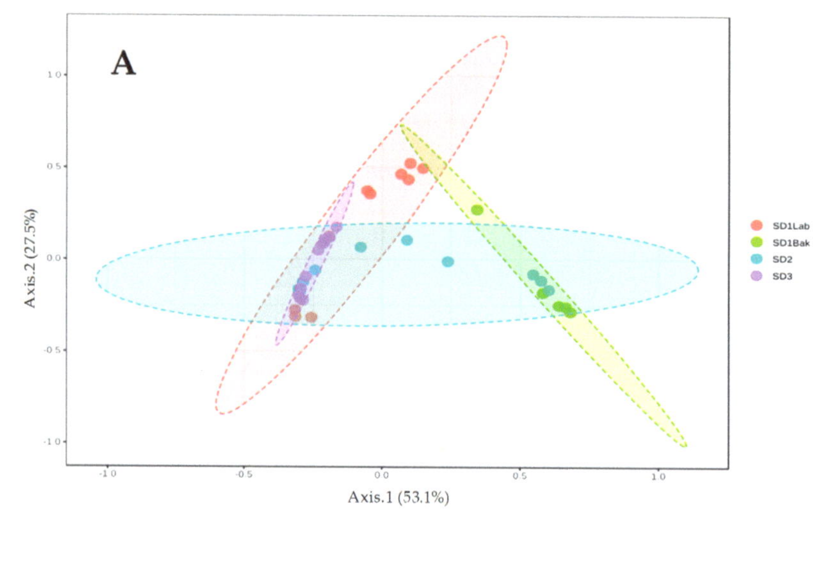

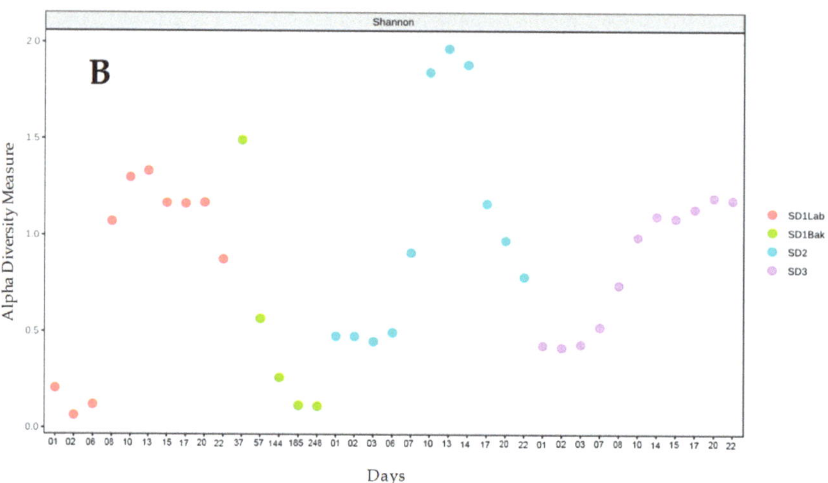

Figure 6. (**A**) Principal component analysis (PCoA) of the bacterial community. The values of axes 1 and 2 are the percentages that can be explained by the corresponding axis. Each color represents one group sample obtained from the same sourdough. (**B**) Shannon index values (alpha-diversity) for each sample, computed on taxonomic information, according to 16S data. Each color represents one group sample obtained from the same sourdough. SD1Lab: laboratory-propagated sourdough. SD1Bak: bakery-propagated sourdough.

To measure the similarity between fungi communities, statistically significant (*p*-value < 0.001 PERMANOVA between sourdoughs) PCoA plots, were performed (Figure 7A). The two axis, explain the 60.9% of variation among samples (PC1 36%

and PC2 24.9%). The plot showed that SD1Lab and SD2 samples clustered together and a clear separation between SD3 and SD1Bak samples according to the first component. Shannon index analysis indicated that the microbial diversity was consistent in the three sourdoughs, with the exception of samples obtained at final points that showed a lower alpha-diversity compared to the initial points (Figure 7B).

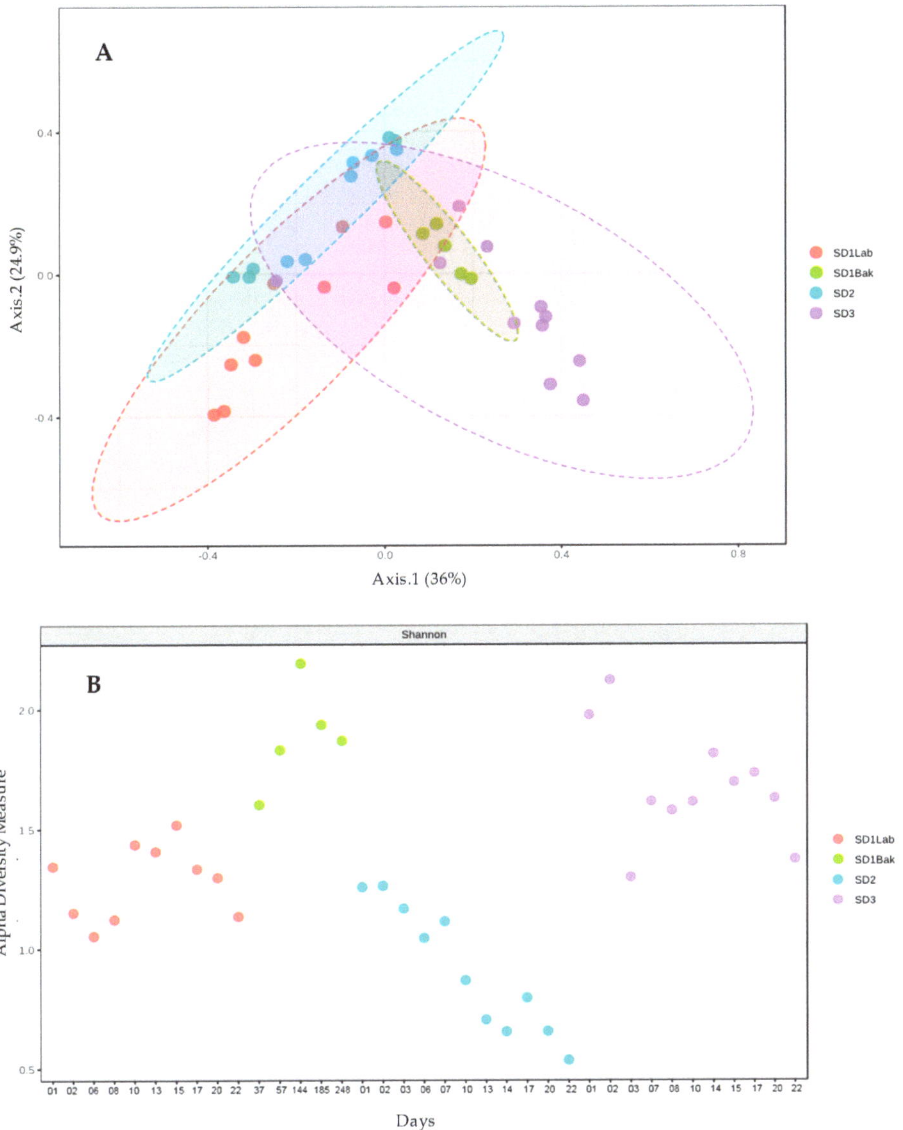

Figure 7. (**A**) Principal coordinate analysis (PCoA) of the fungi community. The values of axes 1 and 2 are the percentages that can be explained by the corresponding axis. Each color represents one group sample obtained from the same sourdough. (**B**) Shannon index values (alpha-diversity) for each sample, computed on taxonomic information, according to ITS data. Each color represents one group sample obtained from the same sourdough. SD1Lab: laboratory-propagated sourdough. SD1Bak: bakery-propagated sourdough.

4. Discussion

The stability of sourdough microflora, in terms of microbial species, is an important factor from an industrial point of view, in order to obtain the standardization of baking processes and of products. The stability and the predominance of specific strains in a sourdough ecosystem is dependent on several factors, such as the metabolic properties of the strains, the microbial interactions, and the technological and ecological parameters [2,16]. In the current work, strains of *L. plantarum*, *S. cerevisiae* and *C. lambica*, isolated from spontaneous sourdoughs, were used to ferment semolina doughs, with the aim of developing a liquid sourdough that had a stable microflora over time. Two fermentation temperatures were investigated and the impact on the microbial strains was observed. The metataxonomic analysis revealed the complexity of the fungi microflora over time, which was characterized by the presence of numerous species; most of them were pathogens of plants, likely suggesting the flour as the origin, and just a few of them were yeasts. With regard to the bacteria, most of them were lactic acid bacteria and their heterogeneity was very low; as a matter of fact, alongside the *L. plantarum* species, inoculated as starter strain, few species developed in all the sourdoughs examined. Above all, *L. brevis* and *Leuc. citreum* were commonly detected from day one to the end of the experiments, except when the *F. sanfranciscensis* became dominant and all the other species disappeared.

The metataxonomic method used in this work does not allow for the identification of a specific microbial strain, so we cannot assert with absolute certainty the persistence of the starter strains during sourdough propagation. However, we could reasonably believe that they were present due to the following reasons: (a) the number of cells inoculated with the starter strains, about 10^7 CFU g^{-1} for bacteria and 10^5 CFU g^{-1} for yeasts, (Figure 1) were close to the values found for the dominant microflora in a mature sourdough (10^6–10^9 CFU g^{-1} for bacteria, 10^5–10^8 CFU g^{-1} for yeasts) [16]; (b) the relative abundance of *L. plantarum* after one day was ca. 90% (Figure 4) in all sourdoughs, whereas it was quite low for *S. cerevisiae*, which tend to increase over time (Figure 5); (c) the first source of sourdough contamination is the flour, where the number of bacteria ranged from 10^4 to 10^6 CFU g^{-1} [18], a value lower than the number of starter cells. The *S. cerevisiae* is one of the most encountered yeast species in spontaneous sourdough [19] whereas *C. lambica*, synonymous with *Pichia fermentans*, is a maltose-negative microorganism that is not considered a typical sourdough microorganism, but it was isolated to a lesser extent from sourdough [20–22]. *Candida humilis* (syn. *Candida milleri*) and *Candida krusei* are the most frequently isolated *Candida* species in sourdough [5]. In this work, more than 10^5 CFU g^{-1} of *S. cerevisiae* and of *C. lambica* were inoculated as starters in sourdoughs. The cells of *C. lambica* decreased over time in all sourdoughs, as showed in Figures 1 and 5, but to a lesser extent in SD1, where the relative abundance was the highest (34.6%) and the decrease over time was slow. In fact, the relative abundance was 1.5% at day 22, while in SD2 it was 1.4% at day 17 and in SD3 it was 1.8% at day 10. The *S. cerevisiae* was found at high relative abundance both in SD1 and SD2 (Table 1), and the increase of the abundance seems to be associated with the decrease of *L. plantarum* and the growth of *F. sanfranciscensis* in both samples, as observed in Figures 4 and 5. The *L. plantarum* is the most employed LAB species in sourdough prepared with starter strains [4], and, despite this, not so many papers studied the permanence in sourdough of *L. plantarum* used as starter. Minervini et al. [23], studying the robustness of seven strains of *L. plantarum*, showed that five of them maintained an elevated number of cells during 10 days of sourdough propagation; nevertheless, new emerging strains were found. The *L. plantarum* is considered a ubiquitous microorganism with a relatively large genome size that allows the expression of important metabolic functions. It has been isolated from different fermented foods including spontaneous sourdough, where it is frequently associated with *L. brevis* [24]. It is noticeable that *L. brevis* was always present in SD1, SD2 and SD3. In SD1, the cell number increased toward the end of laboratory experiment, and in SD3 its relative abundance was quite constant, ranging from 6.4% to 10%. In SD2, the abundance decreased after 10 days, and the same was observed in bakery-propagated SD1, corresponding in both samples to the appearance of the *F. sanfranciscensis* species (Figure 4), which became dominant over the other strains, reaching about 80% of relative abundance at the end of the experiment. The growth of *F. sanfranciscensis* lowered the bacterial diversity, as stated by Comasio et al. [25]. The predominance of *F. sanfranciscensis* in bakery-propagated SD1 is likely due to the power of the "house microbiota", namely the microorganisms contaminating the setting and the equipment of the bakery; the *F. sanfranciscensis* was probably the main bacterial strain in the bakery environment and this led to the replacement of the native strains in the sourdough. The *F. sanfranciscensis* has the smallest genome and the highest density of ribosomal operons within the lactobacilli group, and this feature is retained to favor its predominance in sourdough substrate [16]. Siragusa et al. [26] investigated the predominance of nine strains of *F. sanfranciscensis* inoculated as starters in different type I sourdoughs, and they observed that only three of them were able to dominate during 10 days of continuous propagation. Viiard et al. [27] studied the LAB community of a rye sourdough used in a bakery that was initiated with a commercial starter containing *Limosilactobacillus pontis*; after 28 months of refreshments, the analysis revealed the presence of *F. sanfranciscensis* and *L. pontis*.

As previously reported in the literature [2,28], the dynamics of the microbial community are influenced deeply by the fermentation temperature, but the other process parameters (fermentation time, number of refreshment steps, aeration, dough yield, etc.) are considered important as well. In this work, the effect of different fermentation temperatures (25 °C and 20 °C) on sourdough microflora cannot be easily accounted for. At 25 °C, the total number of viable cells (bacteria and yeast) seems to increase with respect to the sourdoughs fermented at 20 °C (Figure 1); regarding the LAB, three LAB species contributed to increase the number of viable cells, as indicated by metataxonomic analysis in Figure 4; therefore, the high temperature did not favor the growth of *L. plantarum*. Presumptive *Candida* grew up to the 22nd day and the number of presumptive *Saccharomyces* cells is consistent in sourdough fermented at 25 °C. The low fermentation temperature is reported [28] to favor the growth of yeast and heterofermentative LAB species in sourdoughs produced worldwide, whereas homofermentative and facultatively heterofermentative LABs were favored at a high fermentation temperature (>30 °C). In natural sourdoughs, dominated by heterofermentative LABs and the yeast *C. milleri*, the LAB and yeast cells increased when the temperature was raised from 15 °C to 27 °C [29].

The values of pH and TTA observed in sourdough samples were consistent with values found in other papers [4]. The *L. plantarum* species, which is the most abundant and important bacterial species in the laboratory-propagated sourdoughs, showed a similar total relative abundance in SD1, SD2 and SD3 (Table 1), but the decrease over time was more pronounced in SD1 and less pronounced in SD3 (Figure 4). Anyway, the acidifying activity in the first week was greater in SD1 than in SD2 and SD3, and the lowest pH values were observed in SD1. Therefore, the higher fermentation temperature in SD1 could have favored the metabolism of *L. plantarum*; in the first week thereafter, the development of *L. brevis*, a heterofermentative species, could have contributed to the production of organic acids and to the pH decrease [5].

Concerning the sourdough samples fermented at the same temperature (SD2 and SD3) during the first week, the acidification was faster in SD3 than in SD2, likely because of the different method used to inoculate the *L. plantarum*. In fact, the starter strains were inoculated all together in SD2, whereas in SD3 the *L. plantarum* was inoculated alone and left to ferment for 17 h, and then the yeast strains were added in the subsequent backslopping. The analysis of the data reported in Figure 3 indicates that the ΔpH values were more homogeneous in SD3 compared to SD2, and the same phenomenon can be observed for the ΔTTA and bacterial cell density values reported in Figures S2 and S4, respectively. Therefore, in SD3, the growth of lactic acid bacteria and the acidification over the propagation period was more uniform and stable compared to the other samples. Regarding the evolution of starter yeasts, differences can be observed in SD2 and SD3 (Figure 5), unless the fermentation was conducted at the same temperature (20 °C). Indeed, both the *S. cerevisiae* and the *C. lambica* grew well in SD2, where the lowering of dough pH before starter addiction could have favored the growth of yeast, according with that reported by Minervini et al. [16] and to the yeast growth in SD1, where the dough pH was not lowered. On the contrary, in SD3, the lowering of pH did not favor the growth of yeast strains, probably because their growth was conditioned by the rapid growth of *L. plantarum* in the first days. The different methods used for starter addition seem to have affected the behavior of starter strains and the acidification process in sourdoughs.

5. Conclusions

In spontaneous sourdoughs, the first fermentation is commonly carried out by the indigenous microorganism from the flour and/or the environment, and after a few refreshments some species become dominant [16]. In this work, the starter strains were added with the purpose of guiding the first fermentation and to dominate the mature sourdough. The analyses showed that despite the high number of starter cells, other microbial species were able to grow, and sometimes became dominant species. The fungi microbiota were more heterogeneous than the bacteria microbiota, and most species probably originated from flour, being pathogens of plants. For the first time, a strain of *C. lambica* was used as a starter in sourdough fermentation, and the lack of competitiveness towards the other strains was shown. The growth and the stability over time of the starter *L. plantarum* improved when the LAB strain was inoculated alone and left to conduct the first fermentation process for many hours at a low temperature. The high fermentation temperature (25 °C) seems to promote the growth of both bacteria and yeast, but not the growth of the *L. plantarum* starter strain. Therefore, it will be a great challenge to find starter strains that are able to dominate the microflora, grow well at low temperatures for energy saving purposes, and that are competitive enough towards contaminant species, in order to guarantee the persistence of starter strains in sourdough.

Supplementary Materials: The following supporting information can be downloaded at: https://www.mdpi.com/article/10.3390/fermentation8100571/s1. Figure S1: Morphology of the colonies for *S. cerevisiae* and *C. lambica* growing on RB medium. Figure S2: Values of LAB cell density ($\log_{10}$ CFU g^{-1}) The inner line of each box represents the median; the top and bottom of the box represent the 75th and 25th percentiles of the data, respectively. The top and bottom of the bars represent the 100% of the data. The square outside the box plot represents the outliers of the data. Figure S3: Values of yeast cell density ($\log_{10}$ CFU g^{-1}) The inner line of each box represents the median; the top and bottom of the box represent the 75th and 25th percentiles of the data, respectively. The top and bottom of the bars represent the 100% of the data. The square outside the box plot represents the outliers of the data. Figure S4: Values of ΔTTA (difference between the initial TTA value and the values after sourdough refreshment). Data are the means from three independent experiments ($n = 3$). The inner line of each box represents the median; the top and bottom of the box represent the 75th and 25th percentiles of the data, respectively. The top and bottom of the bars represent the 100% of the data. The square outside the box plot represents the outliers of the data.

Author Contributions: Conceptualization, P.C.; formal analysis, V.T., A.C., M.S., S.F. and C.F.; investigation, V.T., A.C. and C.F.; investigation and resources, A.L.; writing—original draft preparation, P.C. and C.F.; writing—review and editing, P.C.; project administration, P.C. and T.R.; funding acquisition, T.R. All authors have read and agreed to the published version of the manuscript.

Funding: This research was funded by Sardegna Ricerche, Science and Technology Park of Sardinia, and Regione Autonoma della Sardegna, under the program POR FESR Sardegna 2014–2020 Azione 1.2.2, and the R&D project "Crunch-Sunalle".

Institutional Review Board Statement: Not applicable.

Informed Consent Statement: Not applicable.

Data Availability Statement: The data presented in this study are available on request from the corresponding author.

Conflicts of Interest: The authors declare no conflict of interest in this research article. Most of the authors are researchers employed by Porto Conte Ricerche Srl, a publicly-owned research company, subsidiary controlled and financed by Sardegna Ricerche, which is an agency of the Regional government of Sardinia.

References

1. Reese, A.T.; Madden, A.A.; Joossens, M.; Lacaze, G.; Dunn, R.R. Influences of ingredients and bakers on the bacteria and fungi in sourdough starters and bread. *mSphere* **2020**, *5*, e00950-19. [CrossRef] [PubMed]
2. Calvert, M.D.; Madden, A.A.; Nichols, L.M.; Haddad, N.M.; Lahne, J.; Dunn, R.R.; McKenney, E.A. A review of sourdough starters: Ecology, practices, and sensory quality with applications for baking and recommendations for future research. *PeerJ* **2021**, *9*, e11389. [CrossRef] [PubMed]
3. Meroth, C.B.; Walter, J.; Hertel, C.; Brandt, M.J.; Hammes, W.P. Monitoring the bacterial population dynamics in sourdough fermentation processes by using pcr-denaturing gradient gel electrophoresis. *Appl. Environ. Microbiol.* **2003**, *69*, 475–482. [CrossRef] [PubMed]
4. Arora, K.; Ameur, H.; Polo, A.; Cagno, R.; Rizzello, C.G.; Gobbetti, M. Thirty years of knowledge on sourdough fermentation: A systematic review. *Trends Food Sci.* **2021**, *108*, 71–83. [CrossRef]
5. De Vuyst, L.; Vrancken, G.; Ravyts, F.; Rimaux, T.; Weckx, S. Biodiversity, ecological determinants, and metabolic exploitation of sourdough microbiota. *Food Microbiol.* **2009**, *26*, 666–675. [CrossRef]
6. De Vuyst, L.; van Kerrebroeck, S.; Leroy, F. Microbial ecology and process technology of sourdough fermentation. *Adv. Appl. Microbiol.* **2017**, *100*, 49–160. [PubMed]
7. De Vuyst, L.; Neysens, P. The sourdough microflora: Biodiversity and metabolic interactions. *Trends Food Sci.* **2005**, *16*, 43–56. [CrossRef]
8. Oshiro, M.; Zendo, T.; Nakayama, J. Diversity and dynamics of sourdough lactic acid bacteriota created by a slow food fermentation system. *J. Biosci. Bioeng.* **2021**, *131*, 333–340. [CrossRef]
9. Landis, E.A.; Oliverio, A.M.; McKenney, E.A.; Nichols, L.M.; Kfoury, N.; Biango-Daniels, M.; Shell, L.K.; Madden, A.A.; Shapiro, L.; Sakunala, S.; et al. The diversity and function of sourdough starter microbiomes. *eLife* **2021**, *10*, e61644. [CrossRef]
10. Fois, S.; Piu, P.P.; Sanna, M.; Roggio, T.; Catzeddu, P. Starch digestibility and properties of fresh pasta made with semolina-based liquid sourdough. *LWT Food Sci. Technol.* **2018**, *89*, 496–502. [CrossRef]
11. Manual DNA Extraction from Food Samples. Available online: https://ita.promega.com/resources/pubhub/applications-notes/an300/an338-manual-dna-extraction-from-food-samples/ (accessed on 24 March 2022).
12. Di Cagno, R.; Pontonio, E.; Buchin, S.; de Angelis, M.; Lattanzi, A.; Valerio, F.; Gobbetti, M.; Calasso, M. Diversity of the lactic acid bacterium and yeast microbiota in the switch from firm-to liquid-sourdough fermentation. *Appl. Environ. Microbiol.* **2014**, *80*, 3161–3172. [CrossRef]
13. Bellemain, E.; Carlsen, T.; Brochmann, C.; Coissac, E.; Taberlet, P.; Kauserud, H. ITS as an environmental DNA barcode for fungi: An in silico approach reveals potential PCR biases. *BMC Microbiol.* **2010**, *10*, 189. [CrossRef] [PubMed]
14. Tanca, A.; Manghina, V.; Fraumene, C.; Palomba, A.; Abbondio, M.; Deligios, M.; Silverman, M.; Uzzau, S. Metaproteogenomics reveals taxonomic and functional changes between cecal and fecal microbiota in mouse. *Front. Microbiol.* **2017**, *8*, 391. [CrossRef] [PubMed]
15. Basic Local Alignment Search Tool. Available online: https://blast.ncbi.nlm.nih.gov/Blast.cgi (accessed on 24 March 2022).
16. Minervini, F.; de Angelis, M.; di Cagno, R.; Gobbetti, M. Ecological parameters influencing microbial diversity and stability of traditional sourdough. *Int. J. Food Microbiol.* **2014**, *171*, 136–146. [CrossRef] [PubMed]
17. Xu, W.-L.; Li, C.-D.; Guo, Y.-S.; Zhang, Y.; Ya, M.; Guo, L. A snapshot study of the microbial community dynamics in naturally fermented cow's milk. *Food Sci. Nutr.* **2021**, *9*, 2053–2065. [CrossRef]
18. Scheirlinck, I.; van der Meulen, R.; de Vuyst, L.; Vandamme, P.; Huys, G. Molecular source tracking of predominant lactic acid bacteria in traditional Belgian sourdoughs and their production environments. *J. Appl. Microbiol.* **2009**, *106*, 1081–1092. [CrossRef]
19. De Vuyst, L.; Harth, H.; van Kerrebroeck, S.; Leroy, F. Yeast diversity of sourdoughs and associated metabolic properties and functionalities. *Int. J. Food Microbiol.* **2016**, *239*, 26–34. [CrossRef]
20. Boyaci-Gunduz, C.P.; Erten, H. Predominant yeasts in the sourdoughs collected from some parts of Turkey. *Yeast* **2020**, *37*, 449–466. [CrossRef]
21. Syrokou, M.K.; Themeli, C.; Paramithiotis, S.; Mataragas, M.; Bosnea, L.; Argyri, A.A.; Chorianopoulos, N.G.; Skandamis, P.N.; Drosinos, E.H. Microbial ecology of greek wheat sourdoughs, identified by a culture-dependent and a culture-independent approach. *Foods* **2020**, *9*, 1603. [CrossRef]
22. Korcari, D.; Ricci, G.; Quattrini, M.; Fortina, M.G. Microbial consortia involved in fermented spelt sourdoughs: Dynamics and characterization of yeasts and lactic acid bacteria. *Lett. Appl. Microbiol.* **2019**, *70*, 48–54. [CrossRef]
23. Minervini, F.; de Angelis, M.; di Cagno, R.; Pinto, D.; Siragusa, S.; Rizzello, C.G.; Gobbetti, M. Robustness of *Lactobacillus plantarum* starters during daily propagation of wheat flour sourdough type I. *Food Microbiol.* **2010**, *27*, 897–908. [CrossRef] [PubMed]
24. Gänzle, M.G.; Zheng, J. Lifestyles of sourdough lactobacilli–Do they matter for microbial ecology and bread quality? *Int. J. Food Microbiol.* **2019**, *302*, 15–23. [CrossRef] [PubMed]

25. Comasio, A.; Verce, M.; van Kerrebroeck, S.; de Vuyst, L. Diverse microbial composition of sourdoughs from different origins. *Front. Microbiol.* **2020**, *11*, 1212. [CrossRef] [PubMed]
26. Siragusa, S.; di Cagno, R.; Ercolini, D.; Minervini, F.; Gobbetti, M.; de Angelis, M. Taxonomic structure and monitoring of the dominant population of lactic acid bacteria during wheat flour sourdough type I propagation using *Lactobacillus sanfranciscensis* starters. *Appl. Environ. Microbiol.* **2009**, *75*, 1099–1109. [CrossRef]
27. Viiard, E.; Bessmeltseva, M.; Simm, J.; Talve, T.; Aaspõllu, A.; Paalme, T.; Sarand, I. Diversity and stability of lactic acid bacteria in rye sourdoughs of four bakeries with different propagation parameters. *PLoS ONE* **2016**, *11*, e0148325. [CrossRef]
28. De Vuyst, L.; van Kerrebroeck, S.; Harth, H.; Huys, G.; Daniel, H.-M.; Weckx, S. Microbial ecology of sourdough fermentations: Diverse or uniform? *Food Microbiol.* **2014**, *37*, 11–29. [CrossRef]
29. Simonson, L.; Salovaara, H.; Korhola, M. Response of wheat sourdough parameters to temperature, NaCl and sucrose variations. *Food Microbiol.* **2003**, *20*, 193–199. [CrossRef]

Review

Achievements of Autochthonous Wine Yeast Isolation and Selection in Romania—A Review

Raluca-Ștefania Rădoi-Encea [1,†], Vasile Pădureanu [2,*,†], Camelia Filofteia Diguță [1], Marian Ion [3], Elena Brîndușe [3] and Florentina Matei [1,2,*]

1 Faculty of Biotechnologies, University of Agronomic Sciences and Veterinary Medicine Bucharest, 59 Mărăști Blvd., District 1, 011464 Bucharest, Romania
2 Faculty of Food Industry and Tourism, Transilvania University of Brașov, 148 Castelului St., 500014 Brașov, Romania
3 Research and Development Institute for Viticulture and Oenology, Valea Călugărească, 107620 Prahova, Romania
* Correspondence: padu@unitbv.ro (V.P.); florentina.matei@biotehnologii.usamv.ro (F.M.)
† These authors contributed equally to this work.

Abstract: Winemaking in Romania has a long-lasting history and traditions and its viticulture dates back centuries. The present work is focused on the development of wine yeast isolation and selection performed in different Romanian winemaking regions during past decades, presenting the advanement of the methods and techniques employed, correlated with the impact on wine quality improvement. Apart from the historical side of such work, the findings will reveal how scientific advancement in the country was correlated with worldwide research in the topic and influenced local wines' typicity. To create an overall picture of the local specificities, the work refers to local grape varieties and the characteristics of the obtained wines by the use of local yeasts as compared to commercial ones. Numerous autochthonous strains of *Saccharomyces* were isolated from Romanian vineyards, of which several demonstrated strong oenological characteristics. Meanwhile, different non-*Saccharomyces* yeast strains were also isolated and are nowadays receiving the attention of researchers seeking to develop new wines according to wine market tendencies and to support wine's national identity.

Keywords: Romania; winemaking; autochthonous yeasts; non-*Saccharomyces* yeast; terroir

Citation: Rădoi-Encea, R.-Ș.; Pădureanu, V.; Diguță, C.F.; Ion, M.; Brîndușe, E.; Matei, F. Achievements of Autochthonous Wine Yeast Isolation and Selection in Romania—A Review. *Fermentation* **2023**, *9*, 407. https://doi.org/10.3390/fermentation9050407

Academic Editors: Agustín Aranda, Roberta Comunian and Luigi Chessa

Received: 15 March 2023
Revised: 13 April 2023
Accepted: 19 April 2023
Published: 23 April 2023

1. Introduction

Winemaking in Romania has a long-lasting history and traditions and its viticulture dates back centuries [1]. With the EU accession in 2007, Romania started a journey with the final goal of putting Romania on the international high-quality wines map. Access to pre- and post-accession funds increased investment in wine making technology, the replacement of low-quality vines, and the replanting vineyards with improved genetic sources [2].

According to OIV (International Organization of Vine and Wine) 2022 statistics [3], Romania is nowadays the sixth largest wine producer in Europe and the thirteenth largest wine producer in the world ranking. The total wine production was estimated at around 4.45 million hl in 2021, increasing from around 3.63 million hl in 2015.

Meanwhile, the total area cultivated with vines decreased from 253.203 ha (1995) to 191.459 ha (2015). Since 2015, when Romania legally declared that wine is considered a food product [4], the area cultivated with vines has still shown some fluctuation, but it remained relatively balanced until 2021, when the number reached 188.891 ha [5].

The delimitation of Romanian viticultural areas was established by the National Office of Vine and Wine Products and is based on the climatic conditions determining the qualitative potential of the grapes and wines, the relief conditions, the applied technologies, the level of the obtained productions, and the qualitative characteristics of the resulting products [6]. Therefore, the Romanian viticultural space consists of 37 vineyards which comprise, in total, 120 viticultural centers and 46 independent viticultural centers, grouped in 8 regions and 3 viticultural areas, as presented in Table 1 and shown in Figure 1.

Table 1. The Romanian viticultural space.

Viticultural Area	Viticultural Region	Vineyards Denominations
Central area, inside the Carpathian arch	The Transylvanian plateau	Târnave, Alba, Sebeș-Apold, Lechința, Aiud
Peri-Carpathian hills	The hills of Moldova	Cotnari, Huși, Iași, Dealu Bujorului, Ivești, Nicorești, Panciu, Odobești, Cotești, Zeletin, Covurlui, Colinele Tutovei
	The hills of Muntenia and Oltenia	Dealu Mare, Sâmburești, Ștefănești, Drăgășani, Dealurile Craiovei, Dealurile Buzăului, Podgoria Severinului, Plaiurile Drancei
	Banat	6 independent centers
	Crișana and Maramureș	Diosig, Miniș-Măderat, Valea lui Mihai, Podgoria Silvaniei
Danube Pontic area	The Dobrogea hills	Murfatlar, Sarica-Niculitel, Istria-Babadag
	The Danube terraces	Ostrov, Greaca
	Region of sands and other favorable lands in the South of the country	Calafat, Sadova-Corabia, Podgoria Dacilor

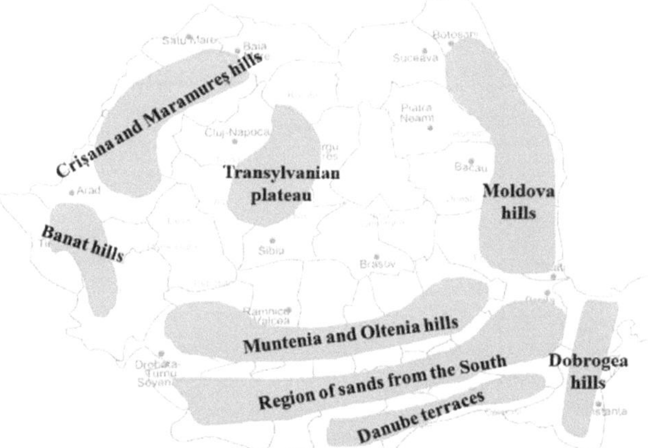

Figure 1. The Romanian viticultural regions and their geographical location.

The trend in Romanian winemaking is to maintain an uprising path in terms of total wine production volume, while also increasing the diversity of local wine types. These goals can be achieved starting from the use of local grape varieties, as well as via the isolation, selection, and then the use of autochthonous yeasts in the production of Romanian wines.

In recent decades, winemakers could choose from a wide variety of commercial yeasts provided by several well-known companies; these are yeasts that display a wide range of special characteristics, adapted to specific needs [7]. In line with the evolution of consumers' preferences and even with climate changes that bring about a higher-than-previous sugar concentration, finding yeasts with special traits was and is a continuous project [8].

Vineyard yeast biodiversity characterization and wine yeast selection are not new entries in wine-making research, but considering the history of wine, these approach can be considerred as young. In the history of winemaking, the use of selected starter cultures did not become widespread practice until the 1970s, and the vast majority of the industrial yeasts belong to *Saccharomyces cerevisiae*; however, currently, it is recongnized that non-*Saccharomyces* species may also be relevant for alcoholic fermentation [9]. It is generally recognized that the current set of the commercial *S. cerevisiae* strains or derived hybrids is not sufficient to provide new technological or organoleptic properties in wine; therefore, new strains are desired, if not essential [10]. Hybrid genomes of *Saccharomyces cerevisiae / Saccharomyces kudriavzevii* yeast strains used for wine making in France (Alsace), Germany, and Hungary have been characterized by the use of microsatellite markers [11]. Autochthonous strains represent alternative genetic resources by which the industry can overcome current challenges. The preservation of spontaneous microflora is essential to obtain the typical flavor and aroma of wines deriving from different grape varieties [12]. Meanwhile, the last two decades, practices of organic vine growing influenced fungi (yeast and molds) biodiversity. This was clearly proven in France, in the Bourgogne region, with respect to the Chardonnay variety [13]. In recent years, on the European level, researchers from different groups and countries have focused on yeast selection and biodiversity issues. Ecological and geographic studies have highlighted that unique strains are associated with particular grape varieties in specific geographical locations [14]. An example of such initiatives was provided by the European project, WILDWINE Project (EU contract 315065), focused on the selection of wild microorganism in five worldwide- recognized wine regions: Nemea and Crete (Greece), Piedmont (Italy), Bordeaux (France), and Priorat (Spain) [15]. In Italy, a wide range of vineyards were examined, covering most of the wine's Italian regions: in the northwest, in the Piedmont region and Monferrato vineyards concerning Barbera grapes [16]; the Barbera variety was also studied in the "Nizza" Barbera d'Asti DOC zone [17]. In Sicily, a wide study was conducted on hundreds of isolates and the superiority of the local strains over the commercial strains was proved [18]. Another focus was on Montepulciano d'Abruzzo "Colline Teramane" premium wine DOCG, produced in Teramo province; the presence of atypical *S. cerevisiae* strains only in a particular vineyard in a restricted area suggests the role of local selective pressure in the origin of distinctive *Saccharomyces* yeast populations [19]. In Spain, several groups conducted similar work, and screening results were reported for wine regions such as Douro, Extremadura, Galicia, La Mancha and Uclés, Ribera del Duero, Rioja, Sherry area, and Valencia [20]. Moreover, in the DOQ Priorat region, isolation was performed on varieties such as Grenache and Carignan [21]; in the northwest, in the Galicia region, biodiversity was studied, comparing organic and conventional culture [22]. Relatively recently, isolates from three appellations of Spanish origin were checked for fingerprinting of interdelta polymorphism; ancient vineyards managed with organic practices showed intermediate to low levels of strain diversity, indicating the existence of stable populations of *S. cerevisiae* strains [23]. In another European area, in the Greek island of Kefalonia, in the Mavrodafni wine region, at the end of the alcoholic fermentation, indigenous yeasts were isolated; selected strains are already in industrial use [24]. In the European eastern neighborhood, in Georgia, a traditional winemaking country, long-term biodiversity studies were conducted in the Dagestan region using various isolation techniques and various substrates [25].

The present work is focused on the development of wine yeast isolation and selection performed in different Romanian winemaking regions during recent decades, presenting the advance of the employed methods and techniques, correlated with the impact on wine quality improvement. Apart from the historical side of the importance of such work, our findings will reveal how scientific advancement in the country is correlated to worldwide research in the topic.

2. Materials and Methods

The current review is based on the available scientific articles that record research regarding the isolation and selection of local wine yeasts from different Romanian vineyards. Most of the sources approached are indexed in different international databases, such as

Google Scholar, ScienceDirect, Web of Knowledge-Clarivate, and CABI. However, the available records in the international databases start from 2005, while records in some native language (Romanian), available in different national libraries, go as far as the beginning of the century, in 1915 [26]. In addition, to create an overall picture of the local specificities, scientific reports were also used in relation to local grape varieties (Fetească regală, Fetească albă, Crâmpoșie, Băbească neagră, Fetească neagră, Grasă de Cotnari, Cadarcă, Tămâioasă românească, etc.) and the characteristics of the wines obtained by the use of local yeasts compared to the commercial ones.

3. Results and Discussion

This review took into account the reported work on wine yeast isolation and selection activities performed in the wine-growing areas of Romania from 1915 to the present.

From the chronological point of view, according to Brîndușe et al. [26], the first report, from 1915, comes from the doctoral thesis of Nițescu M.A. [27]. He made an ample physiological characterization of yeast isolated from different regions and local grape varieties, such as Cotnari (Grasă, Fetească albă), Iași (Fetească neagră), Pietroasele (Grasă), Drăgășani (Tămâioasă românească, Negru moale, Negru vârtos, Crâmpoșie), and Odobești (Tămâioasă românească). This study, conducted in Paris, was positively appreciated by Ribéreau Gayon and Peynaud in 1960, according to the same source [26]. Following this study, in the 1920s–1930s, Dr. Russ and his team (Dr. Moldovan and Dr. Mavromati) founded the national school of wine microbiology and the first Romanian wine yeast collection. In the years 1945–1965, different researchers focused on local wine yeast selection [28–34]. Beginning in the 1970s, isolation and selection work has increased, and the results are detailed below.

In terms of the vineyard region, yeast isolation and selection work was reported in several areas, covering most of the Romanian winemaking regions. For instance, in the Transylvanian plateau, Dănoaie [35] and Stamate et al. [36] focused on the yeast biodiversity in Târnave vineyard, while Oprean [37] studied several Sibiu wine-growing areas. In Moldova, such experiments were conducted by Sandu-Ville et al. [38,39], followed by Viziteu et al. [40] in Cotnari vineyard, by Vasile et al. [41] and by Nechita et al. [42] in the Iași-Copou vineyard, as well as by Găgeanu et al. [43] in Dealurile Bujorului vineyard. In the hills of Muntenia, the research started in Valea Călugărească center by Kontek and Kontek [44,45], followed by Matei Rădoi et al. [46] and Brîndușe et al. [47,48], and in the Buzău vineyard by Bărbulescu et al. [49]. In the Oltenia hills in Tamburești, Banu Mărăcine, Drăgășani, and Târgu Jiu, studies were conducted by Dragomir Tutulescu and Popa [50], while Beleniuc [51] isolated wine yeast from the Murfatlar vineyard in the Dobrogea hills.

3.1. Employed Techniques of Yeast Isolation, Identification and Selection

Different approaches were taken into account during the isolation work, starting from grape washing water [42,43,45,47,52], continuing with the juice from fresh crushed grapes [40,46,52] or must in different fermenting stages: respectively, at the beginning, middle, and end of fermentation [42,48,53]. The employed microbiological media were the classical ones, meaning Sabouraud medium or Yeast Extract Peptone Dextrose (YEPD) supplemented with chloramphenicol. Bărbulescu et al. also made use of a specific medium for yeast isolation (malt extract–peptone yeast extract agar), then another specific medium (yeast extract–malt extract sucrose agar) for the maintenance of the culture [49].

The selection work followed typical steps, i.e., respectively, by monitoring the parameters of the fermentations and the characteristics of the obtained wines. Classically, there were employed tests such as ethanol tolerance [7,42] or the refermentation capacity of the strains [42]. Of the yeasts tested by Nechita et al. from Iași-Copou, five strains proved to be tolerant of high concentrations of ethanol of about 14–15% [42]. Regarding their capacity to restart the stagnated fermentation at 11.5% ethanol and 70 g/L sugars, the strains managed to bring the fermentation to an end and produce dry wines. Dragomir, Tutulescu, and Popa used the standard methods accepted by OIV to isolate, identify, and described their strains' biological, physical, and oenological characteristics from the Oltenia area [50,54]. In the end, most of the authors reported the results of the physicochemical and organoleptical characteristics of the obtained wines after using the selected strains. Following this path, Vasile et al. isolated 86 local yeast strains from the Iași-Copou vineyard, followed by a final selection for the best fermentative characteristics and wine profiles [41,55]. In terms of the killer profile of the isolated yeast, only one report was identified in the databases, in which Matei and Găgeanu reported a killer positive strain isolated in Dealurile Bujorului county [56].

Less conventional methods were used in the characterization and wine yeast selection. For instance, Antoce and Nămoloșanu employed a calorimetric method using a multiplex batch micro-calorimeter (isothermal, conduction type) for the rapid yeast testing for ethanol tolerance in order to select strains that were useful for winemaking [57]. They demonstrated that the method could eliminate labor-intensive cell counting, as well as its high sensitivity and the possibility of measuring cultures grown in intense-colored or high-turbidity media, such as red wine. In addition, this method offers the benefit of simultaneously monitoring a large number of samples in a 48–72-h experiment.

The identification work, hand in hand with yeast biodiversity studies, had a slow evolution in terms of the employed techniques in past decades. Such work rquires know-how and specific tools, and the predominant methods were based on classical morpho-physiological tests, according to Barnett et al. [58,59], Krieger-van Rij [60], and Delfini [61]. Most authors reported studies on the macroscopic features of the colonies, pseudo-mycelium formation, and sporulation on a specific medium [43,44,46,47,52]. Tests such as fermentation and assimilation of different carbohydrates, nitrogen utilization, the use of ethanol as the sole carbon source, and arbutin split were taken into account [37,47,52]. Several authors were using rapid biochemical tests; that is, API galleries [40,46].

Some teams made use of MALDI-TOF mass spectrometry, especially that of Bărbulescu et al., wherein the isolated strains were prepared for the analysis after the extraction of peptides with formic acid, ethanol, and acetonitrile [49]. A similar approach was taken by Corbu and Csutak when studying yeast biodiversity in different traditional fermented foods, including wine [62]. For a more accurate physiological identification of the tested strains, phenotypic phylogeny analyses were also performed using Biolog Microbial ID System according to the manufacturers' specifications [63].

The molecular approach came later on in the country, when PCR-ITS RFLP techniques were employed by Gaspar et al. [64] in Sebeș vineyard (Apold-Blaj centre), followed by Găgeanu et al. [43] in Dealurile Bujorului vineyard, and Dumitrache et al. [53] in Pietroasa center (Dealu Mare vineyard); these results were also coupled with sequencing data. These teams performed conventional DNA extraction, followed by PCR amplification with ITS 1 and ITS 4 primers, continuing with *Hinf*I, *Hae*III, and *Hha*I digestion [43], or *Alu*I and *Taq*I [64], and comparing the obtained profiles with the existent databases.

The first PCR-RAPD approach was taken by Oprean, when different *Saccharomyces* and non-*Saccharomyces* strains, isolated from Sebeș-Apold vineyard, were identified [65]. Relatively recently, apart from using the ITS-RFLP technique of the ITS1-5.8S rDNA-ITS2 region, taking advantage of the restriction enzymes such as *Hinf*I, *Hae*III, *Cfo*I, and *Msp*I, Corbu and Csutak have also employed the RAPD method for the identification of yeast involved in wine spontaneous fermentation [62,63]. In their case, the intraspecific biodiversity (genetic relatedness) of the isolates was detected by analyzing the RAPD profile obtained for each strain and by calculating the similarity index using the Jaccard coefficient (Sij). Similarly, the interspecific biodiversity of the microbial communities from spontaneous fermented products was determined by comparing their profile to the RAPD profile of their co-fermenters; in the end, the dendograms were generated by PyElph, using the UPGAMA (unweighted pair group method with arithmetic mean) method.

3.2. Yeast Biodiversity and Identification Results

The wine yeast studies in Romania followed two different patterns. Most of the authors have isolated and selected different strains, followed by identification only for the strains proving special and/or demonstrating specific winemaking profiles and characteristics. Systematic studies were started only in later 1970s by Kontek et al. (1975–1977). Later on, a few studies took into account the study of the vineyard or fermented grape must yeast biodiversity as a whole [46,62].

A first ample biodiversity report study was performed by Kontek in 1977 [66], in Dealu Mare vineyard (Valea Călugărească centre), adopting the classification proposed by Lodder and Kreger-van Rij [67]. Among 244 isolates, the predominant genus was *Saccharomyces*, with the following species and var.: *S. ellipsoideus* (dominant), *S. bayanus*, *S. carlsbergensis*, *S. cerevisiae*, *S. exiguous*, *S. heterogenicus*, *S. florentinus*, *S. fructuum*, *S. italicus*, *S. oviformis*, *S. rosei*, *S. steinerii*, *S. uvarum*, and *S. logos*. In terms of non-*Saccharomyces* (NS) species, they reported *Candida mycoderma*, *Candida peliculosa*, *Kloeckera apicullata*, *Kloeckera africana*, *Torulopsis stellata*, *Pichia membranaefaciens*, and *Rhodotorula mucilaginosa*.

Later on, Matei Rădoi et al. performed a similar study in the Valea Călugărească center, Dealu Mare vineyard, comparing the data obtained by Kontek team in the 1970s in a double approach: classical morphophysiological study; and by API 20C AUX—Biomerieux [46]. The isolation was performed during 2007–2009 on Cabernet Sauvignon, Merlot, Fetească Neagră, and Pinot Noir varieties. A change in the yeast species profiles was noticed throughout the decades; specifically, the 1970s as compared to the 2000s. Among 262 isolates, the dominant species isolated in the vineyard belonged to the NS species, such as *C. famata*, *K. apiculata*, and *Debaryomyces hansenii*. One year later, a similar study was published in the same area [47], in which the dominant NS species were *C. utilis*, *K. apiculata*, *R. mucilaginosa*, and *D. hansenii*, with the employed method and the results being very close among the two teams. Other reported isolates belonged to *Candida lusitaniae*, *C. stellata*, *C. utilis*, *C. magnoliae*, *C. pelliculosa*, *Pichia anomala*, *P. jadinii*, *Torulaspora delbrueckii*, and *Hanseniaspora uvarum* (Table 2).

Multiple NS species were identified from the Cotnari vineyard by Viziteu et al., namely, *C. mycoderma*, *Hansenula anomala*, *H. uvarum*, *Kluyveromyces* spp., *P. membranafaciens*, and *T. stellata* [40].

Vasile et al. selected three *S. ellipsoideus* strains and determined their influence on the must of three grape varieties from Iași-Copou, namely, Fetească albă, Sauvignon blanc, and Chardonnay [41,55]. Other *Sacharomyces* spp. were reported by Găgeanu et al. in Dealurile Bujorului county (Table 3), such as *S. bayanus*, for instance [43].

The strains isolated and tested in Oltenia county by Dragomir Tutulescu and Popa in 2009–2010 were identified as *K. apiculata*, *P. membranafaciens*, *Rhodotorula glutinis*, *S. ellipsoideus* (the most abundant during must fermentation), and *S. oviformis*, but they also found few representatives of *S. rosei*, *Candida vinaria*, and *Metschnikowia reukaufii* [50,54].

In 2014, Oprean identified in Sebeș-Apold county, by molecular tools, *S. ellipsoideus* and *S. oviformis*, as well as NS yeasts such as *Candida vini* and *K. apiculata* [65]. Similarly, in Blaj centre, Stamate et al. reported as dominant, among 139 isolates, the species of *S. cerevisiae* var. *ellipsoideus*, *K. apiculata*, *S. oviformis*, and *S. bayanus* during must fermentation, while *K. apiculata*, *C. mycoderma*, and *T. stellata* were abundant on the grapes [36].

A general image on the *Saccharomyces* spp. isolated and selected in Romania is presented in Table 3. The main identified *Saccharomyces* species and varieties belong to *S. bayanus*, *S. cerevisiae*, *S. chevalieri*, *S. ellipsoideus*, *S. florentinus*, *S. oviformis* (synonym *S. cerevisiae*), or *S. uvarum*.

Table 2. The non-*Saccharomyces* (NS) yeasts isolated from various winemaking areas in Romania.

Genus	Species	Centre/Vineyard	References
Candida	C. colliculosa	Valea Călugărească, Dealu Mare	[46,47]
		Recaș	[68]
	C. famata	Valea Călugărească, Dealu Mare	[46,47]
	C. lusitaniae	Valea Călugărească, Dealu Mare	[46,47]
	C. magnoliae	Valea Călugărească, Dealu Mare	[46,47]
		Recaș	[68]
	C. mycoderma	Cernavodă, Murfatlar	[52]
		Cotnari vineyard	[40]
	C. pelliculosa	Valea Călugărească, Dealu Mare	[46,47]
	C. sphaerica	Valea Călugărească, Dealu Mare	[46,47]
	C. tropicalis	Recaș	[68]
	C. utilis	Valea Călugărească, Dealu Mare	[46,47]
	C. vini	Drăgășani, Tamburești	[50]
Clavispora	C. lusitaniae	Valea Călugărească, Dealu Mare	[47]
Debaryomyces	D. hansenii	Valea Călugărească, Dealu Mare	[46,47]
Dekkera	D. anomala	Pietroasa vineyard	[53]
Geotrichum	G. penicillatum	Valea Călugărească, Dealu Mare	[47]
Hanseniaspora	H. uvarum	Recaș	[68]
Hansenula	H. anomala	Cernavodă, Murfatlar	[52]
		Cotnari vineyard	[40]
Kloeckera	K. apiculata	Cernavodă, Murfatlar	[52]
		Valea Călugărească, Dealu Mare	[46,47]
		Drăgășani	[33]
		Recaș	[68]
Lachancea	L. kluyveri	Cotnari vineyard	[40]
Metschnikowia	M. pulcherrima	Drăgășani	[50]
		Pietroasa vineyard	[53]
Pichia	P. angusta	Recaș	[68]
	P. anomala	Valea Călugărească, Dealu Mare	[46,47]
	P. fermentans	Cernavodă, Murfatlar	[52]
		Recaș	[68]
	P. jadinii	Valea Călugărească, Dealu Mare	[46,47]
	P. kudriavzevii	Ilfov area	[63]
	P. membranaefaciens	Drăgășani, Tamburești	[50]
		Cotnari vineyard	[40]
	P. ohmeri	Valea Călugărească, Dealu Mare	[47]
Rhodotorula	R. glutinis	Valea Călugărească, Dealu Mare	[46,47]
		Recaș	[68]
	R. minuta	Valea Călugărească, Dealu Mare	[46,47]
	R. mucilaginosa	Cernavodă, Murfatlar	[52]
		Valea Călugărească, Dealu Mare	[47]
		Recaș	[68]
Torulaspora	T. delbrueckii	Valea Călugărească, Dealu Mare	[46,47]

Table 2. *Cont.*

Genus	Species	Centre/Vineyard	References
Torulopsis	*T. stellata*	Cernavodă, Murfatlar	[52]
		Cotnari vineyard	[40]
Zygosaccharomyces	*Z. bailii*	Cotnari vineyard	[40]
	Z. rouxii	Cotnari vineyard	[40]

Table 3. The *Saccharomyces* species and varieties isolated from various winemaking areas in Romania.

Species	Centre/Vineyard	References
S. bayanus	Cernavodă, Murfatlar	[52]
	Dealurile Bujorului	[43]
	Cotnari vineyard	[40]
S. cerevisiae	Buzău vineyard	[49]
	Pietroasa vineyard	[53]
	Recaș	[68]
	Valea Călugărească, Dealu Mare	[46]
	Cotnari vineyard	[40]
S. chevalieri	Cotnari vineyard	[40]
S. ellipsoideus	Cernavodă, Murfatlar	[52]
	Dealurile Bujorului	[43]
	Iași-Copou vineyard	[41]
	Cotnari vineyard	[40]
S. florentinus	Cotnari vineyard	[40]
S. oviformis (synonym *S. cerevisiae*)	Cernavodă, Murfatlar	[52]
	Dealurile Bujorului	[43]
	Cotnari vineyard	[40]
S. uvarum	Cotnari vineyard	[40]

3.3. Selected Yeast Properties and the Final Characteristics of Local Wines

From the available records, a wide range of grape varieties were tested, of which nine are registered as local varieties (Table 4), while the wines' characteristics (Table 5) were assessed for both red wines and white wines, though more attention have been given to the white wines.

In the case of white wines, the local selected yeasts were tested on local varieties (Fetească albă, Fetească regală, Tămâioasă românească), as well as on international varieties (Aligoté, Chardonnay, Sauvignon blanc, Pinot gris, Muscat ottonel).

Regarding Feteasca albă, this type of wine was obtained and tested in Dealu Bujorului, with 13.5% alcohol (v/v) and without residual sugar detected [69], and in Iași, with 11.6% alcohol (v/v) and 0.2 g/L sugars [55]. Colibaba et al. [70], Dobrei et al. [71], and Bora et al. [69] obtained Fetească regală wine from Iași, Miniș-Măderat, and Dealu Bujorului, with an average alcohol content of 13.7% (v/v). The residual sugar content was very different—from 1.9 g/L (Dealu Bujorului) and 3.9 g/L (Miniș-Măderat) to 6.63 g/L (Iași).

Aligoté wines showed some differences in terms of ethanol content from one location to another, but also within the same location. Thus, the Aligoté obtained in Dealu Bujorului had a content of 13.1% ethanol (v/v) with no residual sugars detected [69], while those obtained in Iași had, respectively, 10.08% ethanol (v/v) with 0.72 g/L sugars [70], and 11.33% ethanol without a mention of the residual sugars [72].

Colibaba et al. [73] and Bora et al. [69] also obtained Italian Riesling wines with around 11% ethanol, but the first author obtained a dry wine with 0.77 g/L residual sugar, while the second author obtained a sweet wine with 72 g/L residual sugar.

Table 4. Wine grape varieties from Romanian vineyards fermented with selected autochthonous yeast.

Grape Varieties	Vine Regions	References
Aligoté	Iași	[72]
	Iași	[70]
	Dealu Bujorului	[69]
Băbească gri	Dealu Bujorului	[69]
Cabernet sauvignon	Dealu Mare	[73]
	Miniș-Măderat	[71]
	Dobra (Satu Mare)	[74]
Cadarcă	Miniș-Măderat	[74]
Chardonnay	Iași	[55]
Feteasca albă	Dealu Bujorului	[69]
	Iași	[55]
Feteasca neagră	Miniș-Măderat	[74]
	Panciu	[75]
	Ratești (Satu Mare) and Aliman (Constanța)	[74]
Fetească regală	Dealu Bujorului	[69]
	Iași	[70]
	Miniș-Măderat	[71]
Frâncușă	Iași	[70]
Grasa de Cotnari	Iași	[70]
Italian riesling	Dealu Bujorului	[69]
	Iași	[70]
Merlot	Aliman (Constanța)	[74]
Muscat ottonel	Dealu Bujorului	[69]
	Iași	[70,76,77]
Neuburger	Iași	[70]
Pinot gris	Iași	[70]
	Miniș-Măderat	[71]
Pinot noir	Ratești (Satu Mare)	[74]
Rose traminer	Iași	[70]
Sarba	Dealu Bujorului	[69]
Sauvignon	Dealu Mare	[78]
Sauvignon blanc	Dealu Bujorului	[69]
	Iași	[55,70]
Tamaioasă românească	Iași	[70]
Traminer	Miniș-Măderat	[71]

Muscat Ottonel wines were obtained in two Moldova areas, one from Dealu Bujorului and three from Iași. The wine obtained in Dealu Bujorului was a sweet wine, with 11% ethanol and 30.7 g/L residual sugar [69]. Colibaba et al. [70] and Vararu et al. [76] obtained dry wines from Iași, with less than 2 g/L sugar and 12.2%, respectively, and 13.6% ethanol. The glycerol content of Vararu et al. wine was almost 13 g/L. Focea et al. obtained a sparkling wine with 10.3% ethanol, but without mentioning the sugar content [77].

As for Pinot gris, two wines with an increased ethanol content of about 14% were obtained in Iași [70] and in Miniș-Măderat [71]. Vișan et al. [78] obtained three Sauvignon semi-dry wines from Dealu Mare, with an average of 12.5% ethanol, 11 g/L sugar, and about 8–10 g/L glycerol. Vasile et al. [55] and Colibaba et al. [70] each made a dry Sauvignon blanc from Iași, with 11.2–11.9% ethanol and approx. 1 g/L sugar; wine from 2010 had a content of 7.4 g/L glycerol. The Sauvignon blanc obtained from Dealu Bujorului [69] was semi-dry, with 12 g/L sugar and higher ethanol content of 14.4%.

On red wines' side, Cabernet sauvignon was tested in Dealu Mare [73], Miniș-Măderat [71], and in Dobra, Dealurile Silvaniei [74]. This type of wine had an alcohol content between 12% and 15% (*v/v*); the highest value was obtained in Miniș-Măderat. The residual sugar content was 3.8 g/L in the 2012 study, 10.05 g/L in the 2015 study, and not specified in the 2018 study. Vișan et al. also emphasized that the glycerol content was 9 g/L [73], which contributes to the wine's texture and body [79]. Manolache et al. [74,75] and Dobrei et al. [71] obtained and tested Feteasca neagră wine, with an average of 13.49% ethanol (*v/v*) and 3.48–3.9 g/L residual sugar.

Table 5. Wines obtained in Romanian winemaking areas after fermentation with local yeast and their physicochemical properties.

Grape Varieties	Vine Region	Alcohol Vol. (%)	Residual Sugars (g/L)	Total Acidity (g/L)	Volatile Acidity (g/L)	References
Aligoté	Iași	11.33	*	6.72	0.35	[72]
	Dealu Bujorului	13.1	nd	5.5	0.37	[69]
Băbească gri	Iași	10.08	0.72	9.14	0.33	[70]
	Dealu Bujorului	13.2	12.7	5.9	0.38	[69]
Cabernet Sauvignon	Dealu Mare	13.1	3.8	4.3	0.7	[73]
Cadarcă	Miniș-Măderat	15	10.05	5.5	0.43	[71]
	Dobra (Satu Mare)	12	*	5.42	0.47	[74]
Chardonnay	Miniș-Măderat	13.25	2.44	5.55	0.32	[71]
	Iași	12.4	nd	5.9	0.29	[55]
Fetească albă	Dealu Bujorului	13.5	nd	4	0.39	[69]
	Iași	11.6	0.2	5.6	0.28	[55]
Fetească neagră	Miniș-Măderat	13.97	3.48	5.93	0.42	[71]
	Panciu	13.5	3.9	5.32	0.88	[75]
	Ratești (Satu Mare)	13.06	*	5.98	0.57	[74]
	Aliman (Constanța)	13.43	*	5.41	0.73	[74]
Fetească regală	Dealu Bujorului	13.8	1.9	5.3	0.42	[69]
	Iași	13.94	6.63	6.92	0.43	[70]
	Miniș-Măderat	13.39	3.9	5.7	0.53	[71]
Frâncușă	Iași	11.87	0.63	8.54	0.41	[70]
Grasa de Cotnari	Iași	11.6	1.7	8.55	0.25	[70]
Italian Riesling	Dealu Bujorului	11	72	4.9	0.61	[69]
	Iași	11.83	0.77	7.07	0.29	[70]
Merlot	Aliman (Ostrov)	14.14	*	5.25	0.65	[57]
Muscat ottonel	Dealu Bujorului	11	30.7	4.4	0.54	[69]
	Iași	12.2	1.34	6.43	0.33	[70]
	Iași	13.6	1.67	6.4	0.35	[76]
Sparkling Muscat ottonel	Iași	10.3	*	6.2	0.33	[77]
Neuburger	Iași	12.44	10.63	7.71	0.45	[70]

Table 5. *Cont.*

Grape Varieties	Vine Region	Alcohol Vol. (%)	Residual Sugars (g/L)	Total Acidity (g/L)	Volatile Acidity (g/L)	References
Pinot gris	Iași	14.49	4.81	6.68	0.33	[70]
	Miniș-Măderat	13.39	2.04	5.93	0.47	[71]
Pinot noir	Ratești (Dealurile Silvaniei)	13.47	*	6.01	0.53	[74]
Rose Traminer	Iași	14.1	1.67	6.73	0.25	[70]
Șarba	Dealu Bujorului	14.1	23	5.8	0.54	[69]
Sauvignon	Dealu Mare	12.2	10	5.8	0.3	[78]
		13	12	5.4	0.4	
		12.5	12	5.2	0.4	
Sauvignon blanc	Dealu Bujorului	14.35	12	5.2	0.57	[69]
	Iași	11.24	1.1	5.94	0.29	[70]
	Iași	11.9	0.9	5.95	0.2	[55]
Tămâioasă românească	Iași	11.63	15.47	6.93	0.31	[70]
Traminer	Miniș-Măderat	12.3	50	5.9	0.47	[71]

*: the authors did not mention the residual sugar content in the respective wines; nd: not detected.

Special wines were also obtained in Dealu Mare, Valea Călugărească center by Kontek and Kontek (1976); specifically, Jerez type wines, made of pellicular autochthonous yeast isolates belonging to *S. bayanus* species. These wines reached 15–16% alcohol, a maximum of 4 g H_2SO_4/L acidity, and the most appreciated were the ones with residual sugar of 16–17 g/L. The same authors also reported a cryophilic yeast, identified by classical tools as *S. carslbergensis*, initially isolated from must fermenting at 5 °C; this strain led to rapid wine clarification and produced low volatile content and high glycerol content. Similarly, for the cryophilic property, Tudose et al. selected a *S. ellipsoideus* strain in Iași-Copou centre, which was also resistant to high sulphur hydrogen content [80].

For high-quality sparkling wines, isolates of *S. oviformis* and *S. carlsbergensis* were selected in Blaj county during the 1980s [35]; they were capable of complete sugar consumption, while not stimulating the malolactic fermentation and not producing high volatility.

In the 1980s–1990s, generally, special attention was given to high-alcohol, low-foaming, and high-glycerol wine yeast strains, e.g., in Valea Călugărească center [81] and Iași county [82].

Starting with the 2000s, attention was more focused on the aromatic profile of wines made of local grape varieties and local yeast, while less attention was given to the high alcoholic strength. For instance, Liță et al. reported different local strains of *S. cerevisiae* var. *ellipsoideus* as appropriate candidates for dry white wines made of local varieties, such as Fetească albă and Fetească regală [83]. Moreover, in 2017, Lengyel and Panaitescu reported a local yeast isolated from Gârbova area (Sebeș-Apold vineyard), which was capable of improving the terpene flavor compounds content in Muscat ottonel wines [84]. A deeper study and methodology was reported by Vararu et al. after analyzing the aromatic profile of Muscat ottonel variety fermented with commercial and local yeast from Copou Iași centre [76]; a visual and easy to understand foot-printing was also performed, based on a multiple variable analysis, which established differences in the fermentative volatilome.

3.4. New Selection Directions in the Terroir Concept Context

The conventional practice of producing wines on an industrial scale with the use of *Saccharomyces* species involves controlled fermentation from all points of view. The wines thus obtained can be denominated according to the geographical indication (GI) if certain legislative requirements are followed. However, for an even greater specificity, a possible direction might be the use of local yeasts from each geographical region, in addition to using grapes harvested from those areas.

On another note, one way to obtain local wines is the spontaneous fermentation of grapes, but there are multiple disadvantages. The obtained wines may have different characteristics from one vintage to another, depending on many environmental variables, such as climate (temperature, precipitation, sunlight, wind), biology (microbiota, flora, and fauna), relief (topographic coordinates, geomorphology), and geology (soil types, irrigation, fertilization), as well as human implications, namely, traditions, culture, applied technology, agronomic practices, and legislation [8,85,86]. All these are involved in the concept of *terroir*.

Knight et al. consider the possibility of the existence of the concept of "microbial *terroir*", which implies that the microbial consortia in a certain wine-growing area are specific to that certain area and are producing flavors typical of the area [85]. Their experiments showed that the organoleptic properties of wine are given by *S. cerevisiae* indigenous strains and their origin, which may sustain the microbial aspect of *terroir*; in addition, the biodiversity of the yeast in the vineyards is affected by the micro and macroclimatic conditions of the vine varieties and the geographical location of the vineyard, a fact that would explain why the yeast consortia are different between two different wine-growing regions [87].

A research direction that emerges from the above-mentioned data is the use of autochthonous yeast in the wine industry in order to produce specific wines for certain wine-growing areas. Spontaneous fermentation is an uncontrolled and complex biotechnological process, in which the alteration microorganisms could rapidly multiply and reach too-high levels quickly, which may negatively impact the quality of the finished products [88]; this, even if spontaneous fermentation is correlated with greater complexity, greater wine body, and uncommon flavors [89–91], and it could improve the qualities of the wine by creating unique regional fingerprints [92], it is a process to be avoided. Therefore, one could combine spontaneous fermentation with indigenous yeasts with the safety of controlled processes from the industrial environment [86]. This would imply the use of selected local yeasts as new starter cultures in the winemaking industry, which would be reflected in the specific fingerprint of the finished product [86,93].

The new selection directions regarding the local wine yeasts tend to follow different paths, i.e., obtaining new wines with predetermined properties (high glycerol content, low ethanol content, reduced acidity); creating new and specific technological flows for obtaining certain types of wines, especially in order to avoid the production of certain compounds (biogenic amines, volatile sulfur compounds) in the finished wines; obtaining new wines of controlled origin and with a geographical indication; and completing the oenological practices in the legal specifications.

Thus, the research could be divided into two different directions, namely, that with the use of *Saccharomyces* yeasts, and that with the use of non-conventional (non-*Saccharomyces*) yeasts, in different variations, such as simple cultures, co-fermentation, or in sequential fermentation with *Saccharomyces* yeasts in different proportions. As described above, already, several non-*Saccharomyces* (NS) local yeast were detected during the isolation work and are stored in the owners' collections. In this regard, the usefulness of unconventional yeasts and the need to isolate and select such wine yeasts is further emphasized.

Considering the existence of numerous studies [94–98] which confirm that NS wine yeasts are beneficial a very large proportion, and even essential to obtaining wines with extraordinary organoleptic and sensory properties (Table 6), the selection of these yeast species is desirable in the near future. Among the NS species, only *Dekkera* spp. was reported as having only spoilage impact on wines. In Europe, numerous studies have been registered that argue in favor of non-conventional yeasts for the fermentation of the grape must. It is well-known that numerous NS yeast genera, including, but not limited to, the ones mentioned in Table 6, possess desirable oenological properties, such as the production of glycerol and other higher alcohols [99–101], the decreased ethanol content in the finished wine [102], and also the production of extracellular enzymes [103–105], esters [101,106], or polysaccharides [107].

Table 6. Biotechnological role of some non-*Saccharomyces* yeasts.

Genus	Relevant Species	Initial Technological Significance	Real Biotechnological Role	References
Hanseniaspora/ *Kloeckera*	*H. uvarum/* *H. apiculata*	Contamination /Spoilage	Higher alcohols, acetate, and ethyl esters production	[90,108]
Candida	*C. stellata*	Contamination	Glycerol production, fructophily	[109]
	C. zemplinina/ Starmerella bacillaris	Contamination	Glycerol, succinic acid production; decrease of alcohol content	[94,98]
Metschnikowia	*M. pulcherrima*	Contamination	Esters, terpenes, and thiols production, increase in aroma complexity	[91,98,107]
Pichia	*P. anomala*	Contamination /Spoilage	Increased production of volatile compounds, killer against *Dekkera/Brettanomyces*	[110]
	P. kluyveri		3-mercaptohexan-1-ol and 3-mercaptohexan-1-ol acetate production	[111]
Lachancea/ *Kluyveromyces*	*L. thermotolerans*	Contamination	Glycerol overproduction, reduction of volatile acidity	[112]
Torulaspora	*T. delbrueckii*	Spoilage	Succinic acid, polysaccharides production	[113]
Dekkera/ *Brettanomyces*	*D. bruxellensis*	Spoilage	Spoilage	[8,100]
Schizosaccharomyces	*S. pombe*	Spoilage	Malolactic deacidification; propanol and pyruvic acid production	[98,114]

Taking into account all the properties and real biotechnological roles of these NS yeasts in the production of wine, a new path for their use in grape must fermentation is open, which will avoid the production of certain chemical compounds in the final wines instead of desirable compounds such as esters and glycerol. However, due to the fact that NS yeasts are not able to finish the alcoholic fermentation (they are less efficient in the production of ethanol), the technology should be accmpanied by a sequential inoculation of the grape must [91]. Thus, the NS yeast may be inoculated at the beginning of the fermentation, and, after the fermented must reaches a content of approximatively 10% ethanol, a *Saccharomyces* yeast will be added. In this way, the fermentation will be concluded by the *Saccharomyces* species, while the NS species will produce the necessary metabolites to positively influence the aroma of the wine. A similar alternative involves the simultaneous inoculation of the two types of yeast. Finally, mixed or sequential fermentations with *Saccharomyces* and NS allow the developtment of local wines with a low alcohol content [91].

4. Conclusions

From a historical point of view, the first wine yeast selection work in Romania started in 1915 as part of the international research process started by French teams at the time, and the first local wine yeasts collection was delivered in years 1920s. After the 1970s and until the 1990s, the selection work reached almost all Romanian winemaking regions. The use of novel molecular identification and characterization tools followed the international trend, reaching the country later on (after 2010). The advancement in the past ten years was highly depentend on such techniques, and special selected yeast are nowadays in several local collections. However, their inclusion in international collection was not found in any report, and this is an aspect which should be taken into account in the near future.

Several autochthonous strains of *Saccharomyces* were isolated from Romanian vineyards, grapes, and musts, a part of which demonstrated oenological qualities that are desirable for Romanian local wines.

Moreover, numerous NS yeast strains, belonging to a multitude of different genera, have been isolated and identified from vineyards and wine research stations in Romania, but few Romanian authors have studied and published the use of local NS yeasts in winemaking.

The selection of local yeasts is of great interest for Romanian wine production due to the fact that there is the possibility of expanding the diversity of wines on the market, but also due to the high demand for local, unique products. Actually, it was reported recently [115] that a large majority of Romanian people prefer to consume only local wines. It is also worth mentioning the fact that a larger range of local yeasts used leads to developing a wider range of local wines, which supports Romanian gastronomic identity, culture, and tradition.

Author Contributions: Conceptualization, F.M. and R.-Ș.R.-E.; methodology, R.-Ș.R.-E. and C.F.D.; formal analysis, R.-Ș.R.-E.; investigation, R.-Ș.R.-E., E.B., M.I. and V.P.; resources, R.-Ș.R.-E., V.P., E.B., M.I. and F.M.; writing—original draft preparation, R.-Ș.R.-E. and F.M.; writing—review and editing, R.-Ș.R.-E., V.P., C.F.D. and F.M.; visualization, E.B. and M.I.; supervision, F.M. and V.P.; project administration, F.M. All authors have read and agreed to the published version of the manuscript.

Funding: This research received no external funding.

Institutional Review Board Statement: Not applicable.

Informed Consent Statement: Not applicable.

Data Availability Statement: Not applicable.

Conflicts of Interest: The authors declare no conflict of interest.

References

1. Mărăcineanu, L.; Giugea, N.; Perț, C.; Căpruciu, R. Tradition and Quality of Romanian Viticulture. *Ann. Univ. Craiova* **2018**, *23*, 139–143.
2. Mart, M. Everything You Need to Know about Romanian Wine. *Vincarta* **2017**.
3. International Organisation of Vine and Wine. Country Reports of Romania. 2022. Available online: https://www.oiv.int/what-we-do/country-report?oiv; https://www.oiv.int/what-we-do/data-discovery-report?oiv (accessed on 18 November 2022).
4. Law 164 of 24 June 2015 of Vines and Wine under the System of the Common Organization of the Wine Market. Article 17, Published in Official Journal of Romania no. 472 of 30 June 2015. Available online: https://legislatie.just.ro/Public/DetaliiDocument/169280 (accessed on 4 December 2022).
5. National Institute of Statistics. *Romanian Statistical Yearbook—Time Series (CD-ROM)*; NIS: Bucharest, Romania, 2021.
6. Giugea, N.; Muntean, C.; Mărăcineanu, C.; Călugăru, V. *Resurse Oenoturistice*; Alma: Craiova, Romania, 2020; ISBN 978-606-567-398-4.
7. Antoce, A.O.; Nămoloșanu, I.C.; Matei Rădoi, F. Comparative Study Regarding the Ethanol Resistance of Some Yeast Strains Isolated from Romanian Vineyards. *Rom. Biotechnol. Lett.* **2011**, *16*, 5981–5988.
8. Pretorius, I.S. Tasting the Terroir of Wine Yeast Innovation. *FEMS Yeast Res.* **2020**, *20*, foz084. [CrossRef]
9. Gonzalez, R.; Morales, P. Truth in wine yeast. *Microb. Biotechnol.* **2022**, *15*, 1339–1356. [CrossRef] [PubMed]
10. Molinet, J.; Cubillos, F.A. Wild Yeast for the Future: Exploring the Use of Wild Strains for Wine and Beer Fermentation. *Front. Genet.* **2020**, *11*, 589350. [CrossRef]
11. Erny, C.; Raoult, P.; Alais, A.; Butterlin, G.; Delobel, P.; Matei-Radoi, F.; Casaregola, S.; Legras, J.L. Ecological success of a group of *Saccharomyces cerevisiae/Saccharomyces kudriavzevii* hybrids in the northern european wine-making environment. *Appl. Environ. Microbiol.* **2012**, *78*, 3256–3265. [CrossRef]
12. Pretorius, I.S. Tailoring wine yeast for the new millennium: Novel approaches to the ancient art of winemaking. *Yeast* **2000**, *16*, 675–729. [CrossRef]
13. Grangeteau, C.; Roullier-Gall, C.; Rousseaux, S.; Gougeon, R.D.; Schmitt-Kopplin, P.; Alexandre, H.; Guilloux-Benatier, M. Wine microbiology is driven by vineyard and winery anthropogenic factors. *Microb Biotechnol.* **2017**, *10*, 354–370. [CrossRef]
14. Tofalo, R.; Perpetuini, G.; Schirone, M.; Fasoli, G.; Aguzzi, I.; Corsetti, A.; Suzzi, G. Biogeographical characterisation of *Saccharomyces cerevisiae* wine yeast by molecular methods. *Front. Microbiol.* **2013**, *4*, 166. [CrossRef]
15. Mas, A.; Padilla, B.; Esteve-Zarzoso, B.; Beltran, G.; Reguant, C.; Bordons, A. Taking Advantage of Natural Biodiversity for Wine Making: The WILDWINE Project. *Agric. Agric. Sci. Procedia* **2016**, *8*, 4–9. [CrossRef]
16. Alessandria, V.; Marengo, F.; Englezos, V.; Gerbi, V.; Rantsiou, K.; Cocolin, L. Mycobiota of Barbera grapes from the Piedmont region from a single vintage year. *Am. J. Enol. Vitic.* **2015**, *66*, 244–250. [CrossRef]
17. Costantini, A.; Vaudano, E.; Pulcini, L.; Boatti, L.; Gamalero, E.; Garcia-Moruno, E. Yeast Biodiversity in Vineyard during Grape Ripening: Comparison between Culture Dependent and NGS Analysis. *Processes* **2022**, *10*, 901. [CrossRef]
18. Di Maio, S.; Polizzotto, G.; Di Gangi, E.; Foresta, G.; Genna, G.; Verzera, A.; Scacco, A.; Amore, G.; Oliva, D. Biodiversity of Indigenous Saccharomyces Populations from Old Wineries of South-Eastern Sicily (Italy): Preservation and Economic Potential. *PLoS ONE* **2012**, *7*, e30428. [CrossRef]

19. Tofalo, R.; Perpetuini, G.; Fasoli, G.; Schirone, M.; Corsetti, A.; Suzzi, G. Biodiversity study of wine yeasts belonging to the "terroir" of Montepulciano d'Abruzzo "Colline Teramane" revealed *Saccharomyces cerevisiae* strains exhibiting atypical and unique 5.8S-ITS restriction patterns. *Food Microbiol.* **2014**, *39*, 7–12. [CrossRef] [PubMed]

20. Morata, A.; Arroyob, T.; Bañuelosa, M.A.; Blancoc, P.; Brionesd, A.; Cantorale, J.M.; Castrilloc, D.; Cordero-Buesoe, G.; del Fresnoa, J.M.; Escott, C.; et al. Wine yeast selection in the Iberian Peninsula: *Saccharomyces* and non-*Saccharomyces* as drivers of innovation in Spanish and Portuguese wine industries. *Crit. Rev. Food Sci. Nutr.* **2022**, *10*, 1–29. [CrossRef]

21. Padilla, B.; Garcia- Fernández, D.; González, B.; Izidoro, I.; Esteve-Zarzoso, B.; Beltran, G.; Mas, A. Yeast Biodiversity from DOQ Priorat Uninoculated Fermentations. *Front. Microbiol.* **2016**, *7*, 930. [CrossRef]

22. Castrillo, D.; Rabuñal, E.; Neira, N.; Blanco, P. Yeast diversity on grapes from Galicia, NW Spain: Biogeographical Patterns and the Influence of the Farming System. *OENO One* **2019**, *53*, 573–587. [CrossRef]

23. De Celis, M.; Ruiz, J.; Martín-Santamaría, A.; Alonso, M.; Marquina, D.; Navascués, E.; Gómez-Flechoso, M.Á.; Belda, I.; Santos, A. Diversity of *Saccharomyces cerevisiae* yeasts associated to spontaneous and inoculated fermenting grapes from Spanish vineyards. *Lett. Appl. Microbiol.* **2019**, *68*, 580–588. [CrossRef]

24. Koulougliotis, D.; Eriotou, E. Isolation and identification of endogenous yeast strains in grapes and must solids of Mavrodafni kefalonias and antioxidant activity of the produced red wine. *Ferment Technol.* **2016**, *5*, 1. [CrossRef]

25. Kachalkin, A.V.; Abdullabekova, D.A.; Magomedova, E.S.; Magomedov, G.G.; Chernov, I.Y. Yeasts of the vineyards in Dagestan and other regions. *Microbiology* **2015**, *84*, 425–432. [CrossRef]

26. Brînduşe, E.; Nechita, A.; Ion, M.; Paşa, R.; Fîciu, L.; Ciubucă, A. *Valorificarea Biodiversităţii Drojdiilor Autohtone de Vinificaţie În Scopul Obţinerii Vinurilor Cu Tipicitate de Areal Viticol*; Editura PIM: Iaşi, Romania, 2022; ISBN 978-606-13-7088-7.

27. Niţescu, M.A. Contribution de l'étude Des Levures Roumaines. Ph.D. Thesis, Faculte de Sciences, Universite Paris Cite, Paris, France, 1915.

28. Ringhianu, J.; Septilici, G. Folosirea Drojdiilor Selecţionate La I.A.S. Şi Importanţa Lor În Obţinerea Vinurilor de Calitate. *Rev. I.A.S.* **1958**, *1*, 4–7.

29. Septilici, G. *Contribuţii La Studiul Comparativ al Câtorva Specii de Drojdii*; Institutul de Cercetări Horti-Viticole, Lucrări Ştiinţifice: Paris, France, 1961; Volume 4, pp. 39–47.

30. Septilici, G.; Gherman, M. Elemente noi în pregătirea, livrarea şi folosirea culturilor de drojdii în vinificaţie. *Rev. Grădina Via Şi Livada* **1963**, *3*, 12–19.

31. Septilici, G.; Gherman, M. Rolul drojdiilor specializate în obţinerea vinurilor de calitate. *Rev. Grădina Via Şi Livada* **1963**, *7*, 47–52.

32. Septilici, G.; Sandu-Ville, G. Conducerea fermentaţiei musturilor în toamne cu temperaturi scăzute prin folosirea drojdiilor criofile. *Rev. Grădina Via Şi Livada* **1965**, *11*, 89–94.

33. Gherman, M. Influenţa culturilor de drojdii în amestec asupra calităţii vinurilor. *Rev. Grădina Via Şi Livada* **1963**, *39*, 58–67.

34. Gherman, M. Drojdii de genuri şi specii diferite pentru reglarea acidităţii. *Rev. Grădina Via Şi Livada* **1965**, *38*, 72–75.

35. Dănoaie, F. Cercetări Privind Fiziologia Drojdiilor Izolate Din Podgoria Târnave. Ph.D. Thesis, Babeş-Bolyai University, Cluj-Napoca, Romania, 1989.

36. Stamate, C.; Ţârdea, C.; Burnete, C. Isolation and Identification of Microflora on Grapes for the Varieties in Viticultural Centres Blaj, Târnave Vineyard. *Horticultura* **2006**, *49*, 537–542.

37. Oprean, L. Biotechnological Characteristics of Some *Saccharomyces* Species Isolated from Wine Yeast Culture. *Food Sci. Biotechnol.* **2005**, *14*, 722–726.

38. Sandu-Ville, G. Contribuţii La Studiul Şi Clasificarea Drojdiilor de Vin Din Microflora Vinicolă a Podgoriei Copou-Iaşi. Ph.D. Thesis, Institutul Agronomic Iaşi, Iaşi, Romania, 1974.

39. Sandu-Ville, G.; Savin, C. Conducerea Fermentaţiei Alcoolice a Musturilor Cu Ajutorul Levurilor Criofile. *Cercet. Agron. În Mold.* **1987**, *2*, 68–74.

40. Viziteu, G.A.; Manoliu, A.; Andor, I. Data Concerning the Yeasts Microbiota from Cotnari Vineyard. *Rom. Biotechnol. Lett.* **2008**, *13*, 3673–3680.

41. Vasile, A.; Cotea, V.V.; Mântăluţă, A.; Paşa, R.; Savin, C. The Preliminary Selection of Isolated Yeast Strains from the Indigenous Flora of Iaşi Vineyard. *Sci. Artic. UASVM Iaşi—Hortic. Ser.* **2009**, *52*, 811–816.

42. Nechita, A.; Savin, C.; Paşa, R.; Zamfir, C.I.; Codreanu, M. Isolation of New Types of Yeasts Strains from Indigenous Flora of Iaşi Vineyards. *Sci. Artic. UASVM Iaşi—Hortic. Ser.* **2014**, *57*, 177–182.

43. Găgeanu, A.; Câmpeanu, G.; Diguţă, C.; Matei, F. Isolation and Identification of Local Wine Yeast Strains from Dealurile Bujorului Vineyard. *Sci. Bulletin. Ser. F. Biotechnol.* **2012**, *XVI*, 22–25.

44. Kontek, A.; Kontek, A. Caracteristicile Morfofiziologice Şi Tehnologice Ale Unei Tulpini de *Saccharomyces carlsbergensis* Izolată Din Vin. *An. ICVV* **1975**, *VI*, 543–552.

45. Kontek, A.; Kontek, A. Contribuţii La Studiul Taxonomic al Drojdiilor Din Podgoria Dealu Mare. *An. ICVV* **1976**, *VII*, 597–609.

46. Matei Rădoi, F.; Brînduşe, E.; Nicolae, G.; Tudorache, A.; Teodorescu, R.I. Yeast Biodiversity Evolution over Decades in Dealu Mare—Valea Călugărească Vineyard. *Rom. Biotechnol. Lett.* **2011**, *16*, 113–120.

47. Brînduşe, E.; Tudorache, A.; Fotescu, L. Influence of Ecological Culture System on the Dynamics and Biodiversity of Non-*Saccharomyces* Autochthonous Wine Yeasts. *Sci. Artic. UASVM Iaşi—Hortic. Ser.* **2012**, *55*, 331–336.

48. Brînduşe, E.; Tudorache, A.; Fotescu, L. Selected Autochthonous Yeast Strains with Influence on Wine Quality. *Sci. Artic. UASVM Iaşi—Hortic. Ser.* **2012**, *55*, 365–370.

49. Bărbulescu, I.D.; Begea, M.; Frîncu, M.; Tamaian, R. Identification of Yeasts Tested for Fermentation on Different Carbon Sources. *Prog. Cryog. Isot. Sep.* **2015**, *18*, 51–58.

50. Dragomir Tutulescu, F.; Popa, A. Wine-Growing Areas in Oltenia (Romania) Major Natural Sources for the Isolation, Identification and Selection of Oenological Microorganisms. *Not. Bot. Horti Agrobot. Cluj-Napoca* **2009**, *37*, 139–144.

51. Beleniuc, G. Contribuții La Studiul Drojdiilor Izolate Din Podgoria Murfatlar. Ph.D. Thesis, Al. I. Cuza University: Iași, Romania, 1996.

52. Beleniuc, G. Researches Regarding the Taxonomic Identification of Spontaneous Microflora from Viti-Vinicol Cernavodă Center from Murfatlar Vineyard. *Res. J. Agric. Sci.* **2013**, *45*, 93–99.

53. Dumitrache, C.; Frîncu, M.; Rădoi, T.A.; Bărbulescu, I.D.; Mihai, C.; Matei, F.; Tudor, V.; Teodorescu, R.I. Identification by PCR ITS-RFLP Technique of New Yeast Isolates from Pietroasa Vineyard. *Sci. Bull. Ser. B Hortic.* **2020**, *LXIV*, 287–293.

54. Dragomir Tutulescu, F.; Popa, A. Viticultural Areas from Oltenia—Romania Which Have at Disposal a Rich and Performant Microflora of Oenological Interest. *Agricultura* **2010**, *75*, 33–39.

55. Vasile, A.; Pașa, R.; Savin, C. The Influence of New Yeast Strains from the Indigenous Flora of Iași Vineyard on the Alcoholic Fermentation Process. *Sci. Artic. UASVM Iași—Hortic. Ser.* **2010**, *53*, 459–464.

56. Matei, F.; Găgeanu, A. Killer Profile of Wine Yeast Strains Isolated in Dealurile Bujorului Vineyard. *Rom. Biotechnol. Lett.* **2011**, *16*, 144–147.

57. Antoce, A.O.; Nămoloșanu, I.C. A Rapid Method for Testing Yeast Resistance to Ethanol for the Selection of Strains Suitable for Winemaking. *Rom. Biotechnol. Lett.* **2011**, *16*, 5953–5962.

58. Barnett, J.A.; Payne, R.W.; Yarrow, D. *Yeasts: Characteristics and Identification*, 1st ed.; Cambridge University Press: Cambridge, UK, 1983; ISBN 978-0-521-25296-6.

59. Barnett, J.A. A History of Research on Yeasts 2: Louis Pasteur and His Contemporaries, 1850–1880. *Yeast* **2000**, *16*, 755–771. [CrossRef]

60. Kreger-van Rij, N.J.W. (Ed.) *The Yeasts: A Taxonomic Study*, 3rd ed.; Elsevier Science: Groningen, The Netherlands, 1984; ISBN 978-0-444-80421-1.

61. Delfini, C. Innovative Trends in Oenology and in the Selection of Yeasts and Malolactic Bacteria for Wine Industry. *Riv. Vitic. E Enol.* **1992**, *45*, 17–30.

62. Corbu, V.; Csutak, O. Molecular and Physiological Diversity of Indigenous Yeasts Isolated from Spontaneously Fermented Wine Wort from Ilfov County, Romania. *Microorganisms* **2023**, *11*, 37. [CrossRef]

63. Corbu, V.; Csutak, O. Biodiversity Studies on *Pichia kudriavzevii* from Romanian Spontaneous Fermented Products. *AgroLife Sci. J.* **2020**, *9*, 104–113.

64. Gaspar, E.; Gyöngyvér, M.; Oprean, L.; Iancu, R. Identification and Isolation of Yeast Strains *Saccharomyces bayanus* from Native Sources Using Mollecular Methods. *Ann. Rom. Soc. Cell Biol.* **2011**, *16*, 286–291.

65. Oprean, L. Molecular Identification of Selected Yeast Strains Isolated from Sebeș-Apold Vineyard, Apoldu de Jos Viticultural Area. *Ann. Rom. Soc. Cell Biol.* **2014**, *19*, 45–48.

66. Kontek, A. Studiul Factorilor Care Condiționează Formarea Acidității Volatile a Vinurilor La Vinificarea Primară. Ph.D. Thesis, University of Craiova, Craiova, Romania, 1977.

67. Lodder, J.; Kreger-van Rij, N.J.W. *The Yeasts. A Taxonomic Study*, 2nd ed.; North Holland Publishing Company: Amsterdam, The Netherlands, 1967.

68. Lengyel, E.; Oprean, L.; Tița, O.; Păcală, M.L.; Iancu, R.; Stegărus, D.; Ketney, O. Isolation and Molecular Identifications of Wine Yeast Strains from Autochthonous Aromatic and Semi Aromatic Varieties. *Ann. Rom. Soc. Cell Biol.* **2012**, *17*, 134–138.

69. Bora, F.D.; Donici, A.; Oșlobanu, A.; Fițiu, A.; Babeș, A.C.; Bunea, C.I. Qualitative Assessment of the White Wine Varieties Grown in Dealu Bujorului Vineyard, Romania. *Not. Bot. Horti Agrobot.* **2016**, *44*, 593–602. [CrossRef]

70. Colibaba, L.C.; Cotea, V.V.; Lacureanu, F.G.; Tudose-Sandu-Ville, S.; Rotaru, L.; Niculaua, M.; Luchian, C.E. Studies of Phenolic and Aromatic Profile of Busuioacă de Bohotin Wines. *BIO Web Conf.* **2015**, *5*, 02008. [CrossRef]

71. Dobrei, A.G.; Dobrei, A.; Iordănescu, O.; Nistor, E.; Balla, G.; Mălăescu, M.; Drăgunescu, A. Research Concerning the Correlation of Soil with Wines Quality in Some Varieties of Wine Grapes in Miniș-Măderat Vineyards. *J. Hortic. For. Biotechnol.* **2015**, *19*, 98–105.

72. Vararu, F.; Moreno-Garcia, J.; Moreno, J.; Niculaua, M.; Nechita, B.; Zamfir, C.; Colibaba, C.; Dumitru, G.D.; Cotea, V.V. Minor Volatile Compounds Profiles of "Aligoté" Wines Fermented with Different Yeast Strains. *Not. Sci. Biol.* **2015**, *7*, 123–128. [CrossRef]

73. Vișan, L.; Dobrinoiu, R.; Dumbravă, M. The Chemical and Aromatic Composition in a Romanian Wine Cabernet Sauvignon. *Rom. Biotechnol. Lett.* **2012**, *17*, 6855–6861.

74. Manolache, M.; Pop, T.I.; Babeș, A.C.; Farcaș, I.A.; Muncaciu, M.L.; Călugăr, A.; Gal, E. Volatile Composition of Some Red Wines from Romania Assessed by GC-MS. *Stud. Univ. Babeș-Bolyai Chem.* **2018**, *63*, 125–142. [CrossRef]

75. Manolache, M.; Pop, T.I.; Babeș, A.C.; Farcaș, I.A.; Godoroja, M.; Călugăr, A.; Gal, E. Assessment of Volatile Compounds of Some Red Wine Samples from Republic of Moldova and Romania Using GC-MS Analysis. *Agricultura* **2018**, *105*, 40–47.

76. Vararu, F.; Moreno-Garcia, J.; Niculaua, M.; Cotea, V.V.; Mayen, M.; Moreno, J. Fermentative Volatilome Modulation of Muscat Ottonel Wines by Using Yeast Starter Cultures. *LWT—Food Sci. Technol.* **2020**, *129*, 109575. [CrossRef]

77. Focea, M.C.; Luchian, C.; Zamfir, C.; Niculaua, M.; Moroșanu, A.M.; Nistor, A.; Andrieș, T.; Lacureanu, G.; Cotea, V.V. Organoleptic Characteristics of Experimental Sparkling Wines. *Sci. Artic. UASVM Iași—Hortic. Ser.* **2017**, *60*, 221–226.

78. Vișan, L.; Dobrinoiu, R. Studies on the Aroma of Sauvignon Wine. *Sci. Bulletin. Ser. F. Biotechnol.* **2013**, *17*, 127–131.

79. Scanes, K.T.; Hohmann, S.; Prior, B.A. Glycerol Production by the Yeast *Saccharomyces cerevisiae* and Its Relevance to Wine: A Review. *S. Afr. J. Enol. Vitic.* **1998**, *19*, 17–24. [CrossRef]

80. Tudose, I.; Savin, C.; Stoica, C. Selecția Și Testarea Potențialului Criofil al Unor Tulpini de *Saccharomyces ellipsoideus* Izolate Din Podgoria Iași, Centrul Viticol Copou. *An. ICVV* **2002**, *17*, 256–261.

81. Kontek, A.; Kontek, A.; Rusea, V.; Sandu-Ville, G. Selecția Și Caracterizarea a Două Sușe de Levuri Cu Capacitate Alcooligenă Ridicată. *An. ICVV* **1991**, *13*, 279–292.

82. Sandu-Ville, G.; Popescu, C. Conducerea Fermentației Alcoolice a Musturilor Prin Utilizarea de Drojdii Cu Caracter Nespumant. *Cercet. Agron. În Mold.* **1980**, *1*, 89–92.

83. Liță (Mihai), C.; Antoce, A.O.; Nămoloșanu, I.C. Studiul Influenței Unor Tulpini de Drojdii Selecționate Asupra Fermentației Alcoolice. *Lucr. Științifice USAMV Iași Ser. Hortic.* **2006**, *49*, 701–706.

84. Lengyel, E.; Panaitescu, M. The Management of Selected Yeast Strains in Qualifying Terpene Flavours in Wine. *Manag. Sustain. Dev.* **2017**, *9*, 27–30. [CrossRef]

85. Knight, S.; Klaere, S.; Fedrizzi, B.; Goddard, M.R. Regional Microbial Signatures Positively Correlate with Differential Wine Phenotypes: Evidence for a Microbial Aspect to Terroir. *Sci. Rep.* **2015**, *5*, 14233. [CrossRef]

86. Belda, I.; Zarraonaindia, I.; Perisin, M.; Palacios, A.; Acedo, A. From Vineyard Soil to Wine Fermentation: Microbiome Approximations to Explain the *"terroir"* Concept. *Front. Microbiol.* **2017**, *8*, 821. [CrossRef] [PubMed]

87. Alexandre, H. Wine Yeast Terroir: Separating the Wheat from the Chaff—For an Open Debate. *Microorganisms* **2020**, *8*, 787. [CrossRef]

88. Francesca, N.; Gaglio, R.; Alfonzo, A.; Settani, L.; Corona, O.; Mazzei, P.; Romano, R.; Piccolo, A.; Moschetti, G. The Wine: Typicality or Mere Diversity? The Effect of Spontaneous Fermentations and Biotic Factors on the Characteristics of Wine. *Agric. Agric. Sci. Procedia* **2016**, *8*, 769–773. [CrossRef]

89. Jackson, R.S. *Wine Tasting: A Professional Handbook (Food Science and Technology)*, 2nd ed.; Academic Press: Cambridge, MA, USA, 2009; ISBN 978-0-12-374181-3.

90. Jolly, N.P.; Varela, C.; Pretorius, I.S. Not Your Ordinary Yeast: Non-*Saccharomyces* Yeasts in Wine Production Uncovered. *FEMS Yeast Res.* **2014**, *14*, 215–237. [CrossRef] [PubMed]

91. Carrau, F.; Boido, E.; Ramey, D. Chapter Three—Yeasts for Low Input Winemaking: Microbial Terroir and Flavor Differentiation. *Adv. Appl. Microbiol.* **2020**, *111*, 89–121. [CrossRef]

92. Fleet, G.H. Wine Yeasts for the Future. *FEMS Yeast Res.* **2008**, *8*, 979–995. [CrossRef]

93. Comitini, F.; Capece, A.; Ciani, M.; Romano, P. New Insights on the Use of Wine Yeasts. *Curr. Opin. Food Sci.* **2017**, *13*, 44–49. [CrossRef]

94. Garcia, M.; Esteve-Zarzoso, B.; Arroyo, T. Non-Saccharomyces Yeasts: Biotechnological Role for Wine Production. In *Grape and wine Biotechnology*; IntechOpen: London, UK, 2016; pp. 249–271. ISBN 978-953-51-2693-5.

95. Manzanares, P.; Valles, S.; Viana, F. Chapter 4—Non-*Saccharomyces* Yeasts in the Winemaking Process. In *Molecular Wine Microbiology*; Academic Press: Cambridge, MA, USA, 2011; pp. 85–109. ISBN 978-0-12-375021-1.

96. Englezos, V.; Rantsiou, K.; Torchio, F.; Rolle, L.; Gerbi, V.; Cocolin, L. Exploitation of the Non-*Saccharomyces* Yeast *Starmerella bacillaris* (Synonym *Candida zemplinina*) in Wine Fermentation: Physiological and Molecular Characterization. *Int. J. Food Microbiol.* **2015**, *199*, 33–40. [CrossRef]

97. Sgouros, G.; Chalvantzi, I.; Mallouchos, A.; Paraskevopoulos, Y.; Banilas, G.; Nisiotou, A. Biodiversity and Enological Potential of Non-*Saccharomyces* Yeasts from Nimean Vineyards. *Fermentation* **2018**, *4*, 32. [CrossRef]

98. Benito, A.; Calderon, F.; Benito, S. The Influence of Non-*Saccharomyces* Species on Wine Fermentation Quality Parameters. *Fermentation* **2019**, *5*, 54. [CrossRef]

99. Ciani, M.; Maccarelli, F. Oenological Properties of Non-*Saccharomyces* Yeasts Associated with Wine-Making. *J. Microbiol. Biotechnol.* **1998**, *14*, 199–203. [CrossRef]

100. Belda, I.; Ruiz, J.; Esteban-Fernandez, A.; Navascues, E.; Marquina, D.; Santos, A.; Moreno-Arribas, M.V. Microbial Contribution to Wine Aroma and Its Intended Use for Wine Quality Improvement. *Molecules* **2017**, *22*, 189. [CrossRef] [PubMed]

101. Lee, S.B.; Banda, C.; Park, H.D. Effect of Inoculation Strategy of Non-*Saccharomyces* Yeasts on Fermentation Characteristics and Volatile Higher Alcohol and Esters in Campbell Early Wines. *Aust. J. Grape Wine Res.* **2019**, *25*, 384–395. [CrossRef]

102. Rocker, J.; Strub, S.; Ebert, K.; Grossmann, M. Usage of Different Aerobic Non-*Saccharomyces* Yeasts and Experimental Condition as a Tool for Reducing the Ethanol Content in Wines. *Eur. Food Res. Technol.* **2016**, *242*, 2051–2070. [CrossRef]

103. Maturano, Y.P.; Rodriguez Assaf, L.A.; Toro, M.E.; Nally, M.C.; Vallejo, M.; Castellanos de Figueroa, L.I.; Combina, M.; Vazquez, F. Multi-Enzyme Production by Pure and Mixed Cultures of *Saccharomyces* and Non-*Saccharomyces* Yeasts during Wine Fermentation. *Int. J. Food Microbiol.* **2012**, *155*, 43–50. [CrossRef]

104. Capozzi, V.; Garofalo, C.; Chiriatti, M.A.; Grieco, F.; Spano, G. Microbial Terroir and Food Innovation: The Case of Yeast Biodiversity in Wine. *Microbiol. Res.* **2015**, *181*, 75–83. [CrossRef]

105. Nechita, A.; Filimon, V.R.; Pașa, R.; Damian, D.; Zaldea, G.; Filimon, R.; Zaiț, M. Oenological Characterization of Some Yeast Strains Isolated from the Iași Vineyard Romania. *Rom. J. Hortic.* **2020**, *I*, 141–148. [CrossRef]

106. Lai, Y.T.; Hsieh, C.W.; Lo, Y.C.; Liou, B.K.; Lin, H.W.; Hou, C.Y.; Cheng, K.C. Isolation and Identification of Aroma-Producing Non-*Saccharomyces* Yeast Strains and the Enological Characteristic Comparison in Wine Making. *LWT—Food Sci. Technol.* **2022**, *154*, 112653. [CrossRef]

107. Domizio, P.; Liu, Y.; Bisson, L.F.; Barile, D. Use of Non-*Saccharomyces* Wine Yeasts as Novel Sources of Mannoproteins in Wine. *Food Microbiol.* **2014**, *43*, 5–15. [CrossRef]

108. Andorra, I.; Berradre, M.; Rozes, N.; Mas, A.; Guillamon, J.M.; Esteve-Zarzoso, B. Effect of Pure and Mixed Cultures of the Main Wine Yeast Species on Grape Must Fermentations. *Eur. Food Res. Technol.* **2010**, *231*, 215–224. [CrossRef]

109. Magyar, I.; Toth, T. Comparative Evaluation of Some Oenological Properties in Wine Strains of *Candida stellata, Candida zemplinina, Saccharomyces uvarum* and *Saccharomyces cerevisiae*. *Food Microbiol.* **2011**, *28*, 94–100. [CrossRef]

110. Domizio, P.; Romani, C.; Lencioni, L.; Comitini, F.; Gobbi, M.; Mannazzu, I.; Ciani, M. Outlining a Future for Non-*Saccharomyces* Yeasts: Selection of Putative Spoilage Wine Strains to Be Used in Association with *Saccharomyces cerevisiae* for Grape Juice Fermentation. *Int. J. Food Microbiol.* **2011**, *147*, 170–180. [CrossRef] [PubMed]

111. Anfang, N.; Brajkovich, M.; Goddard, M.R. Co-Fermentation with *Pichia kluyveri* Increases Varietal Thiol Concentrations in Sauvignon Blanc. *Aust. J. Grape Wine Res.* **2009**, *15*, 1–8. [CrossRef]

112. Comitini, F.; Gobbi, M.; Domizio, P.; Romani, C.; Lencioni, L.; Mannazzu, I.; Ciani, M. Selected Non-*Saccharomyces* Wine Yeasts in Controlled Multistarter Fermentations with *Saccharomyces cerevisiae*. *Food Microbiol.* **2011**, *28*, 873–882. [CrossRef] [PubMed]

113. Sadoudi, M.; Tourdot-Marechal, R.; Rousseaux, S.; Steyer, D.; Gallardo-Chacon, J.J.; Ballester, J.; Vichi, S.; Guerin-Schneider, R.; Caixach, J.; Alexandre, H. Yeast-Yeast Interactions Revealed by Aromatic Profile Analysis of Sauvignon Blanc Wine Fermented by Single of Co-Culture of Non-*Saccharomyces* and *Saccharomyces* Yeasts. *Food Microbiol.* **2012**, *32*, 243–253. [CrossRef]

114. Benito, S.; Palomero, F.; Morata, A.; Calderon, F.; Palomero, D.; Suarez-Lepe, J.A. Selection of Appropriate *Schizosaccharomyces* Strains for Winemaking. *Food Microbiol.* **2014**, *42*, 218–224. [CrossRef]

115. Dumitrescu, R. Romanians Prefer to Drink Local Wines. Available online: https://www.romania-insider.com/romanians-prefer-drink-local-wine-2023 (accessed on 3 February 2023).

Article

Exploring the Inhibitory Activity of Selected Lactic Acid Bacteria against Bread Rope Spoilage Agents

Giovanna Iosca [1,2], Joanna Ivy Irorita Fugaban [2], Süleyman Özmerih [2], Anders Peter Wätjen [2], Rolf Sommer Kaas [2], Quốc Hà [2], Radhakrishna Shetty [2], Andrea Pulvirenti [1], Luciana De Vero [1,*] and Claus Heiner Bang-Berthelsen [2,*]

[1] Department of Life Sciences, University of Modena and Reggio Emilia, 42122 Reggio Emilia, Italy; giovanna.iosca@unimore.it (G.I.); andrea.pulvirenti@unimore.it (A.P.)

[2] National Food Institute, Technical University of Denmark, Kemitovet, DK-2800 Kongens Lyngby, Denmark; jofu@food.dtu.dk (J.I.I.F.); sylomen@biosustain.dtu.dk (S.Ö.); apewat@food.dtu.dk (A.P.W.); rkmo@food.dtu.dk (R.S.K.); larryha95@gmail.com (Q.H.); rads@food.dtu.dk (R.S.)

* Correspondence: luciana.devero@unimore.it (L.D.V.); claban@food.dtu.dk (C.H.B.-B.)

Abstract: In this study, a wide pool of lactic acid bacteria strains deposited in two recognized culture collections was tested against ropy bread spoilage bacteria, specifically belonging to *Bacillus* spp., *Paenibacillus* spp., and *Lysinibacillus* spp. High-throughput and ex vivo screening assays were performed to select the best candidates. They were further investigated to detect the production of active antimicrobial metabolites and bacteriocins. Moreover, technological and safety features were assessed to value their suitability as biocontrol agents for the production of clean-label bakery products. The most prominent inhibitory activities were shown by four strains of *Lactiplantibacillus plantarum* (NFICC19, NFICC 72, NFICC163, and NFICC 293), two strains of *Pediococcus pentosaceus* (NFICC10 and NFICC341), and *Leuconostoc citreum* NFICC28. Moreover, the whole genome sequencing of the selected LAB strains and the in silico analysis showed that some of the strains contain operons for bacteriocins; however, no significant evidence was observed phenotypically.

Keywords: lactic acid bacteria; starter culture; organic acids; rope spoilage; bread spoilage; bakery products

check for updates

Citation: Iosca, G.; Fugaban, J.I.I.; Özmerih, S.; Wätjen, A.P.; Kaas, R.S.; Hà, Q.; Shetty, R.; Pulvirenti, A.; De Vero, L.; Bang-Berthelsen, C.H. Exploring the Inhibitory Activity of Selected Lactic Acid Bacteria against Bread Rope Spoilage Agents. *Fermentation* **2023**, *9*, 290. https://doi.org/10.3390/fermentation9030290

Academic Editor: John Samelis

Received: 28 February 2023
Revised: 13 March 2023
Accepted: 14 March 2023
Published: 16 March 2023

1. Introduction

Bread is considered an essential staple food in many cultures, being a valuable source of proteins, lipids, vitamins, and minerals [1]. Unluckily, bread and most of the bakery products available on the market are characterized by short shelf life. Generally, bread quality can be affected by physicochemical decomposition, known as staling, and microbiological contamination, with daily losses ranging from 9.7% to 14.4% [2]. Among bread-spoilage bacteria involved in bakery product loss, those belonging to *Bacillus* spp., including *B. subtilis*, *B. licheniformis*, *B. cereus*, *B. amyloliquefaciens*, *B. mycoides*, *B. pumilus*, and a few species of the genera *Paenibacillus* and *Lysinibacillus*, contribute to two-thirds of total food waste, along with fruit and vegetables [3–7]. Contamination by these microorganisms leads to "ropy" bread, which is characterized by an unpleasant ripe fruity odor similar to an overripe pineapple, melon, valerian, or honey, with a sticky, soft, and discolored crumb due to enzymatic degradation of starch, proteins, and exopolysaccharides (EPS) [8]. These bacteria are normally present in the bakery environment, surfaces, and atmosphere [9]. Furthermore, raw materials such as wheat, seeds, semolina, and brewer's yeast during harvesting and processing conditions are an optimal colonizing source for bacteria, which can prosper during storage periods. Humidity and heat enable bacteria growth to reach high levels of contamination in the flour after the milling process, leading to a higher speed of the rotting process in bread (24–48 h) [7]. The successful colonizing ability of these bacteria is associated with their endospores, a highly thermoresistant structure, which allows retention of their viability during the baking process [10]. The ubiquitous presence and the spore-forming ability of these microorganisms, in combination with the increasing demand for eco-friendly ways of handling food and food products themselves, make the elimination of these spoilage agents a challenge, both at the artisanal and industrial levels. Furthermore, the increasing trend of "green consumerism" demands to be met particularly by the food industries. With this, chemical preservatives are now being rejected, which results in "new" standards considered for food safety and extended shelf-life [11,12]. Ideally, the need for alternative preservatives should be obtained from naturally occurring sources that can be achievable using microorganisms or/and their metabolites [13]. In this context, selected lactic acid bacteria (LAB) can be exploited to prevent spoilage of bread and contribute both to the production of sensory properties and microbial safety of bakery products [14,15]. The use of LAB as a starter culture has a long history in a variety of fermented foods. Particularly in sourdough, lactic acid fermentation is considered one of the most prominent antispoilage "technologies" due to the production of lactic acid, acetic acid, fatty acids, short peptides, and the pH reduction, which lead to the suppression of several spoilage agents [16–18]. Several authors have already investigated different LAB cultures for sourdough as an additive-free method to avoid rope development in bakery products [19–22]. In this study, a wide pool of LAB strains, provided by two recognized culture collections in Italy and Denmark, were tested against some common bread spoilage bacteria, specifically belonging to *Bacillus* spp., *Paenibacillus* spp., and *Lysinibacillus* spp. High-throughput and ex vivo screening assays were performed to select the best candidates. They were further investigated to detect the production of active antimicrobial metabolites and bacteriocins. Moreover, the technological and safety features of the selected strains were assessed. The final goal was the detection of candidate LAB strains with an inhibitory activity to be potentially used as biocontrol agents for the production of clean-label bakery products.

2. Materials and Methods

2.1. Microorganisms Used in This Study

2.1.1. Lactic Acid Bacteria Strains

A total of 18 LAB strains from the *Lactobacillus*, *Pediococcus*, and *Leuconostoc* groups were selected to be tested against bread spoilage agents (Table 1). Five strains were provided by the Unimore Microbial Culture Collection (UMCC), University of Modena and Reggio Emilia (Italy), and the remaining strains were provided by the National Food Institute Culture Collection (NFICC), Technical University of Denmark (Table 1). UMCC strains were identified and characterized by 16 s RNA gene sequencing in previous works and selected for their antispoilage activity [14,15]. Regarding NFICC strains, they were identified by using the MALDI Biotyper® sirius IVD System (BRUKER, Roskilde, Denmark) or through whole-genome sequencing, and selected for their suitability to ferment plant-based substrates [23]. The original culture of the strains is maintained in their respective collections by cryopreservation at $-80\ ^{\circ}\text{C}$ in cryovials containing De Man, Rogosa, Sharpe (MRS) broth (Oxoid, Milan, Italy) mixed with 25% (*v/v*) glycerol. An active culture of each strain was used for all the reported screenings.

2.1.2. Spoilage Bacteria Strains

According to previous works by Saranraj and Gheeta [24] and Valerio et al. [6], a total of 29 different bread spoilage bacterial strains were chosen for the screenings (Table 2). They were provided by the NFICC collection and the German Collection of Microorganisms and Cell Cultures (DSMZ).

Table 1. Lactic acid bacteria strains tested in the present study. They were provided by the National Food Institute Culture Collection (NFICC), Technical University of Denmark and by Unimore Microbial Culture Collection (UMCC), University of Modena and Reggio Emilia (Italy).

Strain Code	Species	Isolation Source
NFICC10	*Pediococcus pentosaceus*	Sourdough
NFICC19	*Lactiplantibacillus plantarum*	Dill
NFICC27	*Lactiplantibacillus plantarum*	Sourdough
NFICC28	*Leuconostoc citreum*	Sourdough
NFICC58	*Pediococcus pentosaceus*	Sourdough
NFICC72	*Lactiplantibacillus plantarum*	Gooseberry
NFICC87	*Leuconostoc citreum*	Beetroot
NFICC94	*Leuconostoc citreum*	Spinach
NFICC103	*Pediococcus pentosaceus*	Pumpkin
NFICC163	*Lactiplantibacillus platarum*	Field pea
NFICC207	*Lactiplantibacillus plantarum*	Glasswort
NFICC293	*Lactiplantibacillus plantarum*	Dragsholm plant
NFICC341	*Pediococcus pentosaceus*	Brewer's spent grain
UMCC 2990	*Fructilactobacillus sanfranciscensis*	Sourdough type I
UMCC 2996	*Lactiplantibacillus plantarum*	Dough for Panettone
UMCC 3002	*Furfurilactobacillus rossiae*	Dough for Panettone
UMCC 3010	*Pediococcus pentosaceus*	Gluten-free sourdough
UMCC 3011	*Leuconostoc citreum*	Dough for Panettone

Table 2. Selected bread spoilage bacteria tested in the present study. They were provided by the German Collection of Microorganisms and Cell Cultures (DSMZ) and the National Food Institute Culture Collection (NFICC).

Strain Code	Species	Isolation Source
DSM 2301	*Bacillus cereus*	Food poisoning incident
DSM 4222	*Bacillus cereus*	-
DSM 4312	*Bacillus cereus*	Vomit
DSM 22905	*Bacillus cytotoxicus*	Vegetable puree
NFICC119	*Lysinibacillus fusiformis*	Beetroot
NFICC432	*Paenibacillus polymyxa*	Walnut
NFICC503	*Bacillus mycoides*	Beech leaves
NFICC510	*Bacillus altitudinis*	Plant
NFICC526	*Bacillus mycoides*	Red fir
NFICC528	*Bacillus subtilis*	Sourdough
NFICC529	*Lysinibacillus sphaericus*	Common Juniper
NFICC530	*Bacillus pumilus*	Common Juniper
NFICC531	*Bacillus simplex*	Common Juniper
NFICC532	*Lysinibacillus fusiformis*	Common Juniper
NFICC740	*Bacillus cereus*	Plant
NFICC781	*Bacillus cereus*	Kombucha
NFICC816	*Bacillus thuringiensis*	Animal feces
NFICC855	*Bacillus weihenstephanensis*	Potato
NFICC869	*Bacillus amyloliquefaciens*	Pasteurized BSG
NFICC871	*Lysinibacillus sphaericus*	Pasteurized BSG
NFICC879	*Lysinibacillus boronitolerans*	Potato
NFICC882	*Lysinibacillus fusiformis*	Potato
NFICC889	*Lysinibacillus boronitolerans*	Potato
NFICC906	*Bacillus simplex*	Potato
NFICC1127	*Bacillus amyloliquefaciens*	Pasteurized BSG *
NFICC1130	*Bacillus amyloliquefaciens*	Pasteurized BSG
NFICC1525	*Bacillus subtilis*	Herring Garum
NFICC1534	*Bacillus subtilis*	Miso
NFICC1549	*Bacillus licheniformis*	Apple pulp

* BSG: brewers' spent grains.

2.2. Screenings for Antibacterial Activity of the LAB Strains

2.2.1. High-Throughput Screening Assay with LAB Cell-Free Supernatants (CFS)

The preliminary screening of the selected LAB activity against 29 bread spoilage strains was performed by using the LAB cell-free supernatants (CFS), following the protocol of Inglin et al. [25] with some modifications. The different spoilage strains were inoculated in a definite medium, specifically, brain heart infusion (BHI) broth (Oxoid, Milan, Italy) (*Bacillus* spp.), M17 broth (Oxoid, Milan, Italy) (*Lysinobacillus* spp.), and potato dextrose (PD) broth (Oxoid, Milan, Italy) (*Paenibacillus* spp.), and incubated at 30 °C for 24 h. The LAB strains were incubated in MRS broth at 30 °C for 48 h. LAB supernatants were obtained by centrifuge at $6000\times g$ 15 min following sterile filtration (0.20 μm) in new Eppendorfs (ThermoFisher, Waltham, MA USA). After 50 μL of BHI/M17/PD containing 0.5% of an overnight culture of the chosen spoilage agent was transferred to a 200 μL clear-glass, flat-bottom 96-well microtiter plate (ThermoFisher, Waltham, MA, USA) using a multichannel pipette. Then, 30 μL of LAB supernatant was transferred to each well. Optical density at 600 nm (OD_{600}) was measured at time zero (t0) using a plate reader infinite M200PRO (Tecan, Männedorf, Switzerland), and results were analyzed using the formula $1.5 \times OD_{600}t0$. Then, the plates were incubated at 30 °C (optimal condition for the spoilage strains), and the OD_{600} was subsequently controlled after 24 h and 48 h. For each strain, OD_{600} value below the formula threshold indicated an inhibition activity.

2.2.2. Double-Agar-Layer Screening Assay

The double-agar-layer screening assay was performed by adapting the protocol described by Iosca et al. [15]. Briefly, spoilage strains were incubated in their respective optimal media mentioned previously (BHI broth, M17 broth, and PD broth) at 30 °C for 24 h. After incubation, initial OD_{600} was measured and adjusted to a final concentration of 10^5 CFU/mL. LAB strains were grown in MRS broth at 30 °C for 24 h, then OD_{600} was measured and adjusted as needed to obtain 10^8 CFU/mL. After, 1 mL of MRS agar was poured into 24-well plates (ThermoFisher, Waltham, MA, USA) and overlaid with 500 μL of the specific growth medium of the spoilage strains (BHI agar for *Bacillus* spp.; M17 agar for *Lysinobacillus* spp.; and PD agar for *Paenibacillus* spp.). After solidification, 10 μL of the spoilage agents was spread on each well and left to dry. Subsequently, a hole of 4 mm was made and 5 μL of each LAB strain was inoculated in the wells. Results were taken after 48 h of incubation at 30 °C, wherein growth inhibitions were noted. Scores were designated as follows: complete inhibition of the spoilage agents was scored 3, strong inhibition was scored 2, and moderate to weak inhibition was scored 1, while no inhibition was scored 0.

2.3. Confirmatory Assay of Antibacterial Activity in Bread Medium (BM)

To evaluate the ex vivo antibacterial activity of the LAB strains selected from the previous screenings, a culture medium from plain wheat bread was simulated following the protocol designed by Verni et al. [26] and Iosca et al. [17] with some modifications. Briefly, 600 g of wheat bread was homogenized using a blender to facilitate the enzyme treatment, and water was added in a 1:4 ratio (*w/w*). After the blended mixture, Neutrase, an endoprotease (1.5 AU/g) from *Bacillus amyloliquefaciens* (Novozymes, Bagsværd, Denmark) was used at 1.5 mL/L, while Amylase (Novozymes, Bagsværd, Denmark) was added at 0.5 g/L. Proteolytic and amylolytic enzymes were then added to facilitate nutrient compound availability. Bread suspensions were then incubated at 55 °C for 18 h, considered the enzymes optimal conditions. After the incubation, mixtures were centrifuged (6000 rpm for 20 min) and supernatants were collected. The medium pH was adjusted to around 5.7, filtered, and sterilized at 121 °C for 15 min. Preselected lactic acid bacteria CFS, obtained as previously described, were then tested in ex vivo conditions using BM to detect the best performers. As above, Inglin et al.'s [25] protocol was performed. LAB and selected *Bacillus* spp. spoilage agents were grown in BM. All screening assays were carried out in triplicate. Data were analyzed and compared using a one-tailed *t*-test. Strains with significant inhibitory activities were selected and further analyzed.

2.4. Assessment of the LAB Bioactive Compounds

2.4.1. High-Performance Liquid Chromatography (HPLC) Detection of the Main Compounds Produced in MRS Fermentation

To detect and identify the main compounds produced by the most prominent inhibitory LAB strains, the CFS of the evaluated strains were examined by HPLC after 14 h, 18 h, and 24 h of fermentation in MRS at 30 °C. HPLC was equipped with an Aminex HPX-87H column (300 × 7.8 mm column) (Bio-Rad, Hercules, CA, USA), and a Shodex RI-101 refractive index detector was used. The flow rate of the mobile phase (5 mM H_2SO_4) was 0.5 mL/min, and the column oven temperature was maintained at 60 °C. All reagents were analytically pure, standard curves were first identified individually, and retention times were calculated. Oxalic acid, tartaric acid, formic acid, lactic acid, citric acid, acetic acid, and succinic acid were the key acids chosen for the analysis, as suggested by Hui-Hu et al. [27]. All the samples were loaded in triplicate. MRS broth was also analyzed as a control. Statistical analysis was performed using Kruskal–Wallis nonparametric analysis. A *p*-value of <0.05 was considered statistically significant.

2.4.2. Bacteriocins Production

The potential bacteriocin production from the various LAB was evaluated by treating LAB's CFS with proteolytic enzymes, including proteinase K, trypsin, and α-chymotrypsin at a final concentration of 0.1 mg/mL. The preparation of CFS was performed as previously described. Subsequently, after 1 h incubation at 37 °C, heat treatment at 95 °C × 10 min was performed to terminate the enzymatic processes before spotting it on a plate [28].

Moreover, another assay of CFS spotted on a BHI was performed according to the method described by Fugaban et al. [28]. Briefly, the selected LAB were cultured in MRS broth at 30 °C for 18 h, and the CFS was obtained by centrifugation (8000× *g*, 10 min). Using sterile 1 M NaOH, supernatant's pH was adjusted to 6.5 and heat treated at 80 °C for 10 min to inactivate putative proteolytic enzymes and eliminated hydrogen peroxide. The supernatant was then filtered with 0.2 μm syringe filters (Sartorius Ministart Syringe hydrophilic Filter, Göttingen, Germany). Ten microliters of the CFS were then spotted on a BHI plate with 1% agar seeded with appropriative test organisms at a final viable cell count of ~10^5 CFU/mL and left to dry. Plates were then incubated for 24 h at 30 °C to observe the formation of inhibition zones. The experiment was conducted in triplicate.

2.4.3. Kinetic Screening Assay

To further confirm that the antibacterial activity of the selected LAB inhibits spoilage growth, the method proposed by Fugaban et al. [28] was followed. Test organisms, *B. cereus* DSM 2301, *B. thuringensis* NFICC816, *B. weihenstephanensis* NFICC855, *B. amyloliquefaciens* NFICC1130 and NFICC1127, and *B. licheniformis* NFICC1549, were grown individually in sterile 96-well flat-bottom plates. After 3 h, LAB-CFS, obtained as previously described, was added to the appropriate wells. Sterile BHI inoculated with 10% test organisms was dispensed in the first 10 columns of the plate, leaving the last two for sterility control and growth control. Experiments were performed in two independent set-ups for each LAB-CFS, with the test organism being treated. Plates were incubated at 30 °C for 18 h and, simultaneously, the OD_{600} nm were measured for 24 h.

2.5. Phenotypical Characterization of the LAB Strains

2.5.1. API Test

Carbohydrate utilization was characterized by using API 50 CHL panel test (bioMérieux, Marcy l'Etoile, France) according to the protocol suggested by the manufacturers. Briefly, all the LAB strains were initially grown in MRS broth for 18 h at 30 °C. Subsequently, cells were collected by centrifugation and washed twice in 0.85% NaCl saline solution. Cell concentrations were adjusted to 0.5 McFarland. Suggested volumes of inoculum were added to each corresponding well. Set-ups were incubated at 30 °C, and color changes were monitored after 24 h and 48 h. Results were interpreted based on the manufacturers' recommendations.

2.5.2. Determination of Acidification Ability in BM USING ICINAC

Acidification profiles were generated for the antibacterial strain in BM. Exponentially growing LAB, previously incubated in MRS at 30 °C for 18 h, were washed as mentioned before, and 1 mL was inoculated in 40 mL of sterile BM, obtained as already described in the previous paragraph. The fermentation was performed in 40 mL of liquid Bread Media and monitored with the *iCinac* system (AMS alliance, Frepillon, France), which allows monitoring of acidification kinetics during fermentation [29,30]. The cultures were incubated in a water bath at 30 °C during all procedures, and sampling was recorded every 30 min for 48 h.

2.6. Safety Assessment
2.6.1. Hemolytic Activity

The hemolytic activity of the LAB present in this study was evaluated using Columbia Blood Agar (Oxoid LTD, Basingstoke, UK) with 5% defibrinized horse blood according to Fugaban et al. [31]. Strains grown for 18 h in MRS at 30 °C were spot-plated (10 µL) on the agar surface. Positive hemolytic activity was indicated by clear yellow zones around the bacterial growth (β-hemolysis). The reference strains used were *Staphylococcus aureus* NFICC1477, *B. cereus* DMS 2301, and *Lc. citreum* NFICC88 as controls for α-, β-, and γ-hemolysis, respectively. All experiments were performed in triplicate.

2.6.2. LAB Antimicrobial Susceptibility

The antimicrobial susceptibility testing (AST) was performed according to the suggestions set by the European Food Safety Authority (EFSA) in the set guidelines for the assessment of AST of human and veterinary significant microorganisms [32]. The assay was performed in a 96-well microplate using microbroth dilution with specific antibiotics from Sigma-Aldrich (ampicillin, chloramphenicol, ciprofloxacin, gentamycin, kanamycin, streptomycin, and vancomycin) on cation-adjusted Mueller–Hinton broth supplemented with MRS (5.0 g/L). The assay included 10 antibiotic dilutions in two-fold and controls (growth and sterility controls). Inocula were adjusted to 0.5 McFarland units and disseminated appropriately to obtain a final concentration of 10^5 CFU/mL. The plates were incubated following EFSA guidance (35 ± 1 °C for 18 h). The lowest concentration with complete bacterial inhibition was recorded as the MIC and analyzed according to the standards set for LAB.

2.7. Whole-Genome Sequencing Analysis: In Silico Screening for Functional Genes and Virulence Genes

Whole-genome sequencing of the selected LAB strains was performed using Illumina technology. Libraries for paired-end sequencing were constructed using the Nextera XT kit (Illumina, CA, USA) guide 15031942v01. The pooled Nextera XT libraries were loaded onto an Illumina NextSeq reagent cartridge using the NextSeq 500/550 Mid Output Kit v2.5 (300 Cycles) with a standard flow cell. Raw FASTQ files were trimmed using standard Trimmomatic settings, and genome assembly was performed using Unicycler [33,34]. Subsequently, annotation and conversion of the genome in protein sequence were obtained using the Bacterial and Viral Bioinformatics Resource Center (BV-BRCbeta, https://www.bv-brc.org/, accessed on 13 January 2023) [35,36]. To detect putative bacteriocins' sequence and ribosomally synthesized and post-translationally modified peptides (RiPPs), BAGEL4 (http://bagel4.molgenrug.nl, accessed on 15 December 2022) was used in combination with UniProt Consortium and the National Center for Biotechnology Information (NCBI, https://www.ncbi.nlm.nih.gov/, accessed on 1 December 2022) using BLAST protein [37,38].

3. Results

3.1. Screening Assays

3.1.1. High-Throughput Preliminary Screening

The high-throughput screening assay suggested by Inglin et al. [25], which aims for an efficient screening for putative antispoilage LAB, was conducted using 96-well plates. The obtained results were analyzed based on the formula: $1.5 \times OD_{600}$ t0. Results are reported in Table S1. All the strains in which the OD_{600} value after 24 h and 48 h resulted to be under the cut-off derived from the formula were considered to be putative inhibitory candidates. All results obtained in this preliminary screening were further evaluated. In this assay, the best antagonistic strains were *L. plantarum* UMCC 2996 and NFICC 19, 27, 72, 163, 207, and 293. Additionally, *P. pentosaceus* UMCC 3010 and NFICC 10, 58, 103, and 341, along with *Lc. citreum* NFICC28 and NFICC94, demonstrated having the highest inhibitory activity in the typical growth media used.

3.1.2. Antibacterial Activity of LAB Assessed by Double-Agar-Layer Screening Assay

Additional assessment of the antibacterial activity of the LAB strains was assessed against *Bacillus* spp. using well diffusion assay in 24-well microtiter plates. Inhibitions were analyzed and graded based on the scores previously mentioned, and results are indicated in Table S2. Assessments of activities showed that the strain *P. pentosaceus* NFICC341 had the broadest inhibition (66% of the spoilage agents), followed by 66% for *P. pentosaceus* UMCC 3010 and 57% for *Lc. citreum* UMCC 3011 and *F. sanfranciscensis* UMCC 2990. A strong inhibition was also detected for *F. rossiae* UMCC 3002 and *L. plantarum* NFICC207, with a percentage inhibition of 52% for both strains. For the other evaluated strains, even though inhibitory abilities were detected, due to a percentage under 50%, no significant antagonistic activity was observed.

3.2. Antibacterial Activity of LAB by Confirmatory Assay in BM

To confirm the activities of the CFS of the selected LAB candidates, an ex vivo experiment was performed using liquid BM. This is to mimic the nutritional bread conditions to allow the evaluation of the actual antimicrobial potential of the candidates. Results obtained after 48 h were investigated with the use of one-tailed *t*-test as reported in Figures 1–6. Based on the results obtained by the first screenings, *Bacillus cereus* DSM 2301 was tested against the CFS of five LAB (Figure 1). The set-up treated with *L. plantarum* NFICC19 showed the strongest inhibitory activity, with a significant difference compared with the control DSM 2301 in BM.

Similar action was detected for *L. plantarum* UMCC 2996, *Lc. citreum* NFICC28, and *F. sanfranciscensis* UMCC 2990. The lowest activity was shown by *L. plantarum* NFICC293. *B. thuringiensis* NFICC816 was challenged with the CFS of six LAB strains, specifically, *L. plantarum* NFICC19, 27, 293, *Lc. citreum* NFICC28, and *P. pentosaceus* NFICC341 (Figure 2). Here, only *Lc. citreum* NFICC28 was able to significantly prevent the growth of the spoilage organism. Additionally, *B. weihenstephanensis* NFICC855 was challenged with various LAB-CFS, including *L. plantarum* strains (NFICC19, 27, 72, 207, and 293), *F. sanfranciscensis* UMCC 2990, *Lc. citreum* NFICC28, and *P. pentosaceus* strains (UMCC 3010, NFICC58, and 341). As reported in Figure 3, significant results were obtained. Candidates from all four species were able to inhibit pathogen growth, particularly *L. plantarum* NFICC72, NFICC163, and *P. pentosaceus* NFICC58.

Two strains of *B. amyloliquefaciens*, NFICC1130 and NFICC1127, were used in these experiments (Figures 4 and 5), recording completely different results even if challenged with the same LAB-CFS strains. The majority of the LAB were able to inhibit *B. amyloliquefaciens* NFICC1130; the highest inhibitory activities were recorded for *P. pentosaceus* NFICC10 and *L. plantarum* NFICC19. On the contrary, *B. amyloliquefaciens* NFICC1127 showed no inhibition with the same LAB-CFS. The CFS of different LAB strains, including *L. plantarum*, *Lc. Citreum*, and *P. pentosaceus*, were tested against *B. licheniformis* NFICC1549, as reported in Figure 6. *L. plantarum* NFICC72 and NFICC293 appear to be the best inhibitory agents, along with *Lc. citreum* NFICC94.

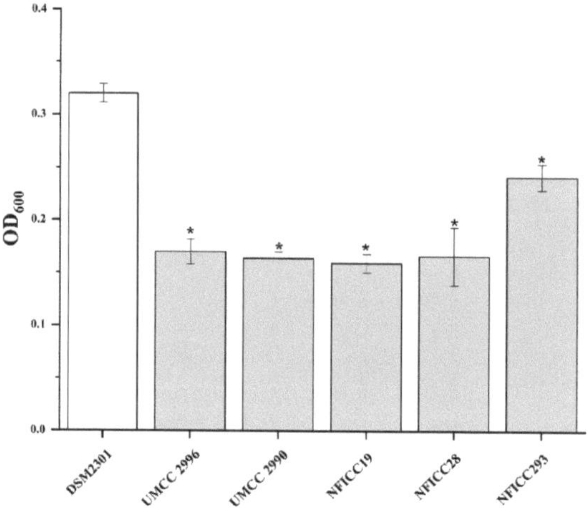

Figure 1. Antibacterial activity of the selected LAB' cell-free supernatant in Bread Media (BM) against *B. cereus* DSM 2301. One-tail *t*-test was performed comparing OD_{600} value after 48 h growth of *B. cereus* DSM 2301 in BM and the same pathogens challenged with CFS of *L. plantarum* (UMCC 2996, NFICC19, and 293), *F. sanfranciscensis* (UMCC 2990), and *Lc. citreum* (NFICC28). The mean value of all the treated samples was significantly lower than the control sample (untreated sample); t(2) = −2.92 and *p* = 0.05. Bars with * are significantly different.

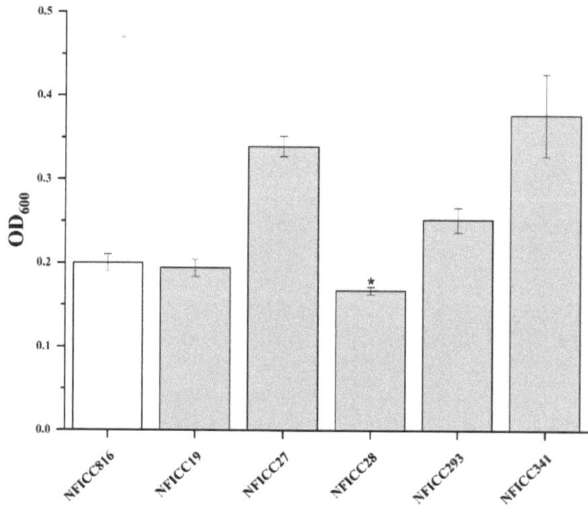

Figure 2. Antibacterial activity of the selected LAB's cell-free supernatant in Bread Media (BM) against spoilage agents *B. thuringensis* NFICC816. One-tail *t*-test was performed on the obtained triplicates comparing OD_{600} value after 48 h growth of NFICC816 in BM and the same pathogens challenged with CFS of *L. plantarum* (NFICC19, 27, and 293) and *P. pentosaceus* (NFICC341). The mean value of one treated sample of *Lc. citreum* (NFICC28) was significantly lower than the control sample (untreated sample), while the other samples respected the null hypothesis; t(2) = −2.92 and *p* = 0.05. Bars with * are significantly different.

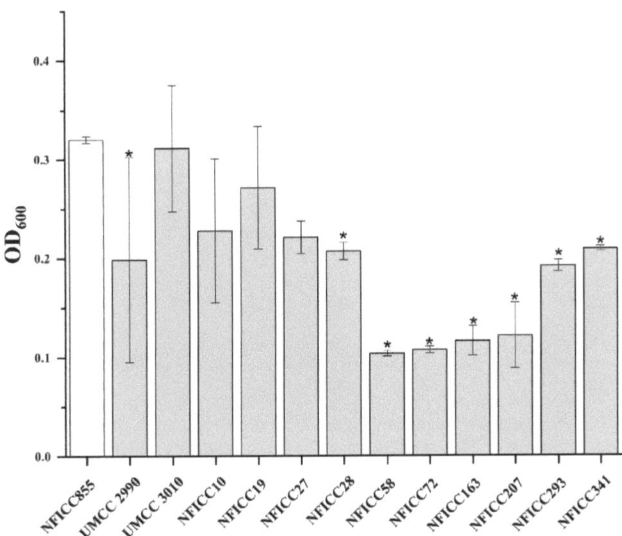

Figure 3. Antibacterial activities of the selected LAB's cell-free supernatant in Bread Media (BM) against spoilage agents *B. weihenstephanensis* NFICC855. One-tail t test was performed on the obtained triplicates comparing OD_{600} value after 48 h growth of NFICC855 in BM and the same pathogens challenged with CFS of *L. plantarum* (NFICC19, 27, 72, 207, and 293), *F. sanfranciscensis* (UMCC 2990), *Lc. citreum* (NFICC28), and *P. pentosaceus* (UMCC 3010, NFICC 58, and 341). The mean value of the samples represented by * was significantly lower than the control sample (untreated sample); $t(2) = -2.92$ and $p = 0.05$.

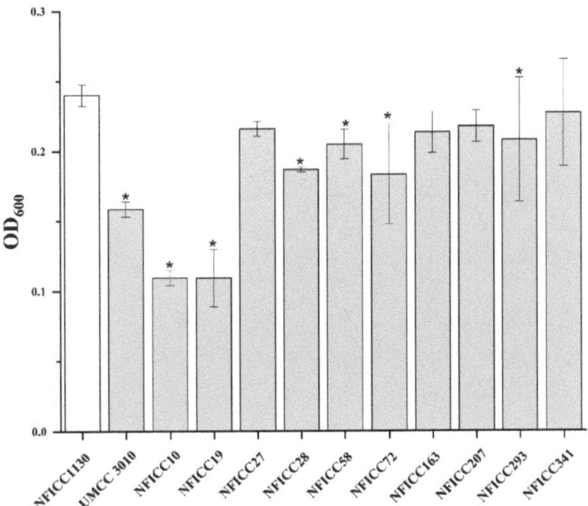

Figure 4. Antibacterial activity of the selected LAB's cell-free supernatant in Bread Media (BM) on spoilage agent *B. amyloliquefaciens* NFICC1130. One-tail *t*-test was performed on the obtained triplicates comparing OD_{600} value after 48 h growth of NFICC1130 in BM and the same pathogens challenged with CFS of *L. plantarum* (NFICC9, 27, 72, 163, 207, and 293), *Lc. citreum* (NFICC28), and *P. pentosaceus* (UMCC 3010, NFICC10, 58, and 341). Bars with * are significantly different, with a mean value significantly lower than the control sample (untreated sample), while samples NFICC 27, 163, 207, 293, and 341 respect the null hypothesis; $t(2) = -2.92$ and $p = 0.05$.

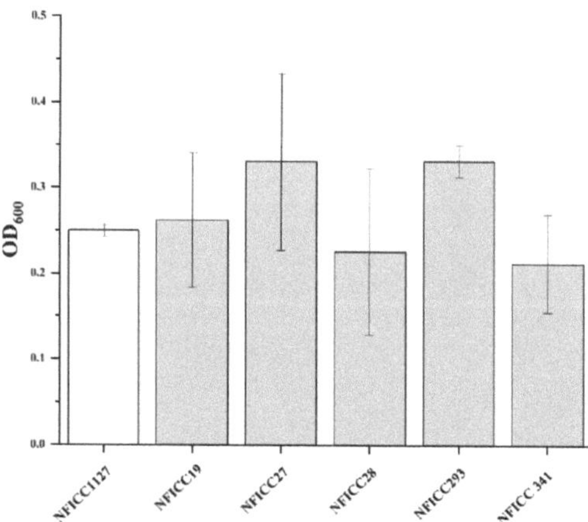

Figure 5. Antibacterial activity of the selected LAB's cell-free supernatant in Bread Media (BM) on spoilage agent *B. amyloquiefaciens* NFICC 1127. One-tail t test was performed on the obtained triplicates comparing OD600 value after 48 h growth of NFICC1127 in BM and the same pathogens challenged with CFS of *L. plantarum* (NFICC19, 27, and 293), *Lc. citreum* (NFICC28), and *P. pentosaceus* (NFICC341). No significant differences between the control and treated samples were observed at t(2) = −2.92 and p = 0.05.

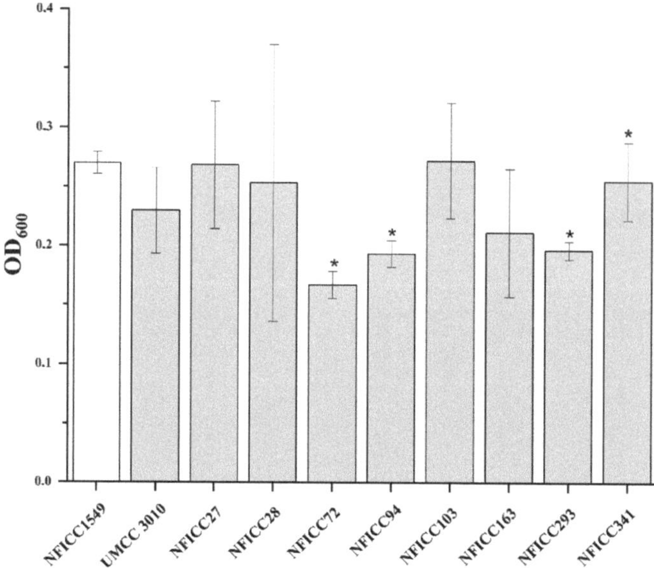

Figure 6. Antibacterial activity of the selected LAB's cell-free supernatant in Bread Medium (BM) on spoilage agent *B. licheniformis* NFICC1549. One-tail *t*-test was performed on the obtained triplicates comparing OD_{600} value after 48 h growth of NFICC1549 in BM and the same spoilage agent challenged with CFS of *L. plantarum* (NFICC27, 72, 163, and 293), *Lc. citreum* (NFICC28 and 94), and *P. pentosaceus* (UMCC 3010 and NFICC341). The * indicate measurements that are significantly different, with a mean lower than the control samples; t(2) = −2.92, p = 0.05.

3.3. Detection of Bioactive Compounds

3.3.1. Metabolites Detected by HPLC during Fermentation in MRS

LAB's ability to produce organic acids, other potentially antimicrobial compounds, and sugar consumption was assessed (Table 3). The CFS of the evaluated strains were examined by HPLC after 14 h, 18 h, and 24 h of fermentation in MRS. Observation of the results shows a corresponding ratio between amounts of lactic acid and acetic acid after 24 h. In congruence with the assessment for acidification, *L. plantarum* strains NFICC72, 163, 293, and 19 showed a strong and fast acidification activity. The present results also show that the ethanol detected is too low or not produced at all. Other compounds such as citric acid, oxalate, tartaric acid, and succinic were detected in trace amounts only.

3.3.2. Evaluation of the Bacteriocins Production and Kinetic Screening Assay

The screening of potential bacteriocin production revealed no antibacterial protein-associated inhibitions. Moreover, the assay of CFS spotted on a BHI confirmed the inhibitory activity of acids compounds and the absence of produced bacteriocins due to the nonformed inhibition zones. The results obtained from the kinetics assay highlight the bacteriostatic activity of the selected LAB. In fact, observations of the results demonstrates the presence of an extended lag phase, supporting the presence of organic acids in the cell-free supernatants and the absence of bacteriocins.

3.4. Assessment of Technological Features of the Candidate LAB Strains

3.4.1. Sugar Fermentation Profiling (API Test)

Sugar fermentation profiles obtained by API 50 CHL galleries were assessed for all the selected LAB, as indicated in Table 4, wherein the most common sugars in bakery products are highlighted (complete API profiles after 48 h are shown in Table S3). Results show that the most utilized sugars by the strains include glucose, fructose, maltose, and cellobiose.

3.4.2. Acidification Ability in BM

To further assess the strains' performance for dough fermentation, their acidification ability was tested in BM during a period of 24 h. In Figure 7, only the best performers are reported. A decrease in pH $\approx$ 4.5, which is considered to be safe, was noted. The strains considered to be the fastest acidifiers were *L. plantarum* (UMCC 2996, NFICC19, and 163), *Lc. citreum* NFICC28, and *P. pentosaceus* UMCC 3010, which reached a pH $\approx$ 4.5 with at least 10 h of fermentation.

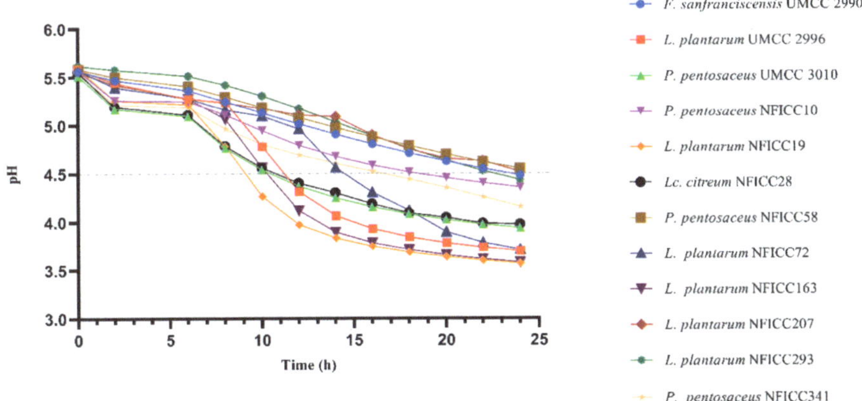

Figure 7. Acidification performance in Bread Medium showed by the best LAB candidates against bread rope-producing strains of *Bacillus* spp.

Table 3. Different compounds (g/L) detected by HPLC after 14, 18, and 24 h of fermentation in MRS broth. The creation of the heat map is based upon the detected MRS broth concentration: green color highlights the production, red color the consumption, and yellow color was used when the detected amount was equal to MRS broth. The measurement were with three independent experiments, and Kruskal–Wallis statistical analysis was performed to detect the significant strains with respect to plain MRS. The significance is marked with an *.

	Sample/time	MRS	F. sanfranciscensis UMCC 2990	L. plantarum UMCC2 996	P. pentosaceus UMCC 3010	P. pentosaceus NFICC10	L. plantarum NFICC19	Lc. citreum NFICC28	P. pentosaceus NFICC58	L. plantarum NFICC72	L. plantarum NFICC163	L. plantarum NFICC207	L. plantarum NFICC293	P. pentosaceus NFICC341
OXALATE	14H	0.31	0.37	0.36	0.38	0.39	0.95 *	0.34	0.32	0.39	0.57 *	0.39	0.30	0.34
	18H		0.37	0.33	0.34	0.40 *	0.93 *	0.39 *	0.38	0.38	0.29	0.36	0.63 *	0.39 *
	24H		0.37	0.44	0.44	0.48 *	1.38 *	0.57 *	0.41	0.70 *	0.69 *	0.32	0.55 *	0.37
CITRIC ACID	14H	1.86	0.00 *	0.00 *	0.00 *	0.00 *	1.77	1.04	1.18	0.00 *	0.00 *	0.00 *	0.00 *	0.38 *
	18H		1.13	0.00 *	1.06	1.12	1.05	1.05	1.86	0.92 *	0.09 *	1.04	0.00 *	0.00 *
	24H		0.99	0.90 *	1.04	0.96	1.60	0.40 *	2.15	0.00 *	0.00 *	0.81 *	0.00 *	1.10
TARTARIC ACID	14H	0.00	1.12 *	1.08	1.13 *	1.12 *	1.17	1.47 *	1.00	1.10 *	1.07	1.19 *	0.97	1.08
	18H		1.12 *	1.00	1.03	1.08 *	0.81	1.32 *	0.00	1.02	0.61	1.17 *	0.00	0.00
	24H		0.00	0.00	0.00	0.00	1.22 *	0.00	0.00	0.00	0.00	0.00	0.00	0.00
GLUCOSE	14H	14.62	14.33	9.34 *	12.25	11.56	6.32 *	6.97 *	8.00 *	13.39	7.02 *	13.62	6.04 *	10.13
	18H		14.99	7.06 *	11.19 *	11.13 *	0.95 *	5.37 *	5.67 *	11.28	4.87 *	11.74	5.23 *	8.59 *
	24H		14.02	11.86	9.04 *	8.98 *	0.98 *	2.78 *	4.03 *	0.00 *	0.00 *	9.52	4.07 *	7.61 *
SUCROSE	14H	1.49	1.79	0.76	1.33	1.57	0.00 *	0.00 *	0.00 *	1.57	0.93	1.03	0.51	1.14
	18H		1.68	0.75	1.26	1.25	0.71	0.00 *	0.00 *	0.49 *	0.51 *	0.92	0.49 *	0.95
	24H		1.72	1.33	1.22	1.34	0.70 *	0.00 *	0.00	0.87 *	0.83 *	1.06	1.11	1.08
GLUTAMIC ACID	14H	0.23	0.08 *	0.25	0.06 *	0.07 *	0.00 *	0.23	0.15	0.08 *	0.10	0.00 *	0.17	0.18
	18H		0.13	0.13	0.11 *	0.12 *	0.21	0.10 *	0.12 *	0.13	0.11 *	0.09 *	0.13	0.18
	24H		0.09	0.00 *	0.00 *	0.00 *	0.23	0.00 *	0.00 *	0.20	0.20	0.00 *	0.00 *	0.00 *
SUCCINIC	14H	0.00	0.00	0.00	0.00	0.00	0.00	0.00	0.00	0.00	0.18	2.07 *	0.00	0.00
	18H		0.75 *	0.59	0.70 *	0.74 *	0.00	0.00	0.07	0.44	0.48	1.10 *	0.44	0.00
	24H		0.00	0.00	0.00	0.00	1.22 *	0.00	0.08	0.00	0.00	1.15 *	0.00	0.00
FORMIC ACID	14H	0.00	0.03	0.00	0.00	0.00	0.00	0.00	0.07 *	0.00	0.00	0.09 *	0.07	0.15 *
	18H		3.76 *	0.01	0.13 *	0.15 *	0.00	0.00	0.11 *	0.03	0.00	0.03	0.00	0.08
	24H		1.44 *	0.01	0.23 *	0.18 *	0.00	0.00	0.12	2.02 *	0.00	0.01	0.00	0.14
ACETIC ACID	14H	0.00	4.36	4.77 *	4.73 *	4.48	4.90 *	5.44 *	0.16	4.34	4.73 *	5.14 *	4.41	4.56
	18H		4.56 *	4.45 *	4.38 *	4.50 *	4.86 *	5.52 *	2.23	4.35	4.49 *	4.89 *	4.49 *	4.47 *
	24H		4.39	4.41	4.40	4.55 *	4.79 *	4.77 *	4.29	4.60 *	4.65 *	4.84 *	4.38	4.51 *
1,3 PROPANDIOL	14H	0.00	0.71	0.85 *	0.83 *	0.82 *	0.86 *	0.92 *	0.71	0.75	0.86 *	0.89 *	0.84 *	0.81
	18H		0.79	0.78	0.81	0.82	0.89 *	0.86 *	0.81	0.53	0.85 *	0.83 *	0.88 *	0.75
	24H		0.67	0.74	0.79 *	0.80 *	0.84 *	0.67	0.84 *	0.70	0.84 *	0.83 *	0.86 *	0.82 *
2,3 BUTANDIOL	14H	0.00	0.05	0.06	0.21 *	0.20 *	0.00	0.00	0.10	0.18 *	0.00	0.29 *	0.00	0.18 *
	18H		0.18 *	0.00	0.16 *	0.18 *	0.00	0.00	0.13	0.14	0.00	0.26 *	0.00	0.06
	24H		0.00	0.06	0.19 *	0.17 *	0.00	0.00	0.15	0.00	0.00	0.26 *	0.00	0.08

Table 3. *Cont.*

	Sample/time	MRS	*F. sanfranciscensis* UMCC 2990	*L. plantarum* UMCC2 996	*P. pentosaceus* UMCC 3010	*P. pentosaceus* NFICC10	*L. plantarum* NFICC19	*Lc. citreum* NFICC28	*P. pentosaceus* NFICC58	*L. plantarum* NFICC72	*L. plantarum* NFICC163	*L. plantarum* NFICC207	*L. plantarum* NFICC293	*P. pentosaceus* NFICC341
LACTIC ACID	14H	0.00	0.59	12.05 *	6.14	5.50	14.73 *	5.47	4.97	2.88	14.25 *	2.95	13.33 *	7.98 *
	18H		0.86	13.08 *	5.59	6.34	19.78 *	6.16	7.91 *	6.56	16.14 *	3.13	14.77 *	9.18 *
	24H		0.78	5.21	9.38	10.08	19.90 *	16.91 *	12.27 *	20.32 *	20.78 *	4.69	16.11 *	11.43
ETOH	14H	0.00	0.00	0.00	0.00	0.00	0.00	1.32 *	0.00	0.00	0.00	0.00	0.00	0.00
	18H		0.00	0.00	0.00	0.00	0.00	1.58 *	0.00	0.00	0.00	0.00	0.00	0.00
	24H		0.00	0.00	0.00	0.00	0.00	0.37 *	0.00	0.00	0.00	1.74 *	0.00	0.00

Table 4. Sugar fermentation profiles of the tested LAB strains obtained by API 50 CHL after 48 h of incubation at 30 °C.

LAB	D(+)-Glucose	D(−)-Fructose	D(+)-Cellobiose	Maltose	Lactose	D(+)-Melibiose	Saccharose	D(+)-Raffinose	Starch	D(+)-Xylose
F. sanfranciscensis UMCC 2990	+	+	+	+	−	−	−	−	−	−
L. plantarum UMCC 2996	+	+	+	+	+	+	+	+	−	−
F. rossiae UMCC 3002	+	+/−	−	+	−	+/−	−	−	−	+
P. pentosaceus UMCC 3010	+	+	+	+	−	−	−	−	−	−
Lc. citreum UMCC 3011	+	+/−	−	+	−	−	−	−	−	+
P. pentosaceus NFICC10	+	+	+	+	+	+	+	+	−	−
L. plantarum NFICC19	+	+	+	+	+	+	+	+	−	−
L. plantarum NFICC27	+	+	+	+	+	+	+	+	−	−
Lc. citreum NFICC28	+	+	+	+	+	+	+	+	−	+
L. plantarum NFICC58	+	+	+	+	+	+	+	+	−	+
L. plantarum NFICC72	+	+	+	+	+	+	+	+	−	−
P. pentosaceus NFICC87	+	+	+	+	−	−	+	−	−	+
Lc. citreum NFICC94	+	+	+	+	−	−	+	−	−	+
P. pentosaceus NFICC103	+	+	+	+	+	+	+	+/−	−	−
P. pentosaceus NFICC163	+	+	+	+	+	+	+	+	−	+/−
L. plantarum NFICC207	+	+	+	+	+	+	+	+	−	+
L. plantarum NFICC293	+	+	+	+	+	+	+	+	−	−
P. pentosaceus NFICC341	+	+	+	+	+	+	+	+	−	+/−

3.5. Safety Profile of the Selected LAB Strains

3.5.1. Hemolytic Activity Profile

Nonhemolytic activity is considered a safe prerequisite for the selection of new strains for food starter culture [39,40]. Results indicate that all the examined strains did not show signs of β-hemolytic activity when grown in Columbia Blood Agar (Table 5). Eleven LAB strains exhibited green-hued zones around colonies showing α-hemolytic activity (partial hemolysis), while seven strains including *F. rossiae* UMCC 3002, *Lc. citreum* UMCC 3011, NFICC28, NFICC94, and *P. pentosaceus* NFICC58 and NFICC341 were γ- hemolytic (nonhemolysis).

Table 5. Hemolytic capacity * of the tested strains.

Tested Strains	γ-Hemolysis	α-Hemolysis	β-Hemolysis
F. sanfranciscensis UMCC 2990		x	
L. plantarum UMCC2996		x	
F. rossiae UMCC 3002	x		
P. pentosaceus UMCC 3010		x	
Lc. citreum UMCC 3011	x		
P. pentosaceus NFICC10		x	
L. plantarum NFICC19		x	
L. plantarum NFICC27		x	
Lc. citreum NFICC28	x		
L. plantarum NFICC58	x		
L. plantarum NFICC72		x	
P. pentosaceus NFICC87	x		
Lc. citreum NFICC94	x		
P. pentosaceus NFICC103		x	
P. pentosaceus NFICC163		x	
L. plantarum NFICC207		x	
L. plantarum NFICC293		x	
P. pentosaceus NFICC341	x		
S. aureus NFICC1477	x		
Lc. citreum NFICC88		x	
B. cereus DMS 2301			x

* The occurrence of the specific hemolytic activity is marked by "x".

3.5.2. Antimicrobial Susceptibility Profiles (AST)

The sensitivity of LAB strains was determined against various antibiotics (streptomycin, ampicillin, vancomycin, gentamicin, kanamycin, and chloramphenicol), and the obtained results were compared based on the cut-offs, as specified in the guidance on the assessment of bacterial susceptibility to antimicrobials of human and veterinary importance from EFSA [32]. Sensitivity profiles are shown in Table 6. Briefly, almost all of the tested LAB show resistance against streptomycin (99%), except *P. pentosaceus* NFICC341 and vancomycin (100%), as already reported by different studies. However, against ampicillin (81%), kanamycin (76%), chloramphenicol (99%), and gentamycin (57%), the majority of them were in accordance with the EFSA cut-off and could be considered susceptible.

3.5.3. In Silico Screening for Resistance Genes

All 18 isolates were analyzed for the presence of acquired resistance genes using ResFinder version 4.2.3 and ResFinder database version 2.0.1 [41] using genome assemblies. Parameters used for a matching identity on the database were required to be at least 80%, whereas the coverage of a matching gene in the database was required to be at least 60%.

3.6. Whole-Genome Sequencing Analysis: Functional Gene and Potential Virulence In Silico Screening

The presence of bacteriocins was initially tested following the protocol by Fugaban et al. [28] with no positive results. In addition, to further confirm the possible absence of bacteriocins, sequences of the whole genome obtained from all the evaluated LAB were analyzed using the web server BAGEL4. Although phenotypically no bacteriocin-associated inhibitions were observed, in silico analysis showed the presence of bacteriocins belonging to class IIb and Iic (particularly penocin and plantaricins A, E, F, K, J, and N) (Table 7). The structures of the putative operons are shown in Figure S1.

Table 6. Minimum inhibitory concentration (MIC) (µg/mL) of the antibiotic susceptibility of the selected LAB. The values reported in bold and blue color are under or equal to recommended EFSA cut-offs and underline the susceptibility of the strains.

Tested LAB	Tested Antibiotics					
	Streptomycin	Ampicillin	Kanamycin	Vancomycin	Chloramphenicol	Gentamycin
F. sanfranciscensis UMCC 2990	≤128	≤16	≤32	512	≤4	≤4
L. plantarum UMCC 2996	≤128	≤32	≤64	512	Resistant	≤256
F. rossiae UMCC 3002	≤128	≤4	≤32	512	≤4	≤4
P. pentosaceus UMCC 3010	≤128	≤2	≤16	512	≤4	≤4
Lc. citreum UMCC 3011	≤128	≤4	≤32	512	≤8	≤8
P. pentosaceus NFICC10	≤128	≤1	≤16	512	≤4	≤4
L. plantarum NFICC19	≤64	≤1	≤16	512	≤4	≤4
L. plantarum NFICC27	≤64	≤1	≤16	512	≤4	≤4
Lc. citreum NFICC28	≤128	≤1	≤32	512	≤4	≤4
L. plantarum NFICC58	≤128	≤1	≤32	512	≤4	≤4
L. plantarum NFICC72	≤64	≤1	≤16	512	≤4	≤4
Lc. citreum NFICC87	≤128	≤1	≤32	512	≤4	≤4
Lc. citreum NFICC94	≤128	≤1	≤32	512	≤4	≤8
P. pentosaceus NFICC103	≤128	≤2	≤32	512	≤4	≤4
P. pentosaceus NFICC163	≤64	≤1	≤16	512	≤4	≤4
L. plantarum NFICC207	≤64	≤1	≤16	512	≤4	≤4
L. plantarum NFICC293	≤128	≤1	≤8	≤8	≤4	≤4
P. pentosaceus NFICC341	≤64	≤2	≤32	512	≤4	≤4

Table 7. Bacteriocins produced by the LAB strains displaying the highest inhibitory activity against ropy agents.

Strain	Species	Bacteriocins Predicted by BAGEL4
NFICC28	*Lc. citreum*	None
NFICC10	*P. pentosaceus*	Penocin A
NFICC58	*P. pentosaceus*	None
NFICC19	*L. plantarum*	Plantaricin E, Plantaricin F Plantaricin K, putative class IIc bacteriocin, putative class IIb bacteriocin
NFICC72	*L. plantarum*	Plantaricin E, Plantaricin F Plantaricin K, putative class IIc bacteriocin, putative class IIb bacteriocin
NFICC163	*L. plantarum*	Plantaricin E, Plantaricin F Plantaricin K, putative class IIc bacteriocin, putative class IIb bacteriocin
NFICC293	*L. plantarum*	Plantaricin A, Plantaricin E, Plantaricin F, Plantaricin J, Plantaricin K, Plantaricin N
UMCC 2996	*L. plantarum*	Plantaricin A, Plantaricin E, Plantaricin F, Plantaricin J, Plantaricin K, Plantaricin N

4. Discussion

Bacillus spp. are well associated with ropy bread and bakery product spoilage [6,7,42]. Clean-label strategies to control microbial spoilage in bakery industries are important, accordingly, some starter cultures, including LAB, are promising alternative biocontrol agents thanks to their ability to produce bacteriocins and organic acids [22,43]. To screen candidates efficiently, we adapted the high-throughput screening assay developed by Inglin et al. [25], which was identified as a fast, low-cost, and accurate primary screening. Based on the results, most of the LAB strains found to be active against *Bacillus* spp. strains originated from fruits and vegetables (NFICC strains) or sourdoughs (UMCC strains). Their ability to adapt to these highly variable and stressful environments has aided in their capacity to grow into different niches, which can be exploited for their beneficial features, such as acidification ability and competitiveness [11,44,45].

To further assess the ability of the strains, a well diffusion assay in 24-well microtiter plates was conducted using live cells, confirming 18 active strains comprised of representatives from *Lactobacillus*, *Pediococcus*, and *Leuconostoc* genera, from both screening assays. Observations in this study were similar to those of Adesulu-Dahunsi et al. [46] against *Bacillus* spp., *Listeria monocytogenes*, and *Escherichia coli*.

The capacity of the *P. pentosaceus* strain NFICC341 to inhibit the growth of the majority of the spoilage microorganisms can be traced back to its origin, which was Brewers' spent grain. The potential of LAB strains isolated from the vegetable matrix was already demonstrated by Puntillo et al. [47]. In accordance with previous studies, similar behavior can be associated with strains of *Lc. citreum*, *L. plantarum*, and *F. sanfranciscensis*, isolated from plants and sourdoughs, which exhibited promising inhibitory activity [48,49]. Furthermore, it was noted that *L. plantarum* and *P. pentosaceus* are well-known species that are able to produce antimicrobial compounds and organic acids, which can influence the safety of food [44,45,50,51]. Differences between the activities of the strains from the same species might be related to the isolation matrix of the strains and their adaptability to different stress conditions [15,52]. While strains UMCC 2996, UMCC 2990, and NFICC28 were isolated from sourdough and shared a similar antimicrobial activity, sample NFICC19 resulted to be the best acidifier.

A confirmation assay was conducted to further assess the activity of the bioactive metabolites produced by the LAB; however, contrary to the initial observations obtained on the high-throughput screening employed, only bacteriostatic activity was observed, in accordance with the previous literature findings [53]. Thus, identification of the nature of the bioactive compound was performed.

The evaluation of the metabolites in the CFS samples analyzed by using HPLC revealed the presence of organic acids in the supernatant, and therefore suggests they may play a significant role in the inhibition activity of the LAB strains tested [19,54]. Organic acids, in particular lactic acid, reduce the matrix's pH without causing a strong sour taste. A fast acidification of the product is responsible for the inhibition of *B. cereus*, as showed by Yang et al. [45], where a reduction in the initial CFU/mL of the pathogen was detected after LAB fermentation. Treatment of proteolytic enzymes on the CFS showed no significant changes in the inhibitory activity of the bioactive strains—ruling out the possible involvement of protein-based bioactive compounds, including bacteriocins and bacteriocin-like inhibitory substances. Additionally, the elimination of heat-labile antimicrobials was excluded by heat treatment (80 °C) of the CFS.

Even though no bacteriocinogenic inhibitions were detected based on the assays performed, the necessity for further investigation of which other possible metabolites can be further exploited, for future potential application of these studied LAB against different types of Gram-positive pathogens and spoilage bacteria, is needed [55]. Therefore, in silico search for possible antibacterial metabolites was conducted, identifying the presence of Class IIb and Class IIc bacteriocin operons on some of the studied LAB strains.

Class II bacteriocins are nonlantibiotics, heat-stable, and small ($\leq$10 kDa) hydrophobic peptides divided into four subclasses. Albeit these strains are detected on the genome, one of the possibilities for not observing bacteriocins from the CFS might be due to the presence of nonsatisfactory environmental conditions or an incomplete expression of all operons required for bacteriocin synthesis, transport, and regulation [56].

The effective biocontrol activity of the LAB strains evaluated would be associated with their rapid rate of acidification, as this is fundamental in food processing to prevent and avoid spoilage contaminants. Sourdough's pH varies depending on the state of fermentation; however, it typically ranges between 3.5 and 5. Harmful microorganisms such as botulism bacteria, *E. coli*, and spoilage fungi cannot develop in an environment with a pH below 4.6, as this acidity keeps them away [5,57]. We observed that the majority of the strains were able to lower the pH around 4–4.5 (considered a safe pH to avoid spoilage) after 10 h, exhibiting a high acidification ability and underlining the massive production of acids (lactic and acetic acid). In fact, low pH and high acidity are the major factors for avoiding/delaying the growth of ropy bacterial agents, thus making the rate of pH reduction during the early stage of fermentation crucial for optimum inhibition [45,53,58]. Organic acids were the key metabolites that were identified in this study. The significance of pH reduction in food stability and preservation has been widely researched and recognized by the scientific community [59,60]. Furthermore, pH value also impacts the pKa of the various acids, leading them to a lower or higher dissociation, affecting the safety and taste of the final product [61].

Before the application of any microorganisms in food systems, an assessment of their safety and technological features is important. In this study, we profiled bioactive strains, identifying their sugar fermentations and their ability to be functional for their intended application in simulated ex vivo models. As observed, these strains can use a wide array of carbohydrates primarily found in bakery products, allowing them to be applied efficiently in this system. Consequently, acidification profiles highlight their ability to thrive in bread and produce necessary metabolites for possible bio-protection [13,62]. On the other hand, safety features were assessed as follows: antimicrobial susceptibility and hemolytic activity. Findings indicate that the evaluated strains have a broad range of resistance against streptomycin and vancomycin. Although this might be concerning, it should be noted that resistance to antibiotics in the majority of LAB is intrinsic. Additional assessment, particularly the identification of the location of associated resistance genes, should be performed to identify the occurrence of acquired resistance [63,64]. However, an in silico search for the presence of antibiotic resistance genes showed no matches on all the genomes of the 18 strains evaluated. Evaluation of hemolytic activity is one of the key factors that should be considered when assessing the safety of functional strains [39]. In this study, we identified partial hemolysis (α-hemolysis) on 11 out of 18 strains, whereas the rest showed no hemolysis. Thus, further confirmation, particularly the presence of genes involved in these putative virulence factors, should be investigated to assess potential risk for the application of the strains, especially for food consumption.

5. Conclusions

Cereals and bakery goods are a fundamental part of the human diet, and their higher susceptibility to microbial spoilage could lead to economic losses and health issues. Generally, LAB starter cultures and their metabolites are found to be promising for controlling spoilage agents. In this study, several LAB strains able to contrast the growth of some common rope spoilage agents were selected by using high-throughput and ex vivo screening assays. Specifically, *L. plantarum* NFICC19, NFICC72, NFICC293, *Lc. citreum* NFICC28, and *P. pentosaceus* NFICC58 and NFICC341 showed the best inhibitory activity. The assessment of their technological and safety features supported their suitability for fermentation processes and the production of bakery products. Moreover, the whole-genome sequencing of the selected LAB strains and the in silico analysis showed that some of the strains contain operons for bacteriocins, yet no significant evidence was observed phenotypically, suggesting that additional analysis needs to be performed to better understand the inhibitory mechanisms involved and validate the application of the strains as potential biocontrol agents.

Supplementary Materials: The following supporting information can be downloaded at: https://www.mdpi.com/article/10.3390/fermentation9030290/s1, Figure S1: Gene encoding for bacteriocins detected in the genome of the best candidate strains. Table S1: Results of the high-throughput preliminary screening of LAB-CFS against spoilage bacteria in common media. Table S2: Results of the dual-plate agar high-throughput screening. Table S3: Results of the complete API test profile after 48 h of incubation at 30 °C.

Author Contributions: Conceptualization, methodology, data curation, investigation, formal analysis, writing—original draft, G.I.; Conceptualization, methodology, data curation, investigation, writing—review and editing, J.I.I.F.; formal analysis, data curation, S.Ö.; data curation, visualization, A.P.W.; investigation, data curation, R.S.K.; data curation, Q.H.; validation, formal analysis, data curation, R.S.; resources, supervision, A.P.; conceptualization, writing—review and editing, supervision, L.D.V.; conceptualization, resources, writing—review and editing, supervision, funding acquisition, C.H.B.-B. All authors have read and agreed to the published version of the manuscript.

Funding: G.I. acknowledges the award of a PhD scholarship from the Department of Life Sciences, University of Modena and Reggio Emilia. R.S. and S.O. were funded by FBCD FFBI grant Apple Pom (project ID 20080). R.S.K. was funded by the European Union's Horizon 2020 research and innovation program (Grant: 874735). C.H.B.-B. has received funding from MycoProtein (Grant Number 34009-20-1682). Lastly, J.I.I.F. was funded by an industrial PhD scholarship (23 IFD LAB DuPont PhD project number 110063) from the Danish innovation foundation. S.Ö. and Q.H. were working as research assistants. Both L.D.V. and A.P. are employees of the Department of Life Sciences, University of Modena and Reggio Emilia.

Institutional Review Board Statement: Not applicable.

Informed Consent Statement: Not applicable.

Data Availability Statement: Not applicable.

Acknowledgments: The authors would like to thank Christina Aaby Svendsen from the Research Group for Genomic Epidemiology, DTU National Food Institute for the technical assistance with performing full-genome Illumina sequencing.

Conflicts of Interest: The authors declare no conflict of interest.

References

1. Zain, M.Z.M.; Shori, A.B.; Baba, A.S. Potential functional food ingredients in bread and their health benefits. *Biointerface Res. Appl. Chem.* **2022**, *12*, 6533–6542. [CrossRef]
2. Goryńska-Goldmann, E.; Gazdecki, M.; Rejman, K.; Łaba, S.; Kobus-Cisowska, J.; Szczepański, K. Magnitude, causes and scope for reducing food losses in the baking and confectionery industry a multi-method approach. *Agriculture* **2021**, *11*, 936. [CrossRef]
3. Cicatiello, C.; Franco, S.; Pancino, B.; Blasi, E.; Falasconi, L. The dark side of retail food waste: Evidences from in-store data. *Resour. Conserv. Recycl.* **2017**, *125*, 273–281. [CrossRef]
4. Muthappa, D.M.; Lamba, S.; Sivasankaran, S.K.; Naithani, A.; Rogers, N.; Srikumar, S.; Macori, G.; Scannell, A.G.M.; Fanning, S. 16S rRNA Based Profiling of Bacterial Communities Colonizing Bakery-Production Environments. *Foodborne Pathog. Dis.* **2022**, *19*, 485–494. [CrossRef] [PubMed]
5. Pepe, O.; Blaiotta, G.; Moschetti, G.; Greco, T.; Villani, F. Rope-producing strains of *Bacillus* spp. from wheat bread and strategy for their control by lactic acid bacteria. *Appl. Environ. Microbiol.* **2003**, *69*, 2321–2329. [CrossRef] [PubMed]
6. Valerio, F.; De Bellis, P.; Di Biase, M.; Lonigro, S.L.; Giussani, B.; Visconti, A.; Lavermicocca, P.; Sisto, A. Diversity of spore-forming bacteria and identification of *Bacillus amyloliquefaciens* as a species frequently associated with the ropy spoilage of bread. *Int. J. Food Microbiol.* **2012**, *156*, 278–285. [CrossRef] [PubMed]
7. Valerio, F.; Di Biase, M.; Huchet, V.; Desriac, N.; Lonigro, S.L.; Lavermicocca, P.; Sohier, D.; Postollec, F. Comparison of three *Bacillus amyloliquefaciens* strains growth behaviour and evaluation of the spoilage risk during bread shelf-life. *Food Microbiol.* **2015**, *45*, 2–9. [CrossRef] [PubMed]
8. Iurie, R.; Cahul, B.P.H.; Uniersity, S.; Turtoi, M.; Dunarea, U.; Galati, D.J. The occurrence of the rope spoilage of bread at Cahulpan bread-making company. *Sci. J. Cahul State Univ. "Bogdan Petriceicu Hasdeu" Econ. Eng. Stud.* **2021**, *2*, 107–116.
9. Thompson, J.M.; Waites, W.M.; Dodd, C.E.R. Detection of rope spoilage in bread caused by *Bacillus* species. *J. Appl. Microbiol.* **1998**, *85*, 481–486. [CrossRef]
10. Pereira, A.P.M.; Stradiotto, G.C.; Freire, L.; Alvarenga, V.O.; Crucello, A.; Morassi, L.L.P.; Silva, F.P.; Sant'Ana, A.S. Occurrence and enumeration of rope-producing spore forming bacteria in flour and their spoilage potential in different bread formulations. *LWT* **2020**, *133*, 110108. [CrossRef]
11. De Vero, L.; Iosca, G.; Gullo, M.; Pulvirenti, A. Functional and healthy features of conventional and non-conventional sourdoughs. *Appl. Sci.* **2021**, *11*, 3694. [CrossRef]
12. Ibrahim, S.A.; Ayivi, R.D.; Zimmerman, T.; Siddiqui, S.A.; Altemimi, A.B.; Fidan, H.; Esatbeyoglu, T.; Bakhshayesh, R.V. Lactic acid bacteria as antimicrobial agents: Food safety and microbial food spoilage prevention. *Foods* **2021**, *10*, 3131. [CrossRef] [PubMed]

13. Wang, Y.; Wu, J.; Lv, M.; Shao, Z.; Hungwe, M.; Wang, J.; Bai, X.; Xie, J.; Wang, Y.; Geng, W. Metabolism Characteristics of Lactic Acid Bacteria and the Expanding Applications in Food Industry. *Front. Bioeng. Biotechnol.* **2021**, *9*, 612285. [CrossRef] [PubMed]

14. De Vero, L.; Iosca, G.; La China, S.; Licciardello, F.; Gullo, M.; Pulvirenti, A. Yeasts and lactic acid bacteria for panettone production: An assessment of candidate strains. *Microorganisms* **2021**, *9*, 1093. [CrossRef] [PubMed]

15. Iosca, G.; De Vero, L.; Di Rocco, G.; Perrone, G.; Gullo, M.; Pulvirenti, A. Anti-spoilage activity and exopolysaccharides production by selected lactic acid bacteria. *Foods* **2022**, *11*, 1914. [CrossRef]

16. De Vuyst, L.; Leroy, F. Bacteriocins from lactic acid bacteria: Production, purification, and food applications. *J. Mol. Microbiol. Biotechnol.* **2007**, *13*, 194–199. [CrossRef]

17. Iosca, G.; Turetta, M.; De Vero, L.; Bang-berthelsen, C.H.; Gullo, M.; Pulvirenti, A. Valorization of wheat bread waste and cheese whey through cultivation of lactic acid bacteria for bio-preservation of bakery products. *LWT* **2023**, *176*, 114524. [CrossRef]

18. Zotta, T.; Parente, E.; Ricciardi, A. Viability staining and detection of metabolic activity of sourdough lactic acid bacteria under stress conditions. *World J. Microbiol. Biotechnol.* **2009**, *25*, 1119–1124. [CrossRef]

19. Cizeikiene, D.; Juodeikiene, G.; Paskevicius, A.; Bartkiene, E. Antimicrobial activity of lactic acid bacteria against pathogenic and spoilage microorganism isolated from food and their control in wheat bread. *Food Control* **2013**, *31*, 539–545. [CrossRef]

20. Katina, K.; Sauri, M.; Alakomi, H.L.; Mattila-Sandholm, T. Potential of lactic acid bacteria to inhibit rope spoilage in wheat sourdough bread. *LWT* **2002**, *35*, 38–45. [CrossRef]

21. Menteş, Ö.; Ercan, R.; Akçelik, M. Inhibitor activities of two *Lactobacillus* strains, isolated from sourdough, against rope-forming *Bacillus* strains. *Food Control* **2007**, *18*, 359–363. [CrossRef]

22. Valerio, F.; De Bellis, P.; Lonigro, S.L.; Visconti, A.; Lavermicocca, P. Use of *Lactobacillus plantarum* fermentation products in bread-making to prevent *Bacillus subtilis* ropy spoilage. *Int. J. Food Microbiol.* **2008**, *122*, 328–332. [CrossRef] [PubMed]

23. Madsen, S.K.; Priess, C.; Wätjen, A.P.; Øzmerih, S.; Mohammadifar, M.A.; Heiner Bang-Berthelsen, C. Development of a yoghurt alternative, based on plant-adapted lactic acid bacteria, soy drink and the liquid fraction of brewers' spent grain. *FEMS Microbiol. Lett.* **2021**, *368*, fnab093. [CrossRef] [PubMed]

24. Saranraj, P.; Gheeta, M. Microbial Spoilage of Bakery Products and Its Control by Preservatives. *Int. J. Pharm. Biol.* **2012**, *3*, 38–48.

25. Inglin, R.C.; Stevens, M.J.A.; Meile, L.; Lacroix, C.; Meile, L. High-throughput screening assays for antibacterial and antifungal activities of *Lactobacillus* species. *J. Microbiol. Methods* **2015**, *114*, 26–29. [CrossRef]

26. Verni, M.; Minisci, A.; Convertino, S.; Nionelli, L.; Rizzello, C.G. Wasted Bread as Substrate for the Cultivation of Starters for the Food Industry. *Front. Microbiol.* **2020**, *11*, 293. [CrossRef] [PubMed]

27. Hui-Hu, C.; Ren, L.Q.; Zhou, Y.; Ye, B.C. Characterization of antimicrobial activity of three *Lactobacillus plantarum* strains isolated from Chinese traditional dairy food. *Food Sci. Nutr.* **2019**, *7*, 1997–2005. [CrossRef]

28. Fugaban, J.I.I.; Vazquez Bucheli, J.E.; Holzapfel, W.H.; Todorov, S.D. Characterization of Partially Purified Bacteriocins Produced by *Enterococcus faecium* Strains Isolated from Soybean Paste Active Against *Listeria* spp. and Vancomycin-Resistant Enterococci. *Microorganisms* **2021**, *9*, 1085. [CrossRef]

29. Corrieu, G.; Spinnler, H.E.; Picque, D.; Jomier, Y. Automated system to follow up 403 and control the acidification activity of lactic acid starters. *Fr. Pat. FR* **1988**, *2*, 612.

30. Madsen, S.K.; Thulesen, E.T.; Mohammadifar, M.A.; Bang-Berthelsen, C.H. Chufa drink: Potential in developing a new plant-based fermented dessert. *Foods* **2021**, *10*, 3010. [CrossRef]

31. Fugaban, J.I.I.; Vazquez Bucheli, J.E.; Kim, B.; Holzapfel, W.H.; Todorov, S.D. Safety and beneficial properties of bacteriocinogenic *Pediococcus acidilactici* and *Pediococcus pentosaceus* isolated from silage. *Lett. Appl. Microbiol.* **2021**, *73*, 725–734. [CrossRef] [PubMed]

32. European Food Safety Authority (EFSA). Guidance on the assessment of bacterial susceptibility to antimicrobials of human and veterinary importance. *EFSA J.* **2012**, *10*, 2740. [CrossRef]

33. Bolger, A.M.; Lohse, M.; Usadel, B. Trimmomatic: A flexible trimmer for Illumina sequence data. *Bioinformatics* **2014**, *30*, 2114–2120. [CrossRef] [PubMed]

34. Wick, R.R.; Judd, L.M.; Gorrie, C.L.; Holt, K.E. Unicycler: Resolving bacterial genome assemblies from short and long sequencing reads. *PLoS Comput. Biol.* **2017**, *13*, e1005595. [CrossRef]

35. VanOeffelen, M.; Nguyen, M.; Aytan-Aktug, D.; Brettin, T.; Dietrich, E.M.; Kenyon, R.W.; Machi, D.; Mao, C.; Olson, R.; Pusch, G.D.; et al. A genomic data resource for predicting antimicrobial resistance from laboratory-derived antimicrobial susceptibility phenotypes. *Brief. Bioinform.* **2021**, *22*, bbab313. [CrossRef] [PubMed]

36. Wattam, A.R.; Davis, J.J.; Assaf, R.; Boisvert, S.; Brettin, T.; Bun, C.; Conrad, N.; Dietrich, E.M.; Disz, T.; Gabbard, J.L.; et al. Improvements to PATRIC, the all-bacterial bioinformatics database and analysis resource center. *Nucleic Acids Res.* **2017**, *45*, D535–D542. [CrossRef] [PubMed]

37. Bateman, A.; Martin, M.J.; Orchard, S.; Magrane, M.; Agivetova, R.; Ahmad, S.; Alpi, E.; Bowler-Barnett, E.H.; Britto, R.; Bursteinas, B.; et al. UniProt: The universal protein knowledgebase in 2021. *Nucleic Acids Res.* **2020**, *49*, D480–D489. [CrossRef]

38. Van Heel, A.J.; De Jong, A.; Song, C.; Viel, J.H.; Kok, J.; Kuipers, O.P. BAGEL4: A user-friendly web server to thoroughly mine RiPPs and bacteriocins. *Nucleic Acids Res.* **2018**, *46*, W278–W281. [CrossRef]

39. Bindu, A.; Lakshmidevi, N. Identification and in vitro evaluation of probiotic attributes of lactic acid bacteria isolated from fermented food sources. *Arch. Microbiol.* **2021**, *203*, 579–595. [CrossRef]

40. Estifanos, H. Isolation and identification of probiotic lactic acid bacteria from curd and in vitro evaluation of its growth inhibition activities against pathogenic bacteria. *Afr. J. Microbiol. Res.* **2014**, *8*, 1419–1425. [CrossRef]

41. Bortolaia, V.; Kaas, R.S.; Ruppe, E.; Roberts, M.C.; Schwarz, S.; Cattoir, V.; Philippon, A.; Allesoe, R.L.; Rebelo, A.R.; Florensa, A.F.; et al. ResFinder 4.0 for predictions of phenotypes from genotypes. *J. Antimicrob. Chemother.* **2020**, *75*, 3491–3500. [CrossRef]

42. Dubey, S.; Sharma, N.; Thakur, S.; Patel, R.; Reddy, B.M. *Bacillus cereus* food poisoning in Indian perspective: A review. *Pharma Innov. J.* **2021**, *10*, 970–975.

43. Qian, M.; Liu, D.; Zhang, X.; Yin, Z.; Ismail, B.B.; Ye, X.; Guo, M. A review of active packaging in bakery products: Applications and future trends. *Trends Food Sci. Technol.* **2021**, *114*, 459–471. [CrossRef]

44. Corsetti, A.; Gobbetti, M.; Smacchi, E. Antibacterial activity of sourdough lactic acid bacteria: Isolation of a bacteriocin-like inhibitory substance from *Lactobacillus sanfrancisco* C57. *Food Microbiol.* **1996**, *13*, 447–456. [CrossRef]

45. Yang, Y.; Tao, W.Y.; Liu, Y.J.; Zhu, F. Inhibition of *Bacillus cereus* by lactic acid bacteria starter cultures in rice fermentation. *Food Control* **2008**, *19*, 159–161. [CrossRef]

46. Adesulu-Dahunsi, A.T.; Sanni, A.I.; Jeyaram, K. Diversity and technological characterization of *Pediococcus pentosaceus* strains isolated from Nigerian traditional fermented foods. *LWT* **2021**, *140*, 110697. [CrossRef]

47. Puntillo, M.; Gaggiotti, M.; Oteiza, J.M.; Binetti, A.; Massera, A.; Vinderola, G. Potential of Lactic Acid Bacteria Isolated From Different Forages as Silage Inoculants for Improving Fermentation Quality and Aerobic Stability. *Front. Microbiol.* **2020**, *11*, 586716. [CrossRef]

48. Arena, M.P.; Silvain, A.; Normanno, G.; Grieco, F.; Drider, D.; Spano, G.; Fiocco, D. Use of *Lactobacillus plantarum* strains as a bio-control strategy against food-borne pathogenic microorganisms. *Front. Microbiol.* **2016**, *7*, 464. [CrossRef]

49. Messens, W.; De Vuyst, L. Inhibitory substances produced by Lactobacilli isolated from sourdoughs—A review. *Int. J. Food Microbiol.* **2002**, *72*, 31–43. [CrossRef]

50. Lindgren, S.E.; Dobrogosz, W.J. Antagonistic activities of lactic acid bacteria in food and feed fermentations. *FEMS Microbiol. Lett.* **1990**, *87*, 149–163. [CrossRef]

51. Thao, T.T.P.; Thoa, L.T.K.; Ngoc, L.M.T.; Lan, T.T.P.; Phuong, T.V.; Truong, H.T.H.; Khoo, K.S.; Manickam, S.; Hoa, T.T.; Tram, N.D.Q.; et al. Characterization halotolerant lactic acid bacteria *Pediococcus pentosaceus* HN10 and in vivo evaluation for bacterial pathogens inhibition. *Chem. Eng. Process. Process Intensif.* **2021**, *168*, 108576. [CrossRef]

52. Ruiz Rodríguez, L.G.; Mohamed, F.; Bleckwedel, J.; Medina, R.; De Vuyst, L.; Hebert, E.M.; Mozzi, F. Diversity and functional properties of lactic acid bacteria isolated from wild fruits and flowers present in northern Argentina. *Front. Microbiol.* **2019**, *10*, 1091. [CrossRef] [PubMed]

53. Gao, Z. Evaluation of different kinds of organic acids and their antibacterial activity in Japanese Apricot fruits. *Afr. J. Agric. Res.* **2012**, *7*, 4911–4918. [CrossRef]

54. Rishi, L.; Mittal, G.; Agarwal, R.K.; Sharma, T. Melioration in Anti-staphylococcal Activity of Conventional Antibiotic(s) by Organic Acids Present in the Cell Free Supernatant of *Lactobacillus paraplantarum*. *Indian J. Microbiol.* **2017**, *57*, 359–364. [CrossRef]

55. Parada, J.L.; Caron, C.R.; Medeiros, A.B.P.; Soccol, C.R. Bacteriocins from lactic acid bacteria: Purification, properties and use as biopreservatives. *Braz. Arch. Biol. Technol.* **2007**, *50*, 521–542. [CrossRef]

56. Zhang, T.; Zhang, Y.; Li, L.; Jiang, X.; Chen, Z.; Zhao, F.; Yi, Y. Biosynthesis and Production of Class II Bacteriocins of Food-Associated Lactic Acid Bacteria. *Fermentation* **2022**, *8*, 217. [CrossRef]

57. Pacher, N.; Burtscher, J.; Johler, S.; Etter, D.; Bender, D.; Fieseler, L.; Domig, K.J. Ropiness in Bread—A Re-Emerging Spoilage Phenomenon. *Foods* **2022**, *11*, 3021. [CrossRef]

58. Cho, W.-I.; Chung, M.-S. Antimicrobial effect of a combination of herb extract and organic acid against *Bacillus subtilis* spores. *Food Sci. Biotechnol.* **2017**, *26*, 1423–1428. [CrossRef]

59. Amit, S.K.; Uddin, M.M.; Rahman, R.; Islam, S.M.R.; Khan, M.S. A review on mechanisms and commercial aspects of food preservation and processing. *Agric. Food Secur.* **2017**, *6*, 1–22. [CrossRef]

60. Gould, G.W. Methods for preservation and extension of shelf life. *Int. J. Food Microbiol.* **1996**, *33*, 5–64. [CrossRef]

61. Su, X.; Wu, F.; Zhang, Y.; Yang, N.; Chen, F.; Jin, Z.; Xu, X. Effect of organic acids on bread quality improvement. *Food Chem.* **2019**, *278*, 267–275. [CrossRef] [PubMed]

62. Ayivi, R.; Gyawali, R.; Krastanov, A.; Aljaloud, S.; Worku, M.; Tahergorabi, R.; Da Silva, R.; Ibrahim, S. Lactic Acid Bacteria: Food Safety and Human Health Applications. *Dairy* **2020**, *1*, 202–232. [CrossRef]

63. Abriouel, H.; Casado Muñoz, M.D.C.; Lavilla Lerma, M.L.; Pérez Montoro, B.; Bockelmann, W.; Pichner, R.; Kabisch, J.; Cho, G.-S.; Franz, C.M.A.P.; Galvez, A.; et al. New insights in antibiotic resistance of *Lactobacillus* species from fermented foods. *Food Res. Int.* **2015**, *78*, 465–481. [CrossRef]

64. Mathur, S.; Singh, R. Antibiotic resistance in food lactic acid bacteria—A review. *Int. J. Food Microbiol.* **2005**, *105*, 281–295. [CrossRef] [PubMed]

Review

The Microbial Community of Natural Whey Starter: Why Is It a Driver for the Production of the Most Famous Italian Long-Ripened Cheeses?

Erasmo Neviani *, Alessia Levante and Monica Gatti

Department of Food and Drug, University of Parma, Parco Area Delle Scienze 49/A, 43124 Parma, Italy; monica.gatti@unipr.it (M.G.)
* Correspondence: erasmo.neviani@unipr.it

Abstract: The remarkable global diversity in long-ripened cheese production can be attributed to the adaptability of the cheese microbiota. Most cheese types involve intricate microbial ecosystems, primarily represented by lactic acid bacteria (LAB). The present study aims to review the microbial community's diversity in dairy fermentation processes, focusing on two famous Italian cheeses, Grana Padano and Parmigiano Reggiano, produced using natural whey starter (NWS). NWS, created by retaining whey from the previous day's cheese batches, forms a microbiological connection between daily cheese productions. Through this technique, a dynamic microbiota colonizes the curd and influences cheese ripening. The back-slopping method in NWS preparation ensures the survival of diverse biotypes, providing a complex microbial community in which interactions among microorganisms are critical to ensuring its technological functionality. As highlighted in this review, the presence of microbial cells alone does not guarantee technological relevance. Critical microorganisms can grow and colonize the curd and cheese. This complexity enables NWS to adapt to artisanal production technologies while considering variations in raw milk microbiota, inhibitory compounds, and manufacturing conditions. This critical review aims to discuss NWS as a key factor in cheese making, considering microbial communities' ability to evolve under different selective pressures and biotic and abiotic stresses.

Keywords: lactic acid bacteria; cheese fermentation; Italian cheeses; natural whey starter; raw milk cheese; dairy microbial ecology

---

Citation: Neviani, E.; Levante, A.; Gatti, M. The Microbial Community of Natural Whey Starter: Why Is It a Driver for the Production of the Most Famous Italian Long-Ripened Cheeses? *Fermentation* **2024**, *10*, 186. https://doi.org/10.3390/fermentation10040186

Academic Editors: Roberta Comunian and Mutamed Ayyash

Received: 3 February 2024
Revised: 14 March 2024
Accepted: 27 March 2024
Published: 29 March 2024

1. Premise: Cheeses and Their Microbiota

Cheese is a product obtained via the acidic or enzymatic destabilization of casein, or more commonly, through a combination of both processes [1,2]. The earliest historical documentation of cheese production was found in ancient texts recovered in Iraq, dating back to about 3200 BC. In truth, the cheese transformation of milk may have spread to different people and places in the world even earlier as a result of random experiences. Subsequently, the relevant procedures would have been empirically reproduced until they became an industrial process through the development of scientific and technological knowledge that forms the basis of modern cheese processing [3–5].

Most modern cheese types are produced using only milk, lactic acid bacteria (LAB), rennet, and often sodium chloride [1,4]. Moreover, the different microbial communities harbored by different types of cheeses arise from raw milk, starter cultures, and adventitious microorganisms that originate from the equipment and cheese-making plant environment [4].

Raw milk is a rich and very attractive substrate for different microbial species that use lactose as a carbon source. Environmental factors play crucial roles in shaping the composition of the raw milk microbiota and in defining its evolution in cheese during ripening [1,4,6–13].

Thus, the manufacture of most cheese types involves a complex and dynamic microbial ecosystem in which several biochemical reactions occur, largely based on lactic acid fermentation by LAB [1,4,6,7,10,14,15]. However, not only the variability but also the versatility of this microbiota define the great differences that can be obtained in the fermentation of curds to produce very different cheeses.

In summary, the microbial evolution in cheese production is a dynamic process that encompasses adverse and fluctuating conditions for fermenting or not fermenting microorganisms that reach the curd alive in different ways.

The contribution of the cheese microbiota to flavor development characterizes the quality and recognition of the cheese and is thus of critical significance. The final cheese flavor, as with many of the final characteristics of cheese, is due to the interactions between the cheese microorganisms, the growth substrates, proteins in the milk, and the cheese environment [14,16]. During cheese manufacturing, environmental parameters such as temperature, pH, osmolarity, and lactose concentration change significantly. These parameters, particularly LAB, can be stressful to the microbiota of all cheeses. Rapid environmental changes impose limitations on the adaptation and cellular duplication of cheese through alternative secondary metabolism, leading to the production of metabolites with impacts on taste and aroma [17].

The process of transforming milk into cheese can be schematically divided into three blocks of operations (Figure 1). The first block of operations takes place before the milk is placed in the coagulation tank (vat) and corresponds to the milk preparation phase, including any refrigeration, possible skimming and pre-maturation, and any heat treatments. The second block of operations occurs in the vat and consists of the stages of milk processing that lead to the production of the curd. The third part consists of a series of operations that transform the curd into cheese. All these procedures together lead to the formation of the peculiar structures, flavors, and aromas of the different types of products.

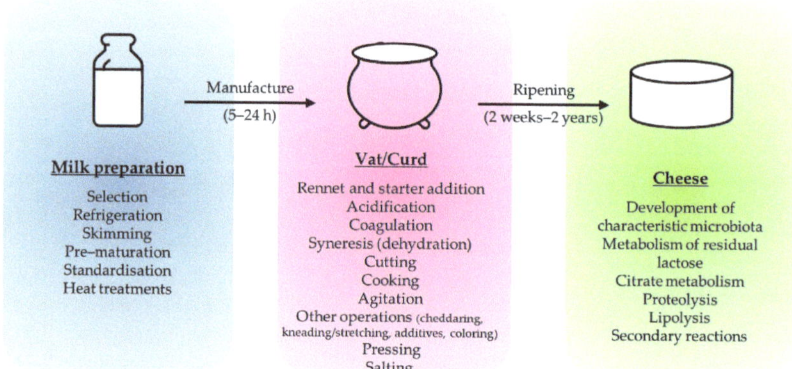

Figure 1. A simplified outline of the general flowsheet for the manufacture of rennet coagulated cheese. Adapted from ref. [4].

Starter LAB first produce mild acidification of the milk, followed, more importantly, by acidification of the curd. These changes define the main characteristics of the three blocks of operations during the first part of the cheese-making process (Figure 1) [4,6,10,18]. The factors related to lactic fermentation and multiplication of the acidifying LAB are critical in transforming the curdled milk during the first hours of maturation. For this reason, these LAB are called starter LAB. Acidification and lactose depletion are the first steps in curd formation. In addition to making the substrate less hospitable to most other microbial species, the acidification resulting from homolactic fermentation induces changes in casein hydration and its ability to remain in a colloidal suspension. In addition to coagulation, cheese curd fermentation plays a central role in defining the rheological structure of the future cheese. Lactic acid is the primary metabolite produced by this type of lactose fermentation. The speed and intensity of acidification induce dramatic changes in the destabilization and structure of casein micelles. Much of the casein is found as a colloidal suspension in raw milk. The stability of this colloidal state is deeply connected to the presence of saline bridges and the availability of calcium phosphate in the state. The acidification caused by the fermentation by LAB modifies this physical–chemical equilibrium, thereby increasing the permeability of the curd [1,4,7,19].

After lactose depletion, bacterial cell death begins. Then, the lysis of the starter LAB, together with other bacteria sensitive to these new adverse conditions, induces the release of a significant amount of intracellular proteolytic enzymes, which constitute the heritage of the cheese microbiota and can participate in the ripening of the cheese. It is well established that cell lysis is a key event for the release of cytoplasmic enzymes into the cheese matrix and is crucial for understanding the contribution of different microbial cells to cheese ripening [17,20–30]. In this way, the proteolysis of caseins should be considered a crucial event in determining the outcome of the process [1,7,16,31].

The number of starter LAB cells in the acidified curd amounts to almost one billion per gram of cheese. This high quantity represents a huge reservoir of enzymatic activity [32]. These enzymes, or at least some of them, can remain active in the curd for very long periods of time and contribute to the different stages of cheese ripening [1,4,7].

From this perspective, starter LAB could be considered responsible for cheese modifications—firstly, as a well-defined cellular entity and secondly, as enzymes released after cell lysis. However, it was recently proposed that cell lysis alone cannot explain the long-term enzymatic activity observed during cheese ripening. From this perspective, bacterial cells that derive from the starter cultures could undergo permeabilization events, allowing for intracellular enzyme activity that might be relevant for prolonged metabolic conversions and thus flavor compound synthesis [17].

In this context, choosing the type of starter LAB for cheese making involves determining, in detail, the microbiota of the vat milk and how it will develop, first in the curd and then in the final cheese. Cheese technology leads to the selection of different bacterial populations. In addition, the complexity of the cheese-making parameters represents an aid or a tool to manage the versatility of the different LAB biotypes. Critically, not all the microorganisms present in the milk and curd that are considered cultivable or detectable based on DNA must fully participate in the dairy transformation. Some microorganisms may simply be present but could be of little interest in cheese production and ripening [17].

The presence of a microbial cell (or, worse, microbial DNA) is not enough to be considered technologically relevant. The most important microorganisms are those capable of multiplying, growing, and colonizing the curd and cheese. Indeed, microorganisms that are technologically relevant and increase the "quality" of the resulting cheese can be relatively rare in the composition of the raw milk and starter microbiome.

2. The Main Factors of the Microbial Ecosystem Involved in Long-Ripened Cheese, Such as Grana Padano and Parmigiano Reggiano

The most commonly studied and famous long-ripened Italian cheeses are Grana Padano (GP) and Parmigiano Reggiano (PR). These two varieties are traditional and long-ripened hard-cooked cheeses produced with raw milk in restricted geographical areas of Northern Italy, delimited by official regulations ("https://www.granapadano.it/wpcontent/uploads/2023/02/SpecificationsGBOct2022-50252.pdf (accessed on 11 January 2024)", "https://www.parmigianoreggiano.com/consortium-specifications-and-legislation/ (accessed on 11 January 2024)") [4,24,33,34]. Both varieties are "Grana cheeses", which refers to a cheese with a grainy structure. Such cheeses have been produced in the Po Valley since the 13th century.

GP and PR cheeses are made from partially skimmed raw milk through lactic acid fermentation and subjected to slow and long ripening for at least 9 and 12 months. LAB from raw milk, commonly called non-starter LAB (NSLAB) and starter LAB, plays a fundamental role in achieving the typical sensory characteristics of these cheeses [4,24].

GP and PR cheeses have many common characteristics and some distinct properties. Although the similarities between these varieties are defined by their cheese-making technologies, the differences are largely determined by the methods of raw milk collection, milk management before coagulation, and ripening conditions. Briefly, to produce PR, the milk is not refrigerated and should be maintained at a temperature no lower than 18 °C. The evening milk is partially skimmed after overnight creaming at about 20 °C in special tanks called "bacinelle". For GP, the feeding of high-quality silage fodder is allowed, and the cheese is produced from two consecutive rounds of milking. This milk is stored at a temperature no lower than 8 °C on the farm. To inhibit the late blowing of cheese, the addition of lysozyme to the vat milk (20–25 ppm) is allowed, as the use of silage favors the contamination of raw milk by spore-forming clostridia. The milk is skimmed via creaming in "bacinelle" for about 12 h at 8–20 °C [24,35].

For both cheeses, the slight microbial acidification that occurs during creaming favors rennet activity in the milk vat. At the same time, slight proteolysis produces short peptides that may favor further growth of the LAB in the natural whey culture (NWS). In both cheeses, calf rennet (powder preparation) is used.

A large amount of the NWS, about 3% (v/v), is added to the vat milk, yielding a total titratable acidity of ca. 28–32 °SH/50 mL [18]. As the use of commercial/selected starters is not allowed, NWS is prepared using whey from the previous cheese-making process. This whey is held under a temperature gradient (from about 50–54 °C to about 30–34 °C for 12–16 h) to reach a final titratable acidity close to about 20–32 °SH/50 mL. It is well known that the microbial composition of the NWS is dominated by thermophilic LAB (about 10^9 CFU/mL), such as *Lactobacillus helveticus* and *Lactobacillus delbrueckii* subsp. *lactis*. [24,26,36,37]. The relative abundance of the species *L. helveticus* and *L. delbrueckii* subsp. *lactis* in the NWS varies according to the dairy used. Recently, Sola et al. [37] proposed a distinction between NWS dominated by the species *L. helveticus* (NWS type-H) and that dominated by the species *L. delbrueckii* (NWS type-D), noting that these ecological differences in starter cultures can influence the early stages of curd acidification [38].

After coagulation, the curd is broken into particles the size of rice grains and cooked at 52–56 °C for 5–15 min under stirring. The time from rennet addition at 32–34 °C to the end of cooking is close to 23 min. The combination of heating and acidifying activity by the NWS allows the formation of curd grains with the right texture along with whey drainage. After cooking, curd grains settle to the bottom of the vat for about 30–50 min with a whey temperature not exceeding 53–55 °C. Then, the curd is removed, cut into two parts, molded, acidified for about 48 h, salted in brine, and ripened for at least 12 months.

The use of the NWS is a practice that started at the beginning of 1900. This process aims to reduce microbiological defects in cheese and has consolidated over time. Acid production at the appropriate rate and time is a key step in the manufacture of high-quality cheese [24,39].

The method used to produce NWS by retaining some of the whey drained from the cheese vat at the end of cheese making leads to the selection of a characteristic microbiota [36]. The different treatments used during cheese making, starting from the addition of NWS to curd removal and the collection of "sweet whey", promote the selection of thermophilic and acid-tolerant lactic bacteria in the acidified whey [18,24]. The most influential parameters are the temperature of curd cooking, the management of the gradient temperature during whey fermentation, and the increase in acidity. Changing one or more of these parameters can lead to the selection of a characteristic microbiota mainly consisting of thermophilic, aciduric, and moderately heat-resistant LAB [24,36,40].

The NWSs obtained in this way demonstrate how the inhabitants of these specific geographical areas empirically learned to use their fermentation capabilities for dairy purposes by exploiting the abilities of specific microbial ecosystems to adapt and evolve. The curd structure in the vat and after breaking and extraction is defined by the acidification of the curd and, consequently, by LAB activity. During the first few hours of cheese making, the thermophilic LAB present in NWS quickly grow in the curd, but the correctness of curd acidification depends on the residual availability of sugar, the pH, the residual moisture, and the temperature. It is known that the LAB in NWS mainly develop in the molded curd within 12 to 24 h and that the growth of these bacteria is coupled with lactic acid production and a decrease in pH to approximately 5.10–5.25. During the first 24 h of cheese molding, the conversion of lactose into lactic acid is the main biochemical process that occurs in cheese [4,24,41]. Moreover, due to the large size of the cheese (a diameter of 35 to 45 cm and a side height of 18 to 25 cm), all these parameters can differ in the different areas of the curd, creating differences in the acidification activity between the central and the external parts of the cheese. Within 48 h, the total LAB count starts to decrease.

The performance of NWS during cheese ripening can also be influenced by its cultivation history, modulating the proteome allocation and metabolic stability in starter cultures, thereby providing novel approaches to influence flavor formation [17,42,43]. Additionally, the presence of NSLAB in raw milk [44,45] can influence starter LAB's growth capacity in vat milk. During the first stage of cheese making, the weak proteolytic activity of NSLAB favors an increase in free amino acids in the milk. These free amino acids, or little peptides, allow the fast growth of starter LAB and, consequently, facilitate acidification kinetics [46]. The metabolic interactions between NSLAB and starter LAB affect the inhibition of spoilage bacteria and curd structuring [1,4,7,19].

The aim for the remainder of this critical review is to discuss the NWS as the driver of the cheese-making process, considering microbial communities and their capability to evolve under the different ecosystems that change during the production cycle of the NWS itself and during that of long-ripened cheeses.

3. Natural Whey Starter—Peculiarly Complex Microbial Ecosystems

According to PDO regulations, to prepare NWS for cheese making the following day, the whey remaining after curd separation, i.e., whey that is cooked and not already acidified (cooked non-acidified whey, often simply called "sweet whey"), is recovered, usually from one vat, and incubated at a decreasing temperature. This back-slopping procedure establishes a microbiological connection between subsequent batches of production.

The composition of these undefined multiple-strain cultures is the sum of the LAB obtained from raw milk and the LAB introduced in the previous batch of cheese with the previous NWS [24,36,39,47,48]. Under this traditional protocol, whey represents the link between these cheeses, which are manufactured each subsequent day. One of the peculiarities of using NWS for cheese making is that it forms a "microbiological bond", which is transferred through whey from one day's milk to that of the following day. In this way, NWS serves as a link between the dairy products manufactured each day. For this reason, following production regulations, cheese must be produced every day.

The success of the NWS is linked to its ability to adapt to the different technological parameters encountered in the cheese production process, thereby maintaining a high level of resilience among thermophilic LAB species. The peculiar adaptive features of the microbiota of NWS allow this undefined culture to retain its technological functions by adapting to a cyclical production process based on back-slopping [24,36,39].

However, this modality of preparation based on the back-slopping principle also enables the survival of many different biotypes, some of which are likely useful for the development of the whey ecosystem itself. A mixture of strains of the same species facilitates the development of a natural starter with a poorly defined composition but a strong ability to self-adapt to variable technological performance, as required by non-standardized cheese-making operations [36,39,49]. Small changes in technological parameters such as the temperature of curd cooking, the temperature and modality of NWS cooling, and differences in the final acidity and pH reached can affect the bacterial consortium present in NWS [24,36,38,39].

Notably, natural whey cultures experience two different thermal gradients. After inoculation into the milk vat, the temperature increases up to about 53–55 °C, exposing the bacteria cells to thermal stress for about one hour, while after separation of the whey from the curd, the temperature decreases, alleviating the severe thermal conditions. It was observed that the two main thermophilic species present in whey, *L. helveticus* and *L. delbrueckii* subsp. *lactis*, respond differently to the different gradients and composition of the environment during the production phases [37]. Indeed, *L. helveticus* is regarded as a more acid-tolerant strain, which might explain its increase in quantity despite the low pH values reached in NWS [36]. Conversely, the inoculated vat milk had higher pH values that, in combination with a possibly higher tolerance to the thermal stress of *L. delbrueckii* subsp. *lactis*, could explain the numerical increase observed after the phase of cooking. This evidence suggests the resilience of this peculiar ecosystem in adapting itself to different stress conditions [36].

Other studies have focused on the intraspecies (i.e., strain) characterization of the most abundant species isolated from NWS, showing how these cultures vary not only by species but also, and primarily, by strains within species, as observed in undefined cultures used for the production of other cheeses [50,51]. Because it is the dominant species in NWS, *L. helveticus* has been the focus of many studies on its phenotypic and genotypic diversity [52].

According to ecological principles, Giraffa et al. evaluated different bacterial interactions involving either stimulatory or inhibitory effects for *L. helveticus*, *L. delbrueckii* ssp. *lactis*, and *L. fermentum* [53]. Certainly, future studies are needed to better understand the role of microbial interactions in the stability and functionality of the NWS ecosystem.

Comparing culture-independent (i.e., microscopy) and culture-dependent (i.e., plate count) quantification [26,39,54] indicates that bacterial viability in NWS cannot be evaluated only based on LAB's capacity to form colonies on MRS or a whey agar medium. The number of total cells, particularly viable cells, is often higher (up to 1 log unit) than the number of colony-forming units. Questions remain about the roles of cells that are viable but not cultivable, which often represent most of the culture. Because these bacteria cannot be cultivated, the role of this population in the whey culture during cheese curd fermentation is not currently known.

Recently, it was demonstrated that *Lactococcus lactis* can form persistent and viable but not culturable (VBNC) cells when exposed to antimicrobial agents [55]. Other reports on the same species suggest that bacteria that enter dormant, low-growth states could be relevant in the microbial ecology of dairy products since they are metabolically active [56]. However, such bacterial strains are challenging to isolate from complex environments such as the NWS.

In general, the molecular mechanisms of nonculturable cells are perplexing, and the condition of VBNC is controversial [57,58].

However, the greatest and most well-known advantage of NWS's biological systems is undoubtedly their wide resistance to lytic bacteriophage attacks [59]. Natural starters are widely considered highly tolerant to phage infection because they are grown in the presence of phages, which leads to the dominance of resistant or tolerant strains [18,60–62]. Although Carminati et al. [62] found that lysogeny occurred in *L. helveticus* cultures isolated from GP NWS, these cultures were found to carry defective phages or killer particles when induced by mitomycin. More recently, a study by Mancini et al. [63] confirmed the prevalence of bacteriophages in NWS cultures used to produce Trentin Grana cheese, despite showing the limited capability of the isolated bacteriophages to form lysis plaques on cultures of *L. helveticus*. The consistent presence of lytic phages in the NWS did not impair their performance. This result could be related to the presence of various bacterial strains of the *L. helveticus* species, each with different phage sensitivity profiles, allowing the species to effectively counteract phage predation [64]. However, when the concentration of bacteriophages increases, adverse effects could be encountered in the sensory profiles of cheeses resulting from such a production process [65]. The presence of phages in cheese might select for resistance traits among bacteria, especially if the bacteria and phage association persists between different production cycles [65].

4. An Ecological Perspective on Natural Whey Starters

Over time, microorganisms have been mutating and evolving to adapt to an ever-changing ecosystem [66]. Microbial populations adapt rapidly when they are introduced to a new environment. However, at the same time, microbial populations could continue to improve indefinitely, albeit slowly, even in a constant environment, through the contributions of individual mutations to fitness improvement [67].

NWS can be considered one complex microbial community among many such communities in nature. The production of fermented products such as NWS is the result of activities not by an individual but by a group of microorganisms. For example, most food fermentation processes depend on mixtures of microorganisms (species and biotypes), which act in concert to produce the desired product characteristics. All fermentation processes are often characterized by the presence of a complex microbiota. Notably, the LAB community of NWS can be discussed while considering the scenario of complex microbial communities.

About 35 years have passed since multicellularity was proposed as a possibility for understanding the growth and development of complex prokaryotic ecosystems. Indeed, the hypothesis that complex microbial ecosystems act like multicellular organisms whose individual components interact and condition each other remains intriguing [68,69].

Intercellular communication and multicellular coordination are known to be widespread among prokaryotes and influence the expression and intensity of multiple phenotypes. Following this approach, the interactions between microorganisms that comprise complex ecosystems represent the decisive factor that influences the development of different microbial cultures [70]. Beyond microbial quantity or the presence of different species, biotypes, and variants, the interactions between microorganisms represent a key factor for understanding the biological functionality of complex microbial ecosystems and their ability to adapt to stress, survive, evolve, and express different phenotypes [68,69].

This evidence highlights the need to deeply explore the diversity of the microbial community involved in natural food fermentation processes and the links between their technological capabilities and product quality [71–74]. The back-slopping principle applied to an environment such as non-thermally treated milk brings the results of NWS very close to those of natural fermentation. Thus, we should explore how the technological choices made by humans can direct growth and microbial metabolism under conditions of stress. It is well known that the technological processes used to produce fermented foods usually involve process conditions that guide fermentation through the imposition of differently selective or elective conditions on the microbiota present. The ability of microorganisms to resist different stress factors enables their resilience under conditions that are hostile to growth and metabolism [75–79]. Microbial selection guides the fermentation process [71,72,80–82].

Concerning this microbial adaptative capacity, Charles Darwin wrote in a letter to Asa Gray, "What a trifling difference must often determine which shall survive, and which perish!" [71]. Therefore, the colonization of food by different microorganisms may also be studied in terms of both ecological strategy and community development [18,83–85].

In natural food systems, the stimulatory and inhibitory effects among microorganisms could support the possibility of maintaining the viability of a crucial part of the microbial population (population stability), even in the presence of continuous changes in the food environment, including those resulting from the metabolic activity of the microorganisms themselves [18,29,53,83–87].

In general, it can be stated that the NWS bacterial consortium is more versatile and robust than pure cultures used in cheese production because it performs more complex activities and can tolerate more variation in the environment. This deduction is based on the concept that bacteria benefit from multicellular cooperation by using the cellular division of labor, accessing resources that cannot effectively be utilized by single cells, collectively defending against antagonists, and optimizing population survival by differentiating into distinct cell types [68]. Even micro-interactions with the environment matrix, including milk, whey, curd, and cheese, can be crucial in defining cellular cooperation and the evolution of microbial communities [69,88–90].

In complex ecosystems, even population heterogeneity can be a determinant in defining resilience against environmental uncertainty [91]. Intraspecies diversity among closely related strains is commonly linked to functionally adaptive traits encoded on genomic islands that are acquired by horizontal gene transfer [92]. The generation of subpopulations with varying plasmid content in natural communities yields selective advantages in the face of environmental uncertainty [50,82,93,94]. Analogously, bacteriophages play a regulatory role in population dynamics through density-dependent predation [50,95].

The presence of isogenic bacteria, populations that are traditionally considered to be composed of identical cells, can also contribute to the survival and evolution of complex microbial ecosystems [96]. Despite containing the same genetic material, protein levels between cells can vary due to stochastic events associated with gene expression and regulation. Thus, cell-to-cell heterogeneity has important implications for allowing populations of cells to diversify and thus survive environmental stress [50,71,82,94,97,98].

It was verified that microbial communities in sourdough microbiota undergo changes in composition that threaten their resilience. To support resilience and good performance, the sourdough metacommunity includes dominant, subdominant, and satellite players, which together ensure gene and transcript redundancy [86].

The microbial consortia offer several advantages for the survival aptitude of the microbiota. These benefits include the increased quality and safety of several food systems, flavor development, and increased stability to improve shelf life and consumer safety. In summary, the interactions that occur within the food ecosystem can play a decisive role in the evolution of all players present in the ecosystem itself. This evidence has increased interest in studying the diversity of the community of fermenting microorganisms and linking the evolution of microbiota to their properties in adapting to technological processes and product quality.

For GP and PR cheese production, NWS represents a large component of the future microbiota of curd and, in collaboration with the raw milk microbiota, the "engine" of the metabolism involved in cheese making and cheese ripening [13,45]. LAB species/biotypes present in vat milk (from NWS and raw milk) can grow, survive, decline, and even become dominant during cheese manufacture. The outcomes depend on metabolic potential, which is species- and even biotype-specific. The environmental conditions that species encounter are first the biochemical composition of the vat milk, followed by the curd matrix modified by acidification and technological parameters.

The large number of NWS LAB cells that develop in the curd, thereby acidifying it, also represent the most important source of the large pool of proteolytic enzymes released after cell lysis; these enzymes are significantly involved in cheese ripening. Enzyme activities are regulated by cheese composition, moisture, NaCl concentration, pH, and temperature, which change not only over time but also according to the different cheese zones [22,25,32,99]. Interactions between SLAB and NSLAB occur starting from the earliest stages of cheese manufacture through cheese ripening. One of the most well-accepted theories indicates that SLAB lysates provide energy sources for NSLAB [18,24,28–30,100–104].

It was previously observed that the bacterial consortia of NWS coevolved during adaptation to whey and curd acidification [36]. The management of this dynamic ecosystem could be considered a superorganism, consisting of the sums of microbial metabolism and the interactions between individual microbes [18,72,105]. This super-organism activity recalls the multicellularity proposed by Shapiro [68,69] as the key to understanding the growth and development of complex prokaryotic ecosystems, whose individual components interact and condition each other.

For simplicity, the different microorganisms of the cheese microbiota can be arbitrarily clustered into three parts (Figure 2). Many of these parts are only occasionally present, likely without any (or minimal) technological significance, or are not yet understood (supporting actors) (Figure 2, area A). The size of this area can vary according to the type of milk used to produce the cheeses. As in the case of raw milk, the more complex the milk microbiota, the larger the area. Another area (Figure 2, area B) includes all microorganisms (biotypes and variants) that are functional in the survival of the whey starter's ecosystem itself and its ability to adapt and survive/evolve. This part of the microbiota determines the resilience of the NWS ecosystem [82]. As suggested by Charles Darwin, this component determines the continuity and survival of the ecosystem [71]. The third part (Figure 2, area C) represents a core microbiota that is not needed for natural whey starter fermentation but remains necessary for curd fermentation. This part represents the microbial core responsible for the cheese-making process and can change during the different moments (or zones) of the cheese-making process. Unlike part B, part C does not represent the biological element of stability and continuity in the whey starter ecosystem itself but only the part that is functional for the transformation of the cheese. This component is capable of adapting to stress factors induced by technology. Starting from this part of the microbiota alone, unlike with part B, there is no certainty that the NWS ecosystem will survive. Therefore, it cannot be excluded that even small variations in the preparation of the starter cultures can significantly influence the dominance of otherwise minority bacterial populations that influence the functional properties of the NWS ecosystem [63].

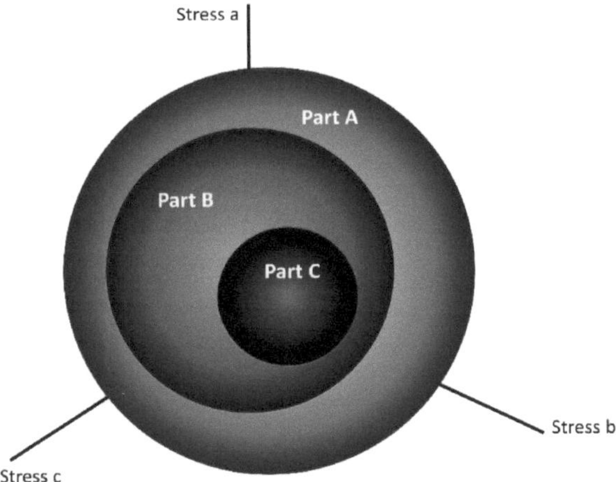

Figure 2. Schematic representation of the microbial complexity observed in the NWS ecosystem. Assuming an environment where different abiotic or biotic stresses (Stress a, b, c) are faced by the NWS microbial community, it is likely that the majority of components (Part A) are superfluous and not relevant to the functionality of the microbial community itself, while different subsets (Part B, Part C) could better adapt to different steps in the cheese-making process. The existence of microbial interactions, within-species biodiversity, and phenotypic variability makes it difficult to disentangle and characterize the contribution of each microbial population to the complex NWS microbiome.

Studying NWS to produce Parmigiano Reggiano cheese, Bertani et al. [36] observed a large number of microorganisms present in the natural culture mixed with raw milk (part A). These microorganisms mainly came from raw milk, a core microbiota useful for the persistence of the ecosystem, and adapted, in turn, to a mixture of raw milk and whey or non-acidified/acidified whey (part B). A minority of the microorganisms adapted to the curd ecosystem (part C). This minority component of the natural whey microbiota (part C) could be considered essential for the cheese-making process. Bertani et al. [36] also showed that it is not possible to develop a natural whey starter useful for cheese making with only part C because part B is able to maintain the complex culture, adapt to vat milk, and produce the natural whey starter fermentation. Even if these bacteria (species and/or biotypes) had found ideal conditions to grow in the NWS, only a minority of them could have better adapted to the curd ecosystem compared to cooked non-acidified whey and the NWS ecosystem [29,38,53,106].

In part A, there are other microorganisms not involved in the persistence or resilience of NWS. Such microorganisms likely come from environmental raw milk contamination. We believe that the interactions between microorganisms belonging to these different parts of the microbiota, not only their presence and amount, are key factors in understanding their technological functionality, ability to adapt to stress, ability to survive, and ability to evolve and express phenotypes in the final cheese.

It can be assumed that a core microbiota would be useful for the persistence of the ecosystem. This microbiota could adapt to the mixture of raw milk and NWS in the vat, to sweet whey after curd separation, or to acidified whey after preparation of the NWS. The majority of LAB present in acidified whey is necessarily adapted to this substrate and capable of reaching the desired level of acidification.

This factor is likely related to the difficulties sometimes encountered when selecting starter cultures with technological and aromatic performance like that achieved using natural starters. Indeed, isolating and using mixtures of strains obtained from milk or a natural starter seems insufficient to obtain starters with good dairy performance. Rather, it is necessary to identify the biotypes, which may represent a minority in the natural starter, with the ability to develop into the curd. Consequently, minority populations or even apparently non-viable strains could be necessary to maintain cell interactions.

5. Conclusions

The biological complexity of the LAB consortium characterizing the NWS is both a strength and a weakness of these natural cultures. Indeed, this biological complexity and biodiversity guarantee the LAB consortium's ability to adapt to artisanal production technologies based on its intrinsic ability to evolve in response to external factors such as the microbiota of raw milk, the presence of any compounds that might inhibit bacterial growth, and the manufacturing conditions. On the other hand, the presence of different species and numerous biotypes makes it more difficult to standardize the daily propagation of the starter. Consequently, the activity of this consortium can vary slightly from day to day.

Despite the findings provided by previous studies, the composition of these natural microbial cultures adapted to the selective pressure of dairy processing is not yet fully understood. It will be necessary for future studies to explore individual players and, especially, the relevant complex ecosystem communities and bacterial interactions.

The ability of all living species to colonize an ecosystem includes their interactions with other species living in the ecosystem. Biotic and abiotic effects determine the evolution of this interaction. NWS seems to represent a good example of this complexity and functionality.

Author Contributions: Conceptualization, E.N.; writing—original draft preparation, E.N., M.G. and A.L.; writing—review and editing, E.N., M.G. and A.L. All authors have read and agreed to the published version of the manuscript.

Funding: This research received no external funding.

Institutional Review Board Statement: Not applicable.

Informed Consent Statement: Not applicable.

Data Availability Statement: Not applicable.

Acknowledgments: This work is dedicated to the memory of Anna Cabrini.

Conflicts of Interest: The authors declare no conflict of interest.

References

1. Alais, C. *Science du Lait: Principes des Techniques Laitières*, 4th ed.; Société D'édition et de Promotion Agro-Alimentaires, Industrielles et Commerciales: Paris, France, 1984.

2. Fox, P.F.; Mcsweeney, P.L.H. *Rennets: Their Role in Milk Coagulation and Cheese Ripening*; Springer: Berlin/Heidelberg, Germany, 1997.
3. Douillard, F.P.; Ribbera, A.; Kant, R.; Pietilä, T.E.; Järvinen, H.M.; Messing, M.; Randazzo, C.L.; Paulin, L.; Laine, P.; Ritari, J.; et al. Comparative Genomic and Functional Analysis of 100 Lactobacillus rhamnosus Strains and Their Comparison with Strain GG. *PLoS Genet.* **2013**, *9*, e1003683. [CrossRef] [PubMed]
4. Gobbetti, M.; Neviani, E.; Fox, P. *The Cheeses of Italy: Science and Technology*; Springer International Publishing: Berlin/Heidelberg, Germany, 2018. [CrossRef]
5. McClure, S.B.; Magill, C.; Podrug, E.; Moore, A.M.T.; Harper, T.K.; Culleton, B.J.; Kennett, D.J.; Freeman, K.H. Fatty acid specific δ^{13}C values reveal earliest Mediterranean cheese production 7200 years ago. *PLoS ONE* **2018**, *13*, e0202807. [CrossRef] [PubMed]
6. Ercolini, D. Secrets of the cheese microbiome. *Nat. Food* **2020**, *1*, 466–467. [CrossRef] [PubMed]
7. Fox, P.F.; Guinee, T.P.; Cogan, T.M.; McSweeney, P.L.H. *Fundamentals of Cheese Science*; Springer: Boston, MA, USA, 2017. [CrossRef]
8. Fox, P.F.; McSweeney, P.L.H. Methods used to study non-starter microorganisms in cheese: A review. *Int. J. Dairy Technol.* **2000**, *53*, 113–119. [CrossRef]
9. Li, N.; Wang, Y.; You, C.; Ren, J.; Chen, W.; Zheng, H.; Liu, Z. Variation in Raw Milk Microbiota Throughout 12 Months and the Impact of Weather Conditions. *Sci. Rep.* **2018**, *8*, 2371. [CrossRef] [PubMed]
10. Parente, E.; Ricciardi, A.; Zotta, T. The microbiota of dairy milk: A review. *Int. Dairy J.* **2020**, *107*, 104714. [CrossRef]
11. Quigley, L.; O'Sullivan, O.; Stanton, C.; Beresford, T.P.; Ross, R.P.; Fitzgerald, G.F.; Cotter, P.D. The complex microbiota of raw milk. *FEMS Microbiol. Rev.* **2013**, *37*, 664–698. [CrossRef] [PubMed]
12. Renoldi, N.; Innocente, N.; Rossi, A.; Brasca, M.; Morandi, S.; Marino, M. Screening of Aroma-Producing Performance of Anticlostridial Lacticaseibacillus casei Strains. *Food Bioprocess Technol.* **2024**. [CrossRef]
13. Bettera, L.; Levante, A.; Bancalari, E.; Bottari, B.; Cirlini, M.; Neviani, E.; Gatti, M. Lacticaseibacillus Strains Isolated from Raw Milk: Screening Strategy for Their Qualification as Adjunct Culture in Cheesemaking. *Foods* **2023**, *12*, 3949. [CrossRef]
14. Afshari, R.; Pillidge, C.J.; Dias, D.A.; Osborn, A.M.; Gill, H. Cheesomics: The future pathway to understanding cheese flavour and quality. *Crit. Rev. Food Sci. Nutr.* **2020**, *60*, 33–47. [CrossRef]
15. Cocolin, L.; Gobbetti, M.; Neviani, E.; Daffonchio, D. Ensuring safety in artisanal food microbiology. *Nat. Microbiol.* **2016**, *1*, 16171. [CrossRef] [PubMed]
16. Christensen, L.F.; Høie, M.H.; Bang-Berthelsen, C.H.; Marcatili, P.; Hansen, E.B. Comparative Structure Analysis of the Multi-Domain, Cell Envelope Proteases of Lactic Acid Bacteria. *Microorganisms* **2023**, *11*, 2256. [CrossRef] [PubMed]
17. Nugroho, A.D.W.; Kleerebezem, M.; Bachmann, H. Growth, dormancy and lysis: The complex relation of starter culture physiology and cheese flavour formation. *Curr. Opin. Food Sci.* **2021**, *39*, 22–30. [CrossRef]
18. Gobbetti, M.; Di Cagno, R.; Calasso, M.; Neviani, E.; Fox, P.F.; De Angelis, M. Drivers that establish and assembly the lactic acid bacteria biota in cheeses. *Trends Food Sci. Technol.* **2018**, *78*, 244–254. [CrossRef]
19. McSweeney, P.L.H.; Fox, P.F.; Cotter, P.D.; Everett, D.W. *Cheese: Chemistry, Physics & Microbiology*; Academic Press: Cambridge, MA, USA, 2017.
20. Chapot-Chartier, M.-P.; Deniel, C.; Rousseau, M.; Vassal, L.; Gripon, J.-C. Autolysis of two strains of Lactococcus lactis during cheese ripening. *Int. Dairy J.* **1994**, *4*, 251–269. [CrossRef]
21. Crow, V.L.; Coolbear, T.; Gopal, P.K.; Martley, F.G.; McKay, L.L.; Riepe, H. The role of autolysis of lactic acid bacteria in the ripening of cheese. *Int. Dairy J.* **1995**, *5*, 855–875. [CrossRef]
22. De Dea Lindner, J.; Bernini, V.; De Lorentiis, A.; Pecorari, A.; Neviani, E.; Gatti, M. Parmigiano Reggiano cheese: Evolution of cultivable and total lactic microflora and peptidase activities during manufacture and ripening. *Dairy Sci. Technol.* **2008**, *88*, 511–523. [CrossRef]
23. El Soda, M.; Farkye, N.; Vuillemard, J.C.; Simard, R.E.; Olson, N.F.; El Kholy, W.; Dako, E.; Medrano, E.; Gaber, M.; Lim, L. Autolysis of Lactic Acid Bacteria: Impact on Flavour Development in Cheese. In *Developments in Food Science*; Elsevier: Amsterdam, The Netherlands, 1995; Volume 37, pp. 2205–2223. [CrossRef]
24. Gatti, M.; Bottari, B.; Lazzi, C.; Neviani, E.; Mucchetti, G. Invited review: Microbial evolution in raw-milk, long-ripened cheese produced using undefined natural whey starters. *J. Dairy Sci.* **2014**, *97*, 573–591. [CrossRef]
25. Gatti, M.; De Dea Lindner, J.; De Lorentiis, A.; Bottari, B.; Santarelli, M.; Bernini, V.; Neviani, E. Dynamics of whole and lysed bacterial cells during Parmigiano-Reggiano cheese production and ripening. *Appl. Environ. Microbiol.* **2008**, *74*, 6161–6167. [CrossRef]
26. Gatti, M.; Bernini, V.; Lazzi, C.; Neviani, E. Fluorescence microscopy for studying the viability of micro-organisms in natural whey starters. *Lett. Appl. Microbiol.* **2006**, *42*, 338–343. [CrossRef]
27. Levante, A.; De Filippis, F.; La Storia, A.; Gatti, M.; Neviani, E.; Ercolini, D.; Lazzi, C. Metabolic gene-targeted monitoring of non-starter lactic acid bacteria during cheese ripening. *Int. J. Food Microbiol.* **2017**, *257*, 276–284. [CrossRef] [PubMed]
28. Lortal, S.; Chapot-Chartier, M.P. Role, mechanisms and control of lactic acid bacteria lysis in cheese. *Int. Dairy J.* **2005**, *15*, 857–871. [CrossRef]
29. Sgarbi, E.; Bottari, B.; Gatti, M.; Neviani, E. Investigation of the ability of dairy nonstarter lactic acid bacteria to grow using cell lysates of other lactic acid bacteria as the exclusive source of nutrients. *Int. J. Dairy Technol.* **2014**, *67*, 342–347. [CrossRef]
30. Wilkinson, M.G.; Guinee, T.P.; O'Callaghan, D.M.; Fox, P.F. Autolysis and proteolysis in different strains of starter bacteria during Cheddar cheese ripening. *J. Dairy Res.* **1994**, *61*, 249–262. [CrossRef]

31. Savijoki, K.; Ingmer, H.; Varmanen, P. Proteolytic systems of lactic acid bacteria. *Appl. Microbiol. Biotechnol.* **2006**, *71*, 394–406. [CrossRef] [PubMed]
32. Gatti, M.; Fornasari, M.E.; Mucchetti, G.; Addeo, F.; Neviani, E. Presence of peptidase activities in different varieties of cheese. *Lett. Appl. Microbiol.* **1999**, *28*, 368–372. [CrossRef] [PubMed]
33. Giraffa, G. The Microbiota of Grana Padano Cheese. A Review. *Foods* **2021**, *10*, 2632. [CrossRef] [PubMed]
34. Zago, M.; Bardelli, T.; Rossetti, L.; Nazzicari, N.; Carminati, D.; Galli, A.; Giraffa, G. Evaluation of bacterial communities of Grana Padano cheese by DNA metabarcoding and DNA fingerprinting analysis. *Food Microbiol.* **2021**, *93*, 103613. [CrossRef]
35. Olivera Rodi, J.; Gonz´ález Ramos, M.J.; Díaz Gadea, P.; Reginensi, S.M. Study of the inhibitory effect of Lactobacillus strains and lysozyme on growth of *Clostridium* spp. responsible for cheese late blowing defect. *Nova Biotechnol. Chim.* **2022**, *22*, e1229. [CrossRef]
36. Bertani, G.; Levante, A.; Lazzi, C.; Bottari, B.; Gatti, M.; Neviani, E. Dynamics of a natural bacterial community under technological and environmental pressures: The case of natural whey starter for Parmigiano Reggiano cheese. *Food Res. Int.* **2020**, *129*, 108860. [CrossRef]
37. Sola, L.; Quadu, E.; Bortolazzo, E.; Bertoldi, L.; Randazzo, C.L.; Pizzamiglio, V.; Solieri, L. Insights on the bacterial composition of Parmigiano Reggiano Natural Whey Starter by a culture-dependent and 16S rRNA metabarcoding portrait. *Sci. Rep.* **2022**, *12*, 17322. [CrossRef] [PubMed]
38. Santarelli, M.; Bottari, B.; Malacarne, M.; Lazzi, C.; Sforza, S.; Summer, A.; Neviani, E.; Gatti, M. Variability of lactic acid production, chemical and microbiological characteristics in 24-hour Parmigiano Reggiano cheese. *Dairy Sci. Technol.* **2013**, *93*, 605–621. [CrossRef]
39. Bottari, B.; Santarelli, M.; Neviani, E.; Gatti, M. Natural whey starter for Parmigiano Reggiano: Culture-independent approach. *J. Appl. Microbiol.* **2010**, *108*, 1676–1684. [CrossRef]
40. Helal, A.; Nasuti, C.; Sola, L.; Sassi, G.; Tagliazucchi, D.; Solieri, L. Impact of Spontaneous Fermentation and Inoculum with Natural Whey Starter on Peptidomic Profile and Biological Activities of Cheese Whey: A Comparative Study. *Fermentation* **2023**, *9*, 270. [CrossRef]
41. Giraffa, G.; Rossetti, L.; Mucchetti, G. Influence of the Temperature Gradient on the Growth of Thermophilic Lactobacilli Used as Natural Starters in Grana Cheese. *J. Dairy Sci.* **1998**, *81*, 31–36. [CrossRef]
42. Teusink, B.; Molenaar, D. Systems biology of lactic acid bacteria: For food and thought. *Curr. Opin. Syst. Biol.* **2017**, *6*, 7–13. [CrossRef] [PubMed]
43. Wels, M.; Siezen, R.; Van Hijum, S.; Kelly, W.J.; Bachmann, H. Comparative Genome Analysis of Lactococcus lactis Indicates Niche Adaptation and Resolves Genotype/Phenotype Disparity. *Front. Microbiol.* **2019**, *10*, 4. [CrossRef]
44. Gobbetti, M.; De Angelis, M.; Di Cagno, R.; Mancini, L.; Fox, P.F. Pros and cons for using non-starter lactic acid bacteria (NSLAB) as secondary/adjunct starters for cheese ripening. *Trends Food Sci. Technol.* **2015**, *45*, 167–178. [CrossRef]
45. Bettera, L.; Levante, A.; Bancalari, E.; Bottari, B.; Gatti, M. Lactic acid bacteria in cow raw milk for cheese production: Which and how many? *Front. Microbiol.* **2023**, *13*, 1092224. [CrossRef]
46. Carminati, D.; Brizzi, A.; Giraffa, G.; Neviani, E. Effect of amino acids on S. salivarius subsp. thermophilus growth in modified milk deprived of non-protein nitrogen fraction. *Milchwissenschaft* **1994**, *49*, 481–540.
47. Rossetti, L.; Carminati, D.; Zago, M.; Giraffa, G. A Qualified Presumption of Safety approach for the safety assessment of Grana Padano whey starters. *Int. J. Food Microbiol.* **2009**, *130*, 70–73. [CrossRef] [PubMed]
48. Fornasari, M.E.; Rossetti, L.; Carminati, D.; Giraffa, G. Cultivability of *Streptococcus thermophilus* in Grana Padano cheese whey starters. *FEMS Microbiol. Lett.* **2006**, *257*, 139–144. [CrossRef] [PubMed]
49. Santarelli, M.; Gatti, M.; Lazzi, C.; Bernini, V.; Zapparoli, G.A.; Neviani, E. Whey Starter for Grana Padano Cheese: Effect of Technological Parameters on Viability and Composition of the Microbial Community. *J. Dairy Sci.* **2008**, *91*, 883–891. [CrossRef] [PubMed]
50. Erkus, O.; De Jager, V.C.L.; Spus, M.; van Alen-Boerrigter, I.J.; Van Rijswijck, I.M.H.; Hazelwood, L.; Janssen, P.W.M.; Hijum, S.A.F.T.v.; Kleerebezem, M.; Smid, E.J. Multifactorial diversity sustains microbial community stability. *ISME J.* **2013**, *7*, 2126–2136. [CrossRef] [PubMed]
51. Somerville, V.; Berthoud, H.; Schmidt, R.S.; Bachmann, H.P.; Meng, Y.H.; Fuchsmann, P.; von Ah, U.; Engel, P. Functional strain redundancy and persistent phage infection in Swiss hard cheese starter cultures. *ISME J.* **2022**, *16*, 388–399. [CrossRef] [PubMed]
52. Gatti, M.; Lazzi, C.; Rossetti, L.; Mucchetti, G.; Neviani, E. Biodiversity in Lactobacillus helveticus strains present in natural whey starter used for Parmigiano Reggiano cheese. *J. Appl. Microbiol.* **2003**, *95*, 463–470. [CrossRef] [PubMed]
53. Giraffa, G.; Mucchetti, G.; Neviani, E. Interactions among thermophilic lactobacilli during growth in cheese whey. *J. Appl. Bacteriol.* **1996**, *80*, 199–202. [CrossRef]
54. Bottari, B.; Ercolini, D.; Gatti, M.; Neviani, E. Application of FISH technology for microbiological analysis: Current state and prospects. *Appl. Microbiol. Biotechnol.* **2006**, *73*, 485–494. [CrossRef]
55. Tatenhove-pel, R.J.; Zwering, E.; Solopova, A.; Kuipers, O.P.; Bachmann, H. Ampicillin-treated Lactococcus lactis MG1363 populations contain persisters as well as viable but non-culturable cells. *Sci. Rep.* **2019**, *9*, 9867. [CrossRef] [PubMed]
56. van Mastrigt, O.; Abee, T.; Lillevang, S.K.; Smid, E.J. Quantitative physiology and aroma formation of a dairy Lactococcus lactis at near-zero growth rates. *Food Microbiol.* **2018**, *73*, 216–226. [CrossRef]

57. Babu, D.; Kushwaha, K.; Juneja, V.K. Viable but Nonculturable. In *Encyclopedia of Food Microbiology*; Elsevier: Amsterdam, The Netherlands, 2014; pp. 686–690. [CrossRef]
58. Emerson, J.B.; Adams, R.I.; Román, C.M.B.; Brooks, B.; Coil, D.A.; Dahlhausen, K.; Ganz, H.H.; Hartmann, E.M.; Hsu, T.; Justice, N.B.; et al. Schrödinger's microbes: Tools for distinguishing the living from the dead in microbial ecosystems. *Microbiome* **2017**, *5*, 86. [CrossRef] [PubMed]
59. Zago, M.; Bonvini, B.; Rossetti, L.; Meucci, A.; Giraffa, G.; Carminati, D. Biodiversity of Lactobacillus helveticus bacteriophages isolated from cheese whey starters. *J. Dairy Res.* **2015**, *82*, 242–247. [CrossRef]
60. Briggiler-Marcó, M.; Capra, M.L.; Quiberoni, A.; Vinderola, G.; Reinheimer, J.; Hynes, E. Nonstarter Lactobacillus strains as adjunct cultures for cheese making: In vitro characterization and performance in two model cheeses. *J. Dairy Sci.* **2007**, *90*, 4532–4542. [CrossRef] [PubMed]
61. Carminati, D.; Giraffa, G.; Quiberoni, A.; Binetti, A.; Suárez, V.; Reinheimer, J. Advances and Trends in Starter Cultures for Dairy Fermentations. In *Biotechnology of Lactic Acid Bacteria*, 1st ed.; Mozzi, F., Raya, R.R., Vignolo, G.M., Eds.; Wiley: Hoboken, NJ, USA, 2010; pp. 177–192. [CrossRef]
62. Carminati, D.; Mazzucotelli, L.; Giraffa, G.; Neviani, E. Incidence of Inducible Bacteriophage in Lactobacillus helveticus Strains Isolated from Natural Whey Starter Cultures. *J. Dairy Sci.* **1997**, *80*, 1505–1511. [CrossRef]
63. Mancini, A.; Rodriguez, M.C.; Zago, M.; Cologna, N.; Goss, A.; Carafa, I.; Tuohy, K.; Merz, A.; Franciosi, E. Massive Survey on Bacterial–Bacteriophages Biodiversity and Quality of Natural Whey Starter Cultures in Trentingrana Cheese Production. *Front. Microbiol.* **2021**, *12*, 678012. [CrossRef] [PubMed]
64. Spus, M.; Li, M.; Alexeeva, S.; Zwietering, M.H.; Abee, T.; Smid, E.J. Strain diversity and phage resistance in complex dairy starter cultures. *J. Dairy Sci.* **2015**, *98*, 5173–5182. [CrossRef] [PubMed]
65. Walsh, A.M.; Macori, G.; Kilcawley, K.N.; Cotter, P.D. Meta-analysis of cheese microbiomes highlights contributions to multiple aspects of quality. *Nat. Food* **2020**, *1*, 500–510. [CrossRef] [PubMed]
66. Bleuven, C.; Landry, C.R. Molecular and cellular bases of adaptation to a changing environment in microorganisms. *Proc. R. Soc. B* **2016**, *283*, 20161458. [CrossRef] [PubMed]
67. Elena, S.F.; Lenski, R.E. Evolution experiments with microorganisms: The dynamics and genetic bases of adaptation. *Nat. Rev. Genet.* **2003**, *4*, 457–469. [CrossRef]
68. Shapiro, J.A. Thinking about bacterial populations as multicellular organisms. *Annu. Rev. Microbiol.* **1998**, *52*, 81–104. [CrossRef]
69. Shapiro, J.A. Bacteria as Multicellular Organisms. *Sci. Am.* **1988**, *258*, 82–89. [CrossRef]
70. De Vos, M.G.J.; Schoustra, S.E.; De Visser, J.A.G.M. Ecology dictates evolution? About the importance of genetic and ecological constraints in adaptation. *EPL* **2018**, *122*, 58002. [CrossRef]
71. Booth, I.R. Stress and the single cell: Intrapopulation diversity is a mechanism to ensure survival upon exposure to stress. *Int. J. Food Microbiol.* **2002**, *78*, 19–30. [CrossRef] [PubMed]
72. Papadimitriou, K.; Pot, B.; Tsakalidou, E. How microbes adapt to a diversity of food niches. *Curr. Opin. Food Sci.* **2015**, *2*, 29–35. [CrossRef]
73. Smid, E.J.; Lacroix, C. Microbe-microbe interactions in mixed culture food fermentations. *Curr. Opin. Biotechnol.* **2013**, *24*, 148–154. [CrossRef]
74. Wolfe, B.E.; Dutton, R.J. Fermented Foods as Experimentally Tractable Microbial Ecosystems. *Cell* **2015**, *161*, 49–55. [CrossRef] [PubMed]
75. Berg, G.; Rybakova, D.; Fischer, D.; Cernava, T.; Vergès, M.-C.C.; Charles, T.; Chen, X.; Cocolin, L.; Eversole, K.; Corral, G.H.; et al. Microbiome definition re-visited: Old concepts and new challenges. *Microbiome* **2020**, *8*, 103. [CrossRef]
76. Jousset, A.; Schmid, B.; Scheu, S.; Eisenhauer, N. Genotypic richness and dissimilarity opposingly affect ecosystem functioning: Genotypic diversity and ecosystem functioning. *Ecol. Lett.* **2011**, *14*, 537–545. [CrossRef]
77. Konopka, A. What is microbial community ecology? *ISME J.* **2009**, *3*, 1223–1230. [CrossRef]
78. Levante, A.; Lazzi, C.; Vatsellas, G.; Chatzopoulos, D.; Dionellis, V.S. Genome Sequencing of five Lacticaseibacillus Strains and Analysis of Type I and II Toxin-Antitoxin System Distribution. *Microorganisms* **2021**, *9*, 648. [CrossRef]
79. Prosser, J.I.; Bohannan, B.J.M.; Curtis, T.P.; Ellis, R.J.; Firestone, M.K.; Freckleton, R.P.; Green, J.L.; Green, L.E.; Killham, K.; Lennon, J.J.; et al. The role of ecological theory in microbial ecology. *Nat. Rev. Microbiol.* **2007**, *5*, 384–392. [CrossRef]
80. De Angelis, M.; Gobbetti, M. Environmental stress responses in *Lactobacillus*: A review. *Proteomics* **2004**, *4*, 106–122. [CrossRef]
81. Hutkins, R.W. (Ed.) *Microbiology and Technology of Fermented Foods*, 1st ed.; Wiley: Hoboken, NJ, USA, 2006. [CrossRef]
82. Ryall, B.; Eydallin, G.; Ferenci, T. Culture History and Population Heterogeneity as Determinants of Bacterial Adaptation: The Adaptomics of a Single Environmental Transition. *Microbiol. Mol. Biol. Rev.* **2012**, *76*, 597–625. [CrossRef] [PubMed]
83. Boddy, L.; Wimpenny, J.W.T. Ecological concepts in food microbiology. *J. Appl. Bacteriol.* **1992**, *73*, 23s–38s. [CrossRef] [PubMed]
84. Juillard, V.; Spinnler, H.E.; Desmazeaud, M.J.; Boquien, C.Y. Phénomènes de coopération et d'inhibition entre les bactéries lactiques utilisées en industrie laitière. *Lait* **1987**, *67*, 149–172. [CrossRef]
85. Mossel, D.A.A.; Struijk, C.B. The contribution of microbial ecology to management and monitoring of the safety, quality and acceptability (SQA) of foods. *J. Appl. Bacteriol.* **1992**, *73*, 1s–22s. [CrossRef] [PubMed]
86. Calabrese, F.M.; Ameur, H.; Nikoloudaki, O.; Celano, G.; Vacca, M.; Junior, W.J.F.L.; Manzari, C.; Vertè, F.; Di Cagno, R.; Pesole, G.; et al. Metabolic framework of spontaneous and synthetic sourdough metacommunities to reveal microbial players responsible for resilience and performance. *Microbiome* **2022**, *10*, 148. [CrossRef]

87. Lynch, C.M.; McSweeney, P.L.H.; Fox, P.F.; Cogan, T.M.; Drinan, F.D. Manufacture of Cheddar cheese with and without adjunct lactobacilli under controlled microbiological conditions. *Int. Dairy J.* **1996**, *6*, 851–867. [CrossRef]
88. Jeanson, S.; Floury, J.; Gagnaire, V.; Lortal, S.; Thierry, A. Bacterial Colonies in Solid Media and Foods: A Review on Their Growth and Interactions with the Micro-Environment. *Front. Microbiol.* **2015**, *6*, 1284. [CrossRef]
89. Skandamis, P.N.; Jeanson, S. Colonial vs. planktonic type of growth: Mathematical modeling of microbial dynamics on surfaces and in liquid, semi-liquid and solid foods. *Front. Microbiol.* **2015**, *6*, 1178. [CrossRef]
90. Skandamis, P.N.; Nychas, G.-J.E. Quorum Sensing in the Context of Food Microbiology. *Appl. Environ. Microbiol.* **2012**, *78*, 5473–5482. [CrossRef] [PubMed]
91. Heuer, H.; Abdo, Z.; Smalla, K. Patchy distribution of flexible genetic elements in bacterial populations mediates robustness to environmental uncertainty: Population-level robustness through genome flexibility. *FEMS Microbiol. Ecol.* **2008**, *65*, 361–371. [CrossRef] [PubMed]
92. Penn, K.; Jenkins, C.; Nett, M.; Udwary, D.W.; Gontang, E.A.; McGlinchey, R.P.; Foster, B.; Lapidus, A.; Podell, S.; Allen, E.E.; et al. Genomic islands link secondary metabolism to functional adaptation in marine Actinobacteria. *ISME J.* **2009**, *3*, 1193–1203. [CrossRef] [PubMed]
93. Avery, S.V. Microbial cell individuality and the underlying sources of heterogeneity. *Nat. Rev. Microbiol.* **2006**, *4*, 577–587. [CrossRef] [PubMed]
94. Koutsoumanis, K.P.; Aspridou, Z. Individual cell heterogeneity in Predictive Food Microbiology: Challenges in predicting a "noisy" world. *Int. J. Food Microbiol.* **2017**, *240*, 3–10. [CrossRef] [PubMed]
95. Rodriguez-Valera, F.; Martin-Cuadrado, A.-B.; Rodriguez-Brito, B.; Pašić, L.; Thingstad, T.F.; Rohwer, F.; Mira, A. Explaining microbial population genomics through phage predation. *Nat. Rev. Microbiol.* **2009**, *7*, 828–836. [CrossRef] [PubMed]
96. Somerville, V.; Schowing, T.; Chabas, H.; Schmidt, R.S.; von Ah, U.; Bruggmann, R.; Engel, P. Extensive diversity and rapid turnover of phage defense repertoires in cheese-associated bacterial communities. *Microbiome* **2022**, *10*, 137. [CrossRef] [PubMed]
97. Meouche, I.E.; Siu, Y.; Dunlop, M.J. Stochastic expression of a multiple antibiotic resistance activator confers transient resistance in single cells. *Sci. Rep.* **2016**, *6*, 19538. [CrossRef]
98. Viney, M.; Reece, S.E. Adaptive noise. *Proc. R. Soc. B Biol. Sci.* **2013**, *280*, 20131104. [CrossRef]
99. Gatti, M.; De Dea Lindner, J.; Gardini, F.; Mucchetti, G.; Bevacqua, D.; Fornasari, M.E.; Neviani, E. A Model to Assess Lactic Acid Bacteria Aminopeptidase Activities in Parmigiano Reggiano Cheese During Ripening. *J. Dairy Sci.* **2008**, *91*, 4129–4137. [CrossRef]
100. Calasso, M.; Mancini, L.; Di Cagno, R.; Cardinali, G.; Gobbetti, M. Microbial cell-free extracts as sources of enzyme activities to be used for enhancement flavor development of ewe milk cheese. *J. Dairy Sci.* **2015**, *98*, 5874–5889. [CrossRef] [PubMed]
101. Lane, C.N.; Fox, P.F.; Walsh, E.M.; Folkertsma, B.; McSweeney, P.L.H. Effect of compositional and environmental factors on the growth of indigenous non-starter lactic acid bacteria in Cheddar cheese. *Lait* **1997**, *77*, 561–573. [CrossRef]
102. Montel, M.C.; Buchin, S.; Mallet, A.; Delbes-Paus, C.; Vuitton, D.A.; Desmasures, N.; Berthier, F. Traditional cheeses: Rich and diverse microbiota with associated benefits. *Int. J. Food Microbiol.* **2014**, *177*, 136–154. [CrossRef] [PubMed]
103. Visser, S. Proteolytic Enzymes and Their Relation to Cheese Ripening and Flavor: An Overview. *J. Dairy Sci.* **1993**, *76*, 329–350. [CrossRef]
104. Wilkinson, M.G.; Kilcawley, K.N. Mechanisms of incorporation and release of enzymes into cheese during ripening. *Int. Dairy J.* **2005**, *15*, 817–830. [CrossRef]
105. De Pasquale, I.; Calasso, M.; Mancini, L.; Ercolini, D.; Storia, A.L.; De Angelis, M.; Di Cagno, R.; Gobbetti, M. Causal Relationship between Microbial Ecology Dynamics and Proteolysis during Manufacture and Ripening of Protected Designation of Origin (PDO) Cheese Canestrato Pugliese. *Appl. Environ. Microbiol.* **2014**, *80*, 4085–4094. [CrossRef]
106. Giraffa, G.; Neviani, E. Different Lactobacillus helveticus strain populations dominate during Grana Padano cheesemaking. *Food Microbiol.* **1999**, *16*, 205–210. [CrossRef]

MDPI AG
Grosspeteranlage 5
4052 Basel
Switzerland
Tel.: +41 61 683 77 34

Fermentation Editorial Office
E-mail: fermentation@mdpi.com
www.mdpi.com/journal/fermentation